Lecture Notes in Computer Science 3448

Commenced Publication in 1973
Founding and Former Series Editors:
Gerhard Goos, Juris Hartmanis, and Jan van Leeuwen

Günther R. Raidl   Jens Gottlieb (Eds.)

Evolutionary Computation in Combinatorial Optimization

5th European Conference, EvoCOP 2005
Lausanne, Switzerland, March 30 - April 1, 2005
Proceedings

 Springer

Volume Editors

Günther R. Raidl
Vienna University of Technology
Institute of Computer Graphics and Algorithms
Algorithms and Data Structures Group
Favoritenstr. 9-11/186, 1040 Vienna, Austria
E-mail: raidl@ads.tuwien.ac.at

Jens Gottlieb
SAP AG
Neurottstr. 16, 69190 Walldorf, Germany
E-mail: jens.gottlieb@sap.com

Cover illustration: *Triangular Urchin* by Chaps
(www.cetoine.com)
Chaps has obtained an MSc in Physics at the Swiss Federal Institute of Technology. He is the developer of the ArtiE-Fract software that was used to create *Triangular Urchin*.
Triangular Urchin (an Iterated Functions System of 2 polar functions) emerged from an urchin structure after a few generations using ArtiE-Fract. The evolutionary process was only based on soft mutations, some of them directly induced by the author.

Library of Congress Control Number: 2005922053

CR Subject Classification (1998): F.1, F.2, G.1.6, G.2.1, G.1

ISSN 0302-9743
ISBN 3-540-25337-8 Springer Berlin Heidelberg New York

Springer is a part of Springer Science+Business Media

springeronline.com

© Springer-Verlag Berlin Heidelberg 2005
Printed in Germany

Typesetting: Camera-ready by author, data conversion by Scientific Publishing Services, Chennai, India
Printed on acid-free paper   SPIN: 11407041   06/3142   5 4 3 2 1 0

Preface

Evolutionary computation (EC) involves the study of problem-solving and optimization techniques inspired by principles of natural evolution and genetics. EC has been able to draw the attention of an increasing number of researchers and practitioners in several fields. Evolutionary algorithms have in particular been shown to be effective for difficult combinatorial optimization problems appearing in various industrial, economics, and scientific domains.

This volume contains the proceedings of EvoCOP 2005, the 5th European Conference on Evolutionary Computation in Combinatorial Optimization. It was held in Lausanne, Switzerland, on 30 March–1 April 2005, jointly with EuroGP 2005, the 8th European Conference on Genetic Programming, and the EvoWorkshops 2005, which consisted of the following six individual workshops: EvoBIO, the 3rd European Workshop on Evolutionary Bioinformatics; Evo-COMNET, the 2nd European Workshop on Evolutionary Computation in Communication, Networks, and Connected Systems; EvoHOT, the 2nd European Workshop on Hardware Optimisation Techniques; EvoIASP, the 7th European Workshop on Evolutionary Computation in Image Analysis and Signal Processing; EvoMUSART, the 3rd European Workshop on Evolutionary Music and Art; and EvoSTOC, the 2nd European Workshop on Evolutionary Algorithms in Stochastic and Dynamic Environments.

EvoCOP, held annually as a workshop since 2001, became a conference in 2004 and it is now the premier European event focusing on evolutionary computation in combinatorial optimization. The events gave researchers an excellent opportunity to present their latest research and to discuss current developments and applications, besides stimulating closer future interaction between members of this scientific community. Accepted papers of previous events were published by Springer in the series Lecture Notes in Computer Science (LNCS volumes 2037, 2279, 2611, and 3004).

The double-blind reviewing process resulted in a strong selection among the submitted papers; the acceptance rate was 36.4%. All accepted papers were presented orally at the conference and are included in this proceedings volume. We would like to give credit to the members of our Program Committee, to whom we are very grateful for their quick and thorough work.

EvoCOP	submitted	accepted	acceptance ratio
2001	31	23	74.2%
2002	32	18	56.3%
2003	39	19	48.7%
2004	86	23	26.7%
2005	66	24	36.4%

EvoCOP 2005 covers evolutionary algorithms as well as related approaches like scatter search, simulated annealing, ant colony optimization, immune algorithms, variable-neighborhood search, hyperheuristics, and estimation of distribution algorithms. The papers deal with representations, analysis of operators and fitness landscapes, and comparison of algorithms. The list of studied combinatorial optimization problems includes prominent examples like graph coloring, quadratic assignment, the knapsack problem, graph matching, packing, scheduling, timetabling, lot-sizing, and the traveling-salesman problem.

For the first time, EvoCOP used a conference management system, VSIS ConfTool 1.1.2, to handle paper submissions and the reviewing process. Harald Weinreich and his team, who developed this software and made it available to us, deserve our gratitude for this open-source project that saved us a lot of time. We would like to thank Philipp Neuner for administrating the conference management system.

Finally, many thanks go to Jennifer Willies, who cared about the administration and coordination of EuroGP 2005, EvoCOP 2005, and the EvoWorkshops 2005, for her tremendous efforts.

March 2005   Günther R. Raidl
                                                                      Jens Gottlieb

Organization

EvoCOP 2005 was organized jointly with EuroGP 2005 and the EvoWorkshops 2005.

Organizing Committee

Chairs: Günther R. Raidl, Vienna University of Technology, Austria
                   Jens Gottlieb, SAP AG, Germany
Local Chair: Marco Tomassini, University of Lausanne, Switzerland
Publicity Chair: Jano van Hemert, Napier University, Edinburgh, UK

Program Committee

Adnan Acan, Eastern Mediterranean University, Turkey
Hernan Aguirre, Shinshu University, Japan
Enrique Alba, University of Málaga, Spain
M. Emin Aydin, London South Bank University, UK
Jean Berger, Defence Research and Development Canada, Canada
Christian Bierwirth, University of Halle-Wittenberg, Germany
Christian Blum, Universitat Politècnica de Catalunya, Spain
Edmund Burke, University of Nottingham, UK
Ernesto Costa, University of Coimbra, Portugal
Carlos Cotta, University of Málaga, Spain
Peter Cowling, University of Bradford, UK
Bart Craenen, Napier University, Edinburgh, UK
David Davis, NuTech Solutions, Inc., USA
Marco Dorigo, Université Libre de Bruxelles, Belgium
Karl Dörner, University of Vienna, Austria
Anton Eremeev, Omsk Branch of the Sobolev Institute of Mathematics, Russia
David Fogel, Natural Selection, Inc., USA
Bernd Freisleben, University of Marburg, Germany
Jens Gottlieb, SAP AG, Germany
Michael Guntsch, Eurobios UK, UK
Walter Gutjahr, University of Vienna, Austria
Jin-Kao Hao, University of Angers, France
Emma Hart, Napier University, Edinburgh, UK
William E. Hart, Sandia National Laboratories, USA
Jano van Hemert, Napier University, Edinburgh, UK

Jörg Homberger, Stuttgart University of Cooperative Education, Germany
Mikkel T. Jensen, Acure, Denmark
Bryant A. Julstrom, St. Cloud State University, USA
Graham Kendall, University of Nottingham, UK
Joshua D. Knowles, University of Manchester, UK
Gabriele Koller, Vienna University of Technology, Austria
Mario Köppen, Fraunhofer IPK, Germany
Jozef J. Kratica, Serbian Academy of Sciences and Arts, Serbia and Montenegro
Ivana Ljubić, Siemens, Austria
Elena Marchiori, Free University Amsterdam, The Netherlands
Dirk C. Mattfeld, TU Braunschweig, Germany
Helmut Mayer, University of Salzburg, Austria
Daniel Merkle, University of Leipzig, Germany
Peter Merz, University of Kaiserslautern, Germany
Zbigniew Michalewicz, University of Adelaide, Australia
Martin Middendorf, University of Leipzig, Germany
Pablo Moscato, University of Newcastle, Australia
Christine L. Mumford, Cardiff University, UK
Francisco J.B. Pereira, University of Coimbra, Portugal
Jakob Puchinger, Vienna University of Technology, Austria
Günther R. Raidl, Vienna University of Technology, Austria
Marcus Randall, Bond University, Australia
Colin Reeves, Coventry University, UK
Marc Reimann, ETH Zurich, Switzerland
Franz Rothlauf, University of Mannheim, Germany
Andreas Sandner, SAP AG, Germany
Marc Schoenauer, INRIA, France
Christine Solnon, University of Lyon I, France
Eric Soubeiga, University of Nottingham, UK
Thomas Stützle, Darmstadt University of Technology, Germany
El-ghazali Talbi, University of Lille, France
Edward Tsang, University of Essex, UK
Ingo Wegener, University of Dortmund, Germany
Takeshi Yamada, NTT Communication Science Laboratories, Japan

Sponsoring Institutions

- EvoNet, the Network of Excellence in Evolutionary Computing
- University of Lausanne, Lausanne, Switzerland

Table of Contents

An External Partial Permutations Memory for Ant Colony Optimization

Adnan Acan

Eastern Mediterranean University, Computer Engineering Department
Gazimagusa, T.R.N.C., Via Mersin 10, Turkey
adnan.acan@emu.edu.tr

Abstract. A novel external memory implementation based on the use of
partially complete sequences of solution components from above-average
quality individuals over a number of previous iterations is introduced.
Elements of such variable-size partial permutation sequences are taken
from randomly selected positions of parental individuals and stored in an
external memory called the partial permutation memory. Partial permutation sequences are associated with lifetimes together with their parent
solutions' fitness values that are used in retrieving and updating the
contents of the memory. When a solution is to be constructed, a partial
permutation sequence is retrieved from the memory based on its age and
associated fitness value, and the remaining components of the partial solution is completed with an ant colony optimization algorithm. Resulting
solutions are also used to update some elements within the memory. The
implemented algorithm is used for the solution of a difficult combinatorial optimization problem, namely the quadratic assignment problem,
for which significant performance achievements are provided in terms of
convergence speed and solution quality.

1 Introduction

Ant colony optimization (ACO) is a nature-inspired general purpose computation method which can be applied to many kinds of optimization problems
[1, 2]. Among many efforts on the development of new variants of ACO algorithms toward improving their efficiency under different circumstances, recently
the idea of knowledge incorporation from previous iterations became attractive
and handled by a number of researchers. Mainly, these population- or memory-
based approaches take their inspiration from studies in genetic algorithms (GAs).
In memory-based GA implementations, information stored within a memory is
used to adapt the GAs behavior either in problematic cases where the solution
quality is not improved over a number of iterations, or a change in the problem
environment is detected, or to provide further directions of exploration and exploitation. Memory in GAs can be provided externally (outside the population)
or internally (within the population) [3].

External memory implementations store specific information within a separate population (memory) and reintroduce that information into the main population at a later moment. In most cases, this means that individuals from memory

G.R. Raidl and J. Gottlieb (Eds.): EvoCOP 2005, LNCS 3448, pp. 1–11, 2005.

are put into the initial population of a new or restarted GA. Case-based memory approaches, which are actually a form of long term elitism, are the most typical form of external memory implemented in practice. In general, there are two kinds of case-based memory implementations: in one kind, case-based memory is used to re-seed the population with the best individuals from previous generations [4], whereas a different kind of case-based memory stores both problems and solutions [5]. Case-based memory aims to increase the diversity by reintroducing individuals from previous generations and achieves exploitation by reusing individuals from case-based memory when a restart from a good initial solution is required.

Other variants of external memory approaches are provided by several researchers for both specific and general purpose implementations. Simoes and Costa introduced an external memory method in which gene segments are stored instead of the complete genomes [6, 7]. Acan et al. proposed a novel external memory approach based on the reuse of insufficiently utilized promising chromosomes from previous generations for the production of current generation offspring individuals [8].

The most common approaches using internal memory are polyploidy structures. Polyploidy structures in combination with dominance mechanisms use redundancy in genetic material by having more than one copy of each gene. When a chromosome is decoded to determine the corresponding phenotype, the dominant copy is chosen. This way, some genes can shield themselves from extinction by becoming recessive. Through switching between copies of genes, a GA can adapt faster to changing environments and recessive genes are used to provide information about fitness values from previous generations [9].

In ACO the first internally implemented memory-based approach is the work of Montgomery et al. [10]. In their work, named as AEAC, they modified the characteristic element selection equations of ACO to incorporate a weighting term for the purpose of accumulated experience. This weighting is based on the characteristics of partial solutions generated within the current iteration. Elements that appear to lead better solutions are valued more highly, while those that lead to poorer solutions are made less desirable. They aim to provide, in addition to normal pheromone and heuristic costs, a more immediate and objective feedback on the quality of the choices made. Basically, considering the TSP, if a link (r, u) has been found to lead to longer paths after it has been incorporated into a solution, then its weight $w(r, u) < 1$. If the reverse is the case, then $w(r, u) > 1$. If the colony as a whole has never used the link (r, u), then its weight is selected as 1. The authors suggested simple weight update procedures and proposed two variations of their algorithm. They claimed that the achieved results for different TSP instances are either equally well or better than those achieved using normal ACS algorithm.

The work of Guntsch et al. [11] is the first example of an external memory implementation within ACO. Their approach, P-ACO, uses a population of previously best solutions from which the pheromone matrix can be derived. Initially the population is empty and, for the first k iteration of ants, the best solutions found in each iteration enters the population. After that, to update the population, the best solution of an iteration enters the population and the oldest one is removed. That is, the population is maintained like a FIFO-queue. This way, each solution in the population influences the decisions of ants over exactly k iterations. For every solution in the population, some amount of pheromone is added to the corresponding edges of the construction graph. The whole pheromone information of P-ACO depends only on the solutions in the population and the pheromone matrix is updated as follows: whenever a solution π enters the population, do a positive update as $\forall i \in [1, n] : \tau_{i\pi(i)} \rightarrow \tau_{i\pi(i)} + \Delta$ and whenever a solution σ leaves the population, do a negative update as $\forall i \in [1, n] : \tau_{i\sigma(i)} \rightarrow \tau_{i\sigma(i)} - \Delta$. These updates are added to the initial pheromone value τ_{init}. The authors also proposed a number of population update strategies in [12] to decide which solutions should enter the population and which should leave. In this respect, only the best solution generated during the past iteration is considered as a candidate to enter the population and the measures used in population update strategies are stated as age, quality, prob, age and prob, and elitism. In age strategy, the oldest solution is removed from the population. In quality strategy, the population keeps the best k solutions found over all past iterations, rather than the best solutions of the last k iterations. Prob strategy probabilistically chooses the element to be removed from the population and the aim is to reduce the number of identical copies that might be caused the quality strategy. Combination of age and prob strategies use prob for removal and age for insertion into the population. In elitist strategy, the best solution found by the algorithm so far is never replaced until a better solution is found.

Recently, Acan proposed two novel external memory strategies for ACO [13]. In this approach, a library of variable size solution segments cut from elite individuals of a number of previous generations is maintained. There is no particular distribution of ants in the problem space and, in order to construct a solution each ant retrieves a segment from the library based on its goodness in its parent solution, takes the end component of the segment as its starting point, and completes the rest to form a complete feasible solution. The proposed approach is used for the solution of traveling salesman problem (TSP) and the quadratic assignment problem (QAP) for which significantly better solutions are achieved compared to conventional ACO algorithms.

This paper introduces another population based external memory approach where the population includes variable-size partial permutation sequences taken from elite individuals of previous iterations. Initially, the memory is empty and an ant colony optimization algorithm runs for a small number of iterations to initialize the memory of partial permutations. Each stored sequence is associated with a lifetime and its parent's objective function value that will be used as measures for retrieving and updating partial solutions within the memory. In order to construct a solution, a particular ant retrieves a partial solution (a partial permutation sequence) from the memory based on a defined performance measure and fills in the unspecified components within it. Constructed solutions are also used to update the memory. The details of the practical implementation are given in the following sections. The proposed ACO strategy is used to solve the well-known quadratic assignment problem for which significant performance improvements are achieved, compared to the well-known Max-Min AS algorithm, in terms of both solution quality and the convergence speed.

This paper is organized as follows. The basics of ACO-based solution construction procedures for the the quadratic assignment problem (QAP) are presented in Section 2. The proposed approach is described with its implementation details in Section 3. Section 4 covers the results and related discussions. Conclusions and future research directions are given in Section 5.

2   ACO for Quadratic Assignment Problem

In this section, the basic solution construction procedure of ACO algorithms for the solution of QAP will be briefly described. Given a set of N facilities, a set of N locations, distances between pairs of locations, and flows between pairs of facilities, QAP is described as the problem of assigning each facility to a particular location so as to minimize the sum of the product between flows and the distances. More formally, if $D = [d_{ij}]$ is the $N \times N$ distance matrix and $F = [f_{pq}]$ is the $N \times N$ flow matrix, where d_{ij} is the distance between locations i and j and f_{pq} is the amount of flow between facilities p and q, QAP can be described by the following equation:

$$\min_{\pi \in \Pi} \sum_{i=1}^{N} \sum_{j=1}^{N} d_{\pi(i)\pi(j)} f_{ij} \tag{1}$$

where Π is the set of all permutations of integers from 1 to N, and $\pi(i)$ gives the location of facility i within the current solution (permutation) $\pi \in \Pi$. The term $d_{\pi(i)\pi(j)} f_{ij}$ is the cost of simultaneously assigning facility i to location $\pi(i)$ and facility j to location $\pi(j)$.

Quadratic assignment problem belongs to the class of NP-hard combinatorial optimization problems and the largest instances that can be solved with exact algorithms are limited to instances of size around 30 [14]. Hence, the only feasible way to deal with the solution of large QAP instances is to use heuristic approaches which guarantee to reach a locally optimal solution in reasonable computation times. Several modern heuristics, mostly evolutionary or nature-inspired, are developed since 1975 and successfully applied for the solution of many provably hard optimization problems including the QAP. In this respect, ant colony optimization is successfully used for the solution of QAP and, compared to other metaheuristic approaches, better results are obtained for many difficult problem instances [15, 16, 17, 18, 19].

The QAP is one of the most widely handled problem for the illustration and testing of ACO strategies. An ACO algorithm uses a number of artificial ants for the construction of solutions such that each ant starts from a particular initial assignment and builds a solution in an iterative manner. An ant builds a solution by randomly selecting an unassigned facility and placing it into one of the remaining free locations, until no unassigned facilities are left. In the construction of a solution, the decision of an ant for its next assignment is affected by two factors: the pheromone information and the heuristic information which both indicate how good it is to assign a facility to the free location under consideration. No heuristic information is used in the implementation of MMAS-QAP and pheromone concentration τ_{ij} refers to the desire of assigning facility of i to location j. Very detailed steps of implementation for MAX-MIN AS on the solution of QAP can be found in [16], that is also considered as the main reference for ACO-strategy and parameter value selections in the proposed approach.

3   The Proposed External Memory Approach

The basic inspiration behind the use of a partial-permutations memory is to store values of some solution components that result in above-average fitness so that they can be reused in future iterations to provide further intensification around potentially promising solutions without destroying diversification capabilities of the underlying ACO algorithm.

In the proposed approach, there are m ants that build the solutions. An external memory of variable-size partial solution sequences from a number of elite solutions of previous iterations is maintained. Initially the memory is empty and a number ACO iterations is performed to fill in the memory. In this respect, after every iteration, the top k-best solutions are considered and a number of randomly positioned solution components are selected from these individuals, one sequence per solution, and stored in the memory. Uniform probability distribution is used in all random number generations. The memory size M is fixed and a number of ACO iterations are repeated until the memory is full. Generation of a partial permutation from a given parent solution is illustrated in Figure 1. Simply, an arbitrary number of solution components are selected from the parent solution and stored as a partially complete permutation within the memory. Since we need the locus of the stored solution components when they will be used, partial permutation sequences are stored as fixed-length sequences where unselected parts are filled with zeros. With this description of individual partial permutation sequences, the external memory is an MxN array where M is the number of elements and N is the length of a solution.

When a new solution is to be created, a partially complete permutation is retrieved from the memory selects a partial permutation sequence from the memory based on its lifetime and fitness values. In this respect, each partial solution sequence i in the memory is assigned with a performance score as $P\_Score(i) = Fitness(i) + Age^2(i)$. With this assignment, higher P_Score values are given to those with higher age attributes in cases of closer fitness values. The main idea behind this selection strategy is to give higher priority of being used to older sequences before they are removed from the memory. This also helps to provide further diversification by trying some of the previously promising search directions.

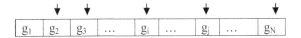

g_1	g_2	g_3	...	g_i	...	g_j	...	g_N

a) Parent solution.

0	g_2	g_3	0...0	g_i	0...0	g_j	0...0	g_N	Fitness	Age

b) Partial permutation sequence stored in memory.

Fig. 1. a) A parent solution and b) the partial permutation sequence containing randomly located solution components stored in the memory

After the initial phase, MAX-MIN AS algorithm works in conjunction with the implemented external memory as follows: there is no particular assignment of ants over the problem space and, in order to construct a solution, each ant selects a partial permutation sequence from memory using a tournament selection strategy and the starting point for the corresponding ant becomes the first unassigned position of the partial solution. That is, the ant builds the rest of the solution by assigning the unassigned components of the retrieved partial solution in a random order. In the selection of partial solution sequences from memory, each ant makes a tournament among Q randomly selected partial permutation sequences and the best one is selected as the seed to start for construction. The solution construction procedure is the same as the one followed in MAX-MIN AS algorithm. After all ants complete their solution construction procedures, pheromone updates are carried in exactly the same way it is done in MAX-MIN AS algorithm. Based on these descriptions, the proposed ACO algorithm can be put into an algorithmic form described in Algorithm 1.

In memory update procedure, the solutions constructed by all ants are sorted in ascending order and the top k-best are considered as candidate parents from which new partial permutation sequences will be extracted and inserted into memory. One random-length and randomly-positioned partial permutation sequence is taken from each elite solution and it replaces either the worst of those memory elements having worse fitness values than of the extracted sequence or the worst of those oldest sequences within the memory.

4   Experimental Results

To study the performance of the proposed external memory ACO approach, it is compared with a well-known classical implementation of ACO algorithms, namely the MAX-MIN AS algorithm, for the solution of a widely known and difficult combinatorial optimization problem, the QAP. The QAP problem instances used in evaluations are taken from the web-site http://www.opt.math.tu-

Algorithm 1: ACO with an external partial permutations memory

1. Initialize the external memory.
2. Repeat
   (a) For each of m ants, find an initial assignment by assigning a randomly selected facility to a randomly selected location.
   (b) Let all ants construct a solution iteratively by randomly selecting an unassigned facility and placing it into one of the free locations.
   (c) Compute the objective function values for all the constructed solutions.
   (d) Use local search to improve the solutions of a number of elite ants.
   (e) Sort the solutions in ascending order of their objective function values.
   (f) Consider the top k-best and extract a randomly-positioned and randomly-sized partial permutation from each solution and insert the partial solution sequences into the memory. Also, store the length of each sequence and the cost of its parent. Meanwhile, set the age of the inserted element to 1.
3. Update the pheromone matrix.
4. Until the memory is full.
5. ITER=1
6. Repeat
   (a) Let all ants select a partial solution sequence from the external memory using tournament selection.
   (b) Let all ants construct a solution starting from the partial permutation (sequence) they retrieved from the memory.
   (c) Compute the objective function values for all the constructed solutions.
   (d) Use local search to improve the solutions of a number of elite ants.
   (e) Update pheromone matrix.
   (f) Increase ages of memory elements by 1.
   (g) Update memory.
   (h) ITER=ITER+1.
7. Until ITER > Max_Iter.

graz.ac.at/qaplib. These selected problems are representative instances of different problem groups, commonly handled by several researchers, and have reasonable problem sizes for experimental evaluations.

The ACO algorithm used with which the external memory is integrated is MMAS-QAP [16] which is a well-known and provably successful algorithm for the solution of QAP instances. All programs are prepared using Matlab 6.5 programming language and the computing platform is a 1 GHz PC with 256 MB of memory. Each experiment is performed 20 times over 1000 iterations. The relative percentage deviation, RPD, is used to express the accuracy of experimental results. RPD is a measure of the percentage difference between the best known solution and an actual experimental result. RPD is simply calculated by the following formula.

$$RPD = 100 \times (\frac{Cost_{Actual} - Cost_{Best}}{Cost_{Best}}) \tag{2}$$

Details of implementations and the sets of parameter values used are given below.

In the implementation of the proposed ACO strategy, the memory size M and the number of ants in ACO are set equal to the number of locations (facilities) in QAP instances. The tournament size $Q = 7$, and the top $k = 0.05 * M$ solutions are used in updating the memory. The local search algorithm used is 2.5-opt and 25 elite solutions are allowed to be processed within the local search. Maximum lifetime of individuals in the shared-memory is taken as 5 iterations.

In the implementation of the ACO algorithms, the two sets of parameter values taken from well-known publications are used as follows [16]: First set of parameters are used in initializing the partial permutations memory and a highly biased probabilistic selection scheme is followed in the selection of solution components. In this respect, the next element for which $\tau^\alpha.\eta^\beta$ is maximal is selected with probability $q_0 = 0.9$. Together with the local search used, memory elements are initialized with partially complete permutations extracted from potentially promising solutions. Other parameter values used in this phase are as follows: $\rho = 0.1$, $\alpha = 1$, $\beta = 2$, $\tau_{init} = 1/n$, $\tau_{max} = 3.0$. Parameters used during the normal course of ACO algorithms are $\rho = 0.02$, $\alpha = 1$, $\beta = 2$, $\tau_{max} = 1/(\rho.L_{Best})$, $\tau_{init} = \tau_{max}$, and $\tau_{min} = 1/(2n)$.

Stagnation in ACO is detected when

$$\sum_{\tau_{ij}} min(\tau_{max} - \tau_{ij}, \tau_{ij} - \tau_{min}) \tag{3}$$

is less than 10^{-5}. In this case, ACO is restarted by setting all elements of the pheromone matrix to τ_{max}.

Experimental results for the solved QAP instances are given in Table 1.

From Table 1, it can be observed that the proposed ACO strategy with an external partial permutations memory performs better than the MAX-MIN AS supported with the same local search algorithm, in all problem instances. In addition, the effectiveness of the external memory and the implemented memory management approaches is observed much more clearly for larger problem instances. This is an expected result because partial solutions from potentially promising solutions provide more efficient intensification for more difficult problems. The main drawback of the proposed approach is the long running times. It is a well-known fact that Matlab is a useful programming language to prepare prototypes, however it can be very slow in the execution of loops, which is also the case in the presented implementations. The CPU times with the used hardware platform change from 15 minutes (for the wil50 QAP instance) to 4 hours (for the tai100a QAP instance). However, these CPU time are smaller than those required for MMAS algorithm within the same computing platform, programming language, and algorithm parameters. For example, CPU times for MMAS change from 22 minutes (for the wil50 QAP instance) to 6 hours (for the tai100a QAP instance). Implementations with a faster programming language is currently an ongoing but yet not completed task.

Table 1. Results for Max-Min AS, and the proposed approach on QAP instances

Strategy	Instance	RPD		
		Min	Average	Max
Max-Min AS	wil50	1.831	3.823	4.412
	wil100	3.512	5.711	7.3
	tai100a	8.910	13.36	18.44
	tai100b	10.117	15.19	19.5
	tai35b	4.218	9.3	14.5
	tai40b	4.887	6.1	11.7
	tai50b	5.628	7.6	13.6
	tai60b	7.917	9.5	17.32
	tai80b	13.226	17.5	28.49
	tai100a	21.11	25.6	33.51
	sko100a	17.6	20.3	24.1
	sko100b	19.44	22.7	26.8
MX-MIN AS	wil50	0.98	2.61	3.73
with a Partial	wil100	2.49	3.26	4.73
Permutations Memory	tai100a	4.91	8.32	11.94
	tai100b	7.96	10.27	14.81
	tai35b	3.21	6.73	10.66
	tai40b	3.16	5.25	8.86
	tai50b	3.83	5.91	8.45
	tai60b	4.92	7.53	11.74
	tai80b	7.74	10.16	15.36
	tai100a	16.49	21.71	27.45
	sko100a	14.51	19.11	22.35
	sko100b	16.43	20.21	24.81

5   Conclusions and Future Research Directions

In this paper a novel ACO strategy using an external memory of partial permutations from elite solutions of previous iterations is introduced. The stored partial permutations are used in the construction of solutions in the current iteration to provide further intensification around potentially promising solutions. Using partially constructed solutions also improve time spent in the construction of solutions since part of the solution is already available. Performance of the proposed external memory approach is tested using several instances of a well-known hard combinatorial optimization problems.

From the results of case studies, it can easily be concluded that the proposed ACO strategy performs better than Max-Min AS algorithm in terms of the convergence speed and the solution quality. It can easily be observed that, even though the proposed strategy better than the MAX-MIN AS strategy, its real effectiveness is seen for larger size problems. This is an expected result because information incorporation from previous iterations should be helpful for more difficult problem instances. The proposed ACO strategy is very simple to implement and it does not bring significantly additional computational or storage cost to the existing ACO algorithms.

Further investigation of the proposed ACO strategy for the solution of other difficult problems, particularly the non-stationary optimization problems, may be considered as a direction for future studies.

References

1. Dorigo, M., Caro, G.D., Gambardella, L.M.: Ant algorithms for distributed discrete optimization, Artificial Life, Vol. 5, (1999), 137-172.
2. Dorigo, M., Caro, G.D.,: The ant colony optimization metaheuristic, In Corne, D., Dorigo, M., Glover, F. (eds.): New ideas in optimization, McGraw-Hill, London, (1999), 11-32.
3. Eggermont, J., Lenaerts, T.: Non-stationary function optimization using evolutionary algorithms with a case-based memory, Technical report, Leiden University Advanced Computer Science (LIACS) Technical Report 2001-11.
4. Ramsey, C.L., Grefenstette, J. J.: Case-based initialization of GAs, in Forest, S., (Editor): Proceedings of the Fifth International Conference on Genetic Algorithms, San Mateo, CA, (1993), 84-91.
5. Louis, S., Li, G.: Augmenting genetic algorithms with memory to solve traveling salesman problem, Proceedings of the Joint Conference on Information Sciences, Duke University, (1997), 108-111.
6. Simoes, A., Costa, E.: Using genetic algorithms to deal with dynamic environments: comparative study of several approaches based on promoting diversity, in W. B. Langton et al. (eds.): Proceedings of the genetic and evolutionary computation conference GECCO'02, Morgan Kaufmann, New York, (2002), 698.
7. Simoes, A., Costa, E.: Using biological inspiration to deal with dynamic environments, Proceedings of the seventh international conference on soft computing MENDEL'2001, Czech Republic, (2001).
8. Acan, A., Tekol, Y.: Chromosome reuse in genetic algorithms, in Cantu-Paz et al. (eds.): Genetic and Evolutionary Computation Conference GECCO 2003, Springer-Verlag, Chicago, (2003), 695-705.
9. Lewis, J., Hart, E., Ritchie, G.: A comparison of dominance mechanisms and simple mutation on non-stationary problems, in Eiben, A. E., Back, T., Schoenauer, M., Schwefel, H. (Editors): Parallel Problem Solving from Nature- PPSN V, Berlin, (1998), 139-148.
10. Montgomery, J., Randall, M: The accumulated experience ant colony for the traveling salesman problem, International Journal of Computational Intelligence and Applications, World Scientific Publishing Company, Vol. 3, No. 2, (2003), 189-198.
11. Guntsch, M., Middendorf, M.: A population based approach for ACO, in S. Cagnoni et al., (eds.): Applications of Evolutionary Computing - EvoWorkshops2002, Lecture Notes in Computer Science, No:2279, Springer Verlag, (2002), 72-81.
12. Guntsch, M., Middendorf, M.: Applying population based ACO for dynamic optimization problems, in M. Dorigo et al., (eds.): Ant Algorithms - Third International Workshop ANTS2002, Lecture Notes in Computer Science, No:2463, Springer Verlag, (2002), 111-122.
13. A. Acan. An External Memory Implementation in Ant Colony Optimization. In *Ant Colony Optimization and Swarm Intelligence - ANTS2004*, page 73-82, Brussels, September 2004.

14. Stützle, T., Fernandes, S.: New benchmark instances for the QAP and the experimental analysis of algorithms. In: Gottlieb, J., Raidl, G. R. (eds.): Evolutionary Computation in combinatorial Optimization - EvoCOP 2004, Springer-Verlag, (2004), 199-209.

15. Stützle, T., Dorigo, M.: ACO Algorithms for the Quadratic Assignment Problem. In: Corne, D., Dorigo, M., Glover, F. (eds.): New Ideas in Optimization, McGraw-Hill, (1999), 33-50.

16. Stützle, T.: MAX-MIN ant system for the quadratic assignment problem. Technical Report AIDA-97-04, Darmstadt University of Technology, Computer Science Dept., Intellectics Group, (1997).

17. Stützle, T.: Iterated local search for the quadratic assignment problem. Technical Report AIDA-99-03, Darmstadt University of Technology, Computer Science Dept., Intellectics Group, (1999).

18. Maniezzo, V.: Exact and approximate nondeterministic tree-search procedures for the quadratic assignment problem. INFORMS Journal on Computing, 11(4), (1999), 358-369.

19. Maniezzo, V., Colorni, A., Dorigo, M.: The ant system applied to the quadratic assignment problem. Technical Report IRIDIA/94-28, Universite Libre de Bruxelles, (1994).

A Novel Application of Evolutionary Computing in Process Systems Engineering

Jessica Andrea Carballido[1,2], Ignacio Ponzoni[1,2], and Nélida Beatriz Brignole[1,2]

[1] Laboratorio de Investigación y Desarrollo en Computación Científica (LIDeCC)
Departamento de Ciencias e Ingeniería de la Computación,
Universidad Nacional del Sur, Av. Alem 1253, 8000, Bahía Blanca, Argentina
[2] Planta Piloto de Ingeniería Química – CONICET,
Complejo CRIBABB, Camino La Carrindanga km.7 CC 717, Bahía Blanca, Argentina
{jac, ip, nbb}@cs.uns.edu.ar

Abstract. In this article we present a Multi-Objective Genetic Algorithm for Initialization (MOGAI) that finds a starting sensor configuration for Observability Analysis (OA), this study being a crucial stage in the design and revamp of process-plant instrumentation. The MOGAI is a binary-coded genetic algorithm with a three-objective fitness function based on cost, reliability and observability metrics. MOGAI's special features are: dynamic adaptive bit-flip mutation and guided generation of the initial population, both giving a special treatment to non-feasible individuals, and an adaptive genotypic convergence criterion to stop the algorithm. The algorithmic behavior was evaluated through the analysis of the mathematical model that represents an ammonia synthesis plant. Its efficacy was assessed by comparing the performance of the OA algorithm with and without MOGAI initialization. The genetic algorithm proved to be advantageous because it led to a significant reduction in the number of iterations required by the OA algorithm.

Keywords: Combinatorial Optimization Problem, PSE, Process-Plant Instrumentation Design, Multi-Objective Genetic Algorithm, Observability Analysis.

1 An Application in the Field of Process Systems Engineering

Process plants are networks of industrial items of equipment physically connected by streams. The instrumentation design problem is a challenging activity in the area of Process Systems Engineering (PSE). It consists in deciding on the most convenient amount, location and type of measuring devices to be incorporated into the industrial process. The objective is to achieve complete knowledge of the plant's operating conditions, while satisfying other goals such as sensor-cost minimization and maximum reliability. Due to the complexity of this task, the development of automatic decision-support tools for this purpose has become a challenge [1].

The computer-aided design of process-plant instrumentation is an iterative procedure that comprises several steps. In the first place, a steady-state mathematical model is built in order to represent plant behavior under stationary operating conditions. This

G.R. Raidl and J. Gottlieb (Eds.): EvoCOP 2005, LNCS 3448, pp. 12–22, 2005.
© Springer-Verlag Berlin Heidelberg 2005

model is a set of algebraic equations that correspond to mass and energy balances, as well as relationships employed to estimate thermodynamic properties like densities, enthalpies, and equilibrium constants. A rigorous model usually involves not only linear functionalities, but also many bilinear and nonlinear equations. Apart from the model, an initial instrument configuration also has to be defined. This preliminary design classifies model variables into measured and unmeasured ones, the former being those whose values will be obtained directly from the sensors.

The next step is to carry out the Observability Analysis (OA), which consists in pinpointing the unmeasured variables that will be observable, i.e. those that can be calculated by means of model equations, regarding the measurements as constants. The OA Algorithm (OAA) used for this purpose [2] analyzes the structural relationships between model equations and unmeasured variables. This analysis is performed by permuting a sparse occurrence matrix built from information about both the model and the measurements in order to obtain a desirable pattern.

It is important to remark that some variables are critical for the industrial process under study because they represent vital information about it (i.e. temperature in a reactor), while others could be considered scarcely relevant. In principle, a careful OA should yield a classification where all the key unmeasured variables are observable. If, after an execution of the OAA, the result contains critical indeterminable variables, the configuration of sensors should be modified and the OA has to be repeated. In this way, the OA normally becomes an iterative procedure.

The last major step required to complete the entire design procedure is the classification of the measurements, also known as redundancy analysis [3]. This task should be carried out only after a satisfactory OA has been achieved.

This paper is focused on the search for an accurate automated OA initialization strategy. The flow diagram of the OA procedure with manual initialization is shown in Fig. 1, where the rectangular boxes represent automated tasks, while the others are associated with expert activities handed over to the decision maker (DM).

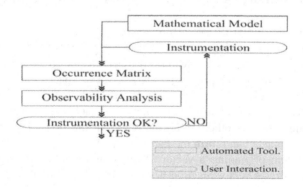

Fig. 1. Iterative process for the OA stage with manual initialization

Both the OA time efficiency and quality of the results depend on the starting point, and will therefore benefit from a careful choice of sensors. Notice that the number of iterations required in order to reach an acceptable result may vary significantly with the initialization. Since the OAA sweeps are very expensive as regards computing time, it is highly advantageous to have as few iterations as possible. This objective can be achieved by choosing an adequate initial instrument configuration. At present, there are no algorithms to make optimal selections in this sense. Therefore, plant engineers choose the sensors exclusively on the basis of their skill and experience. The purpose of this work is to develop an automated tool to tackle this problem, thus supporting them in the making of these complex decisions. The new scheme, shown in Fig. 2, represents the interaction between the DM and the instrumentation design package when the automated tool for initial configuration has been incorporated.

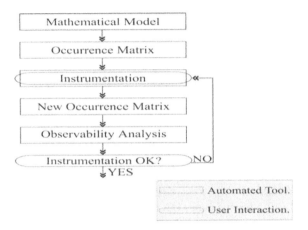

Fig. 2. Iterative process for the OA stage with automated initialization

2   The Sensor Choice as a Combinatorial Optimization Problem

The selection of the initial set of instruments can be classified as a combinatorial optimization problem involving several objectives expressed in different units and in mutual conflict. In particular, these features characterize the so-called multi-objective or multi-criterion optimization problems (MOPs), whose special characteristic is that they have no single solution. There is a set of valid solutions instead, and each one may be considered the solution of the problem. This holds since none of them out-weighs or "dominates" the others in all the objective functions. The valid solutions are called non-dominated and form the Pareto front. The MOP is defined by Osyczka [4] as "the problem of finding a vector of decision variables which satisfies con-straints and optimizes a vector function whose elements represent the objective func-tions. These functions form a mathematical description of performance criteria, which are usually in conflict with each other. Hence, the term 'optimize' means finding such a solution which would give the values of all the objective functions acceptable to the decision maker."

For our particular problem, the conflicting objectives are associated with sensor reliability, purchase and installation costs of the network, and observability level provided by the resulting mathematical model. It is important to remark that the MOGAI is intended as a module for a specialized decision support system, where the DM plays a major role interacting at several points of the procedure as shown in Fig. 2. In this way, the definitive decisions that ultimately determine each configuration are always made by the DM. Then, a multi-objective approach is useful because it not only allows the DM to choose among several feasible alternative solutions, but also helps him weigh various criteria simultaneously. It is interesting to note that, in this case, the DM's expertise can never be totally replaced by an automated tool because there are many subtle, sometimes subjective, aspects impossible to capture with enough detail through mathematical formulations.

There is a wealth of literature about multi-objective optimization techniques [5, 6, 7, *passim*], ranging from the conventional approaches to the evolutionary ones. Traditional methods are very limited [5]. In general, as problem size grows, these strategies are too expensive to allow obtain results in polynomial times. Since Rosenberg [8] pointed out the potential of evolutionary algorithms for MOP solving interest of the evolutionary community in this area has grown enormously. This is justified on the grounds that most real-life problems are MOPs, and also because evolutionary algorithms have the inherent capability of finding the Pareto front in reasonable times [5]. Genetic algorithms (GAs) are particularly suitable for MOPs because they simultaneously deal with a set of possible solutions (population). Thus, several members of the Pareto optimal set can be found in a single run, instead of having to perform various runs, as is the case of the traditional mathematical programming methods. Moreover, in comparison with the typical optimization methods, GAs are less susceptible to shape or continuity, easily dealing with discontinuous or concave Pareto fronts.

3 Main Objective and Proposal

In this work we describe a new automated tool, whose purpose is to find a satisfactory initial sensor network configuration for process plants so that the number of iterations involved in the OA is reduced. In this case, a configuration is considered desirable when it is cheap, reliable and meaningful in the sense that it should provide as much plant information as possible. At the same time, short computing times were required, this being a standard demand for any initialization method. When there are several conflicting objectives, the notion of "optimum" means that we are really trying to find a good trade-off solution that fulfills all the targets as satisfactorily as possible.

Several aspects of our specific application led us to use a GA. First of all, it is not imperative for us to find an optimum, a solution near the Pareto front being good enough. Besides, we need various candidate solutions for the DM to make the final decision. Finally, the tool must be fast and efficacious for huge problem instances.

Founded on the reasons explained above, we decided to implement a multi-objective genetic algorithm based on an aggregative non Pareto method. The technique combines (or "aggregates") all the objectives into a single one, without incorporating the concept of Pareto optima directly. This approach was adopted mainly because it is efficient and works especially well for a small number of objectives [7].

4   The Genetic Algorithm

The input of the MOGAI for OA initialization is the occurrence matrix O built from the steady-state mathematical model of the plant under study. O's rows and columns respectively correspond to model equations and variables. In principle, every process variable, namely temperatures, pressures, flow rates and compositions, is associated to a sensor that could be chosen for its measurement. So, the GA also needs information about the cost and reliability of each potential instrument. In this work, the cost of measuring a variable was calculated as the price of the device plus its installation costs, while the reliability of a variable was considered inversely proportional to the instrument's average error reported by the manufacturer. This information was loaded in **N**-dimensional vectors, where **N** is the total number of model variables.

Representation of Potential Solutions to the Problem - Main Operators. The individuals were represented in the canonical (binary) form. Bit-flip mutation and one-point crossover operators were employed. Each genotype, represented here by the symbol i, should be interpreted as an entire sensor configuration, where a nonzero value on one of the bits means that the variable on that position should be measured. The string length is equal to the total amount of variables (**N**) in the model.

Parameter Control. An excellent review about parameter control was published by Eiben et al. [9]. They support the statement that any static set of parameters, i.e. one whose values remain fixed throughout a run, is in principle inappropriate. For this reason, we decided to explore parameter control techniques as an alternative. In particular, we obtained good results by using an adaptive mutation operator inspired in the one proposed by Bäck [10]. In our case, we defined an initial mutation probability equal to $1/l$, where l is the length of the chromosome. This quantity is decreased during the evolution so as to increase exploitation as the algorithm evolves. When this parameter is applied, it is combined with an adaptive value based on the fitness of the individual to be mutated. The idea behind this operator is to give more chances of mutation to those individuals that are far from the optimum. As we shall see later, there is a utopian optimum for our fitness function, whose value is equal to the number of objectives.

Infeasible Individuals. An individual is not feasible when it contains a non-zero in a position that represents an unmeasurable variable, such as an enthalpy. We have given special treatment to infeasible individuals as follows. First, the initial population was generated with a restriction on the positions of the unmeasurable variables, which were always initialized with a zero. In addition, as new gene data could be introduced only through the mutation operator, those positions were regarded as "non-mutable". In this case, it was essential to implement this policy instead of applying penalties since the first test runs, where the generation of infeasible individuals was allowed, resulted in populations with too few valid individuals, at most 30%. Furthermore, these variables must be coded since they have to be present when the observability term of the fitness function is calculated, as will be discussed later.

Selection and Replacement. The selection method is based on the roulette wheel approach, which picks out the individuals that will constitute the parent pool according to the value of the objective function. The chosen individuals replace the old ones, building a new population that in turn undergoes crossover and mutation.

The Convergence Criterion. From a recent review about stopping criteria for GAs [11] it is clear that, in general terms, it is unadvisable to run a GA for a fixed number of generations. For this reason we decided to implement an adaptive termination condition based on the concept of schemata. A schema is a template that establishes similarities among chromosomes. It is represented through a string of symbols in {0. 1, #}, where # is a wildcard. For example, string 011001 is an instance of the schemata 01##0#. As stated by Radcliffe [12], when two parents are instances of the same schema, the offspring will also be an instance of that schema. In particular, if the schema carries good fitness to its instances, the whole population will tend to converge over the bits defined by that schema. Once convergence has been reached, all the offspring will be instances of that schema. Thus, the solution will also be an instance of that schema. For this reason, our criterion analyzes the genotypes until a high percentage becomes an instance of the same schema. For general information on genotypic termination criteria, see [13].

4.1 The Multi-objective Fitness Function

This algorithm aims at finding the individual that simultaneously exhibits the best trade-off performance with respect to the following three objective functions:

The Cost Term. Given a cost vector cv of length N, the total cost of an individual is the sum of the values of all the elements in cv that correspond to nonzero entries in i.

$$C\,(i) = \sum_{j=1}^{N} (cv[\,j\,]*i[\,j\,]) \ . \tag{1}$$

The Reliability Term. Given a reliability vector rv, and following the same line of reasoning, we have:

$$R\,(i) = \sum_{k=1}^{N} (rv[k]*i[k]) \ . \tag{2}$$

The Observability Term. In contrast with the other two objective functions, this one cannot be calculated in a straightforward way. Its estimation was based on the mathematical operation called Forward Triangularization (FT). Details on the FT procedure can be found in [2]. FT returns estimates on the number of unmeasured variables that can be directly calculated by solving individual equations from the system of algebraic equations, given the measurements defined in i. In short, the value returned by the observability function is:

$$Ob\,(i) = \mathrm{FT}\,(i). \tag{3}$$

The FT algorithm is the basic core of the OAA. The latter also includes other modules with more rigorous analysis tools, whose purpose is to refine the FT results at the expense of much higher computing times. Then, in view of its short run times, the FT constitutes an ideal criterion for an initialization algorithm.

The Merging Approach. The aggregating policy for the construction of the fitness function requires a criterion to reconcile the values of all the individual objectives, judiciously combining them so that none is undervalued. The standard procedure consists in normalizing each of them in the [0,1] range. Therefore, in this paper, the fitness function F was defined in terms of the three normalized objectives as follows:

$$F(i) = NR(i) + NOb(i) + 1 - NC(i) .\tag{4}$$

Our algorithm tends to maximize F, its values always lying between 0 and the total number of individual objectives. Equation 4 can be naturally expanded to meet this requirement for a greater number of objectives in the following way:

$$F(i) = \sum_{p=1}^{n} NOM_p + m - \sum_{q=1}^{m} NOm_q .\tag{5}$$

where n and m are the number of objectives to be maximized or minimized, respectively, $NOM_p \in [0, 1]$ is the pth normalized objective to be maximized, $NOm_q \in [0, 1]$ is the qth normalized objective to be minimized, and $F(i) \in [0, n+m]$.

The optimal (utopian) situation, i.e. $F(i) = n+m$, occurs when all the objectives to be maximized are equal to 1, while those to be minimized become 0. It should be noted that these features are remarkably advantageous. First of all, the expansion to consider additional objectives is straightforward. Besides, F moves within a closed bounded range of values, thus providing a clear threshold to be reached.

Number of Evaluations. The MOGAI evaluates F only when necessary. Whenever an individual remains unchanged from one generation to the other, its fitness value is not recalculated. Implementing this feature led to 10% savings in the number of evaluations, thus proportionally reducing the execution time of a complete GA run.

5 Experimentation

Brief Description of the Plant Under Analysis: The algorithmic performance was assessed by carrying out the instrumentation analysis of an industrial plant whose main features are described in Bike [14]. The plant produces 1500 ton/day of anhydrous liquid ammonia at 240 K and 450kPa with a minimum purity of 99.5%. The product is obtained by means of the Haber-Bosch process, which consists in a medium-pressure synthesis in a catalytic reactor followed by an absorption procedure that removes the ammonia with water. The liquid output from the absorber enters a distillation column that yields pure ammonia as top product. The plant also contains a sector with membranes, where hydrogen is recovered and then recycled to the feed. The rigorous mathematical model of this plant, used to build the occurrence matrix was generated using the ModGen package [15]. The resulting system contained 557 non-linear algebraic equations and 546 process variables.

The MOGAI Parameters: The population size was fixed in 100 individuals. Crossover probability was set at 0.7. The initial mutation probability was 0.0018 and, as explained above, it was forced to decrease as the algorithm evolved, its value being also combined with the fitness of the individual. The genotype length **N**, which amounted to the total number of process variables, was 546.

Some Industrial Results: Both the feasibility and convenience of using the MOGAI as an initialization tool for structural OAAs were evaluated through a detailed study of the ammonia plant. The most promising classifications obtained from a MOGAI run were analyzed. The results were compared in terms of sensor acquisition costs, reliability of the chosen instrumentation, and level of knowledge about the process obtainable both through direct measurements and estimations carried out from the model equations. With these guidelines, the most convenient initialization for the rigorous OA was selected. A complete OA process was executed next, and the results were compared against the configuration without automatic initialization suggested by Ponzoni et al. [16].

For the first stage, the three solutions whose features are summarized in Table 1 were selected. The letters M, O and I indicate the number of measured, observable and indeterminable variables, respectively. It can be observed that all the fitness values are satisfactory since they are close to 3, which is the upper bound for **F**. In all cases, the reliability of the resulting configuration is greater than 99%. With respect to costs, B is significantly cheaper than A or C. However, in terms of observability, the results indicate that A and C are preferable since they have a lower number of indeterminable variables.

Table 1. Three MOGAI solutions

Config.	Fitness Value	Cost	Observability		
			M	O	I
A	2.538	$25,168	105	286	155
B	2.502	$12,642	92	275	179
C	2.512	$24,343	104	298	144

In order to complete the analysis, it is necessary to determine which indeterminable variables are critical, since their values should be known accurately in the final configuration. Table 2 shows the distribution of the critical variables and the incidence of measuring them in the final cost of the instrumentation. The expression C indicates the increment in the cost associated to the purchase and installation of the sensors that wouldmeasure all the indeterminable critical variables.

Table 2. Details on the critical variables for the solutions in Table 1

Config.	Critical Variables			Cost	
	M	O	I	C.	Final C.
A	7	10	12	$ 3,645	$28,813
B	6	10	13	$ 4,364	$17,006
C	7	10	12	$ 3,134	$27,477

From the number of indeterminable variables present in A and B (see Table 1) one could infer that the best configuration is C. However, from Table 2 it is clear that the number of indeterminable variables of interest is, in fact, similar. Hence, the most significant difference lies in the cost, thus favoring configuration B. If it is assumed that the final configuration should have no undeterminable critical variables, it becomes necessary to introduce in the analysis the costs derived from the addition of sensors to monitor all of them. The minimum cost increment associated to the incorporation of these measurements corresponds to configuration C. However, this is not enough to compensate the original difference in costs. Then, all in all, it can be concluded that B is the most convenient alternative.

Finally, in Table 3 the results obtained after carrying out the rigorous OA initialized with configuration B, are compared against the instrumentation reported in Ponzoni et al. [16] (configuration P). The OAA employed in those experiments was a GS-FLCN implementation [2] with manual initialization.

Table 3. Concluding Results

Config.	Cost	Observability		
		M	O	I
P	$ 14,772	52	257	237
B	$ 17,006	105	289	152

In terms of cost, configuration P seems to be more convenient. However, this choice leaves 9 undeterminable critical variables. If we added sensors at those points and considered the corresponding cost increments, the budget would raise to $ 17,922, thus becoming more expensive than B's. Furthermore, the use of the MOGAI also leads to better knowledge about the process, with a reduction over 40% in the total number of non-observable variables.

The use of the MOGAI as an initialization tool implies gains in both time and effort, also improving the reliability of the results by taking into account a higher number of interest factors. For this industrial case, the total amount of time required by the complete OA procedure was reduced in 83% thanks to the automatic initialization. The number of OA iterations diminished and therefore, there was a decrease in the effort the DM had to make for his analysis. More specifically, for the ammonia synthesis plant, the average run times of the MOGAI in a PC Pentium IV (2.8 GHz) amounted to approximately 150 seconds, while a complete iteration of the OA cycle normally takes more than an hour. This shows that the computational effort invested in making an automatic initialization is negligible in comparison with the order of magnitude of the times required by an OA iteration.

6 Conclusions

In this article we tackled the problem of selecting the best configuration of sensors to instrument an ammonia synthesis plant in order to assess the convenience of applying the MOGAI as an automated tool for initialization purposes. The objective function of the GA contemplates terms associated to cost, reliability and observability.

From the comparative analysis of the results achieved with and without automated initialization, it is possible to conclude that the use of the MOGAI makes the design methodology more efficient. Moreover, the automated initialization leads to results of higher quality by directing the search to the simultaneous fulfillment of several objectives.

Acknowledgments

The authors would like to express their acknowledgment to the "Agencia Nacional de Promoción Científica y Tecnológica" from Argentina, for their economic support given through Grant N°11-12778. It was awarded to the research project entitled "Procesamiento paralelo distribuido aplicado a ingeniería de procesos" (ANPCYT Res N°117/2003) as part of the "Programa de Modernización Tecnológica, Contrato de Préstamo BID 1201/OC-AR".

References

1. Vazquez, G.E., Ferraro S.J., Carballido J.A., Ponzoni I, Sánchez M.C., Brignole N.B.: The Software Architecture of a Decision Support System for Process Plant Instrumentation. WSEAS Transactions on Computers, 4, 2, (2003), 1074-1079.
2. Ponzoni I., Sánchez M.C., Brignole N.B.: A New Structural Algorithm for Observability Classification. Ind. Eng. Chem. Res., 38, 8, (1999), 3027-3035.
3. Ferraro S.J., Ponzoni I, Sánchez M.C., Brignole N.B.: A Symbolic Derivation Approach for Redundancy Analysis. Ind. Eng. Chem. Res., 41, 23, (2002), 5692-5701.
4. Osyczka A.: Multicriterion Optimization in Engineering with FORTRAN Programs. Ellis Horwood Limited, (1984).
5. Toscano Pulido G.: Optimización Multiobjetivo Usando Un Micro Algoritmo Genético. Tesis de Maestría en Inteligencia Artificial, Universidad Veracruzana LANIA, (2001).
6. Fonseca C.M.: Multiobjective Genetic Algorithms with Application to Control Engineering Problems. PhD Thesis, Department of Automatic Control and Systems Engineering University of Sheeld, (1995).
7. Coello Coello C.A.: A Comprehensive Survey of Evolutionary-Based Multiobjective Optimization Techniques. Knowledge and Information Systems, An International Journal, 1, 3, (1999), 269–308.

8. Rosenberg R.S.: Simulation of Genetic Populations with Biochemical Properties. PhD thesis, University of Michigan, Ann Harbor, Michigan, (1967).
9. Eiben A.E., Hinterding R., Michalewicz Z.: Parameter Control in Evolutionary Algorithms IEEE Transactions on Evolutionary Computation, 3, 2, (1999), 124-141.
10. Bäck, T., Hammel, U., Schwefel, H.P.: Evolutionary Computation: Comments on the History and Current State, IEEE Transactions on Evolutionary Computation, 1, 1, (1997), 3-17.
11. Safe M., Carballido J., Ponzoni I., Brignole N.B.: On Stopping Criteria for Genetic Algorithms. In: Bazzan, A., Labidi, S. (eds.): Advances in Artificial Intelligence SBIA 2004 Lecture Notes in Artificial Intelligence Vol. 3171 Springer-Verlag, Berlin Heidelberg New York (2004) 405–413.
12. Radcliffe N.J.: Equivalence Class Analysis of Genetic Algorithms. Complex Systems 5, (1991).
13. Michalewicz, Z.: Genetic Algorithms + Data Structures = Evolution Programs. 3rd edn. Springer-Verlag, Berlin Heidelberg New York, (1996).
14. Bike S.: Design of an Ammonia Synthesis Plant, CACHE Case Study. Department of Chemical Engineering, Carnegie Mellon University, (1985).
15. Vazquez G.E., Ponzoni I., Sánchez M.C., Brignole N.B.: ModGen: A Model Generator for Instrumentation Analysis. Advances in Engineering Software, 32, (2001), 37-48.
16. Ponzoni I., Brignole N.B., Bandoni J.A.: Estudio de Instrumentación para una Planta de Producción de Amoníaco empleando un Nuevo Algoritmo de Clasificación. AADECA´98, Argentina, 1, (1998) 59-64.

Choosing the Fittest Subset of Low Level Heuristics in a Hyperheuristic Framework

Konstantin Chakhlevitch and Peter Cowling

MOSAIC Research Centre,
Department of Computing,
University of Bradford,
Bradford BD7 1DP, United Kingdom
{K.Chakhlevitch, P.I.Cowling}@Bradford.ac.uk
http://www.mosaic.brad.ac.uk

Abstract. A hyperheuristic is a high level procedure which searches over a space of low level heuristics rather than directly over the space of problem solutions. The sequence of low level heuristics, applied in an order which is intelligently determined by the hyperheuristic, form a solution method for the problem. In this paper, we consider a hyperheuristic-based methodology where a large set of low level heuristics is constructed by combining simple selection rules. Given sufficient time, this approach is able to achieve high quality results for a real-world personnel scheduling problem. However, some low level heuristics in the set do not make valuable contributions to the search and only slow down the solution process. We introduce learning strategies into hyperheuristics in order to select a fit subset of low level heuristics tailored to a particular problem instance. We compare a range of selection approaches applied to a varied collection of real-world personnel scheduling problem instances.

1   Introduction

Personnel scheduling involves the allocation of the available workforce to timeslots and locations and the assignment of jobs to members of staff. Real-world personnel scheduling problems are NP-hard combinatorial optimisation problems with many complex constraints and a huge number of feasible solutions. Heuristic methods able to find solutions of a good quality in a limited time are often used in personnel scheduling. A recent review of staff scheduling models and methods is given in [1] where the authors identify the development of more general and flexible algorithms as one of the principal research goals. Such algorithms should be robust enough to cope with the changes in problem specifications and scheduling environment [1].

In [2], Burke et al. present an overview of the latest work in the area of *hyperheuristics* – an important direction towards the generalisation of the search methodology. A hyperheuristic is an approach which operates at a higher level of abstraction than a metaheuristic and intelligently chooses the appropriate low level heuristic from a given set depending upon the current state of the

G.R. Raidl and J. Gottlieb (Eds.): EvoCOP 2005, LNCS 3448, pp. 23–33, 2005.
© Springer-Verlag Berlin Heidelberg 2005

problem [3]. Problem-specific information is concentrated in a set of low level heuristics (which represent simple local search neighbourhoods or dispatching rules) and in the objective function(s) of the problem. All that a hyperheuristic requires to make its decisions is the objective value(s) following the application of each low level heuristic and possibly the CPU time used by the low level heuristic. This makes a hyperheuristic robust and highly flexible.

Several hyperheuristic approaches for real-world scheduling problems use metaheuristics as high level procedure. Cowling et al. [8] develop a hyper-GA approach for a personnel scheduling problem. Their approach is based on genetic algorithm (GA) where the chromosome represents a sequence of low level heuristics. Heuristics are applied in the order given by the chromosome. Han et al. [9] further improve the hyper-GA's performance by enabling the chromosome length to change adaptively during the hyperheuristic run. The method of evolving heuristic choice is successfully implemented by Hart et al. in [7] for a complex scheduling problem of chicken catching and transportation. Burke et al. [10] present a tabu search based hyperheuristic applied to nurse rostering and university timetabling problems where a tabu list of low level heuristics with poor performance is maintained.

Another group of hyperheuristics employs learning mechanisms for making decisions. Gratch and Chien [4] consider a statistical approach to adaptively solve the real-world problem of scheduling satellite communications. Nareyek [5] presents a weight adaptation method which learns how to select attractive low level heuristics during the search. Ross et al. [6] use a learning classifier system based on an evolutionary algorithm to learn a solution process for various instances of the bin-packing problem. Their method determines an order in which simple heuristics should be applied to solve particular problem instance. Cowling et al. ([3], [11]) investigate a hyperheuristic based on statistical ranking of low level heuristics. In this method, historical information about the recent performance of low level heuristics is accumulated in a choice function. The selection of low level heuristic at each decision point depends on the current value of the corresponding choice function.

In [13], we present a range of hyperheuristics where greedy and random methods are mixed in order to maintain a good balance between intensification and diversification of the search. We also show how hyperheuristic approaches can be enhanced by adding tabu lists of recently applied low level heuristics or recently modified events. Since for complex real-world problems there is generally no obvious choice of low level heuristics, we introduce a scheme for designing a large set of low level heuristics for a personnel scheduling problem. Low level heuristics represent all possible combinations of simple selection rules for events and resources. The results of hyperheuristic runs for a real-world personnel scheduling problem are promising. The disadvantage of the hyperheuristic methods described in [13] is that they are relatively slow since selection of low level heuristic to apply at each decision point involves examining all heuristics from a large set. In [14], we analyse the behaviour of individual low level heuristics and their contribution to the construction of the solution. We conclude that some low level heuristics from the set are fitted better for the particular problem instance than others. The effectiveness of each low level heuristic may vary a great deal from instance to instance and is hard to predict. These conclusions have motivated us to investigate learning strategies for choosing the subset of the fittest low level heuristics. This is the main contribution of this paper. Learning low level heuristics allows us not only to get rid of a significant amount of redundancy in our approach, but, given similar amount of CPU time, to improve further the results previously achieved for all instances of the trainer scheduling problem.

The paper is organised as follows. In the next three sections we briefly formulate the problem and describe our set of low level heuristics and hyperheuristic algorithms. The learning approaches are introduced in Section 5. In Section 6 we analyse the results of all approaches. Section 7 concludes the paper.

2   The Trainer Scheduling Problem

The trainer scheduling problem arises in a large financial institution which regularly organises training for its staff. The problem involves assigning a number of compound events of three different types (i.e. training courses, meetings and projects) to a limited number of training staff, locations, and timeslots. Each timeslot represents a day of the week. We must provide a schedule for 3 months activity. A numerical priority is specified for each event which reflects management's view of the event's utility. The travel of each trainer is penalised depending on the time taken to travel from the home location of the trainer to the location where the event is conducted. The objective is to maximise the total priority for scheduled events while minimising the total travel penalty for the training staff.

The problem is heavily constrained due to a number of limitations related to possible trainers and locations for the events, availability of trainers and rooms, room types and capacities, time windows and durations for the events, workloads for trainers etc (see [13] for details). The integer programming formulation of a much simplified version of a similar problem is given in [8].

3   Low Level Heuristics

We divide events into two subsets: already scheduled and not yet scheduled. For each category of events we develop separate lists of event selection rules and resource selection rules. Selection rules are chosen in such a way as to reflect the subgoals of the objective function and the constraints with some random selection for diversification. For example, the criteria for selection events which are not yet scheduled can be highest priority, smallest number of possible trainers or locations etc. For already scheduled events we employ such criteria as highest travel penalty, widest time window, largest number of possible trainers or locations and others. The resource selection rules concern the selection of alternative locations, trainers, timeslots or their combinations. Selecting different resources for the event, we intend to reduce the travel penalty for the event and free up resources which can be used later for scheduling other events. We refer to [13] for further details and complete lists of selection rules.

Combining ("multiplying") event selection rules with resource selection rules for each category of events, we construct a set of 95 low level heuristics (25 heuristics for not scheduled events and 70 heuristics for scheduled events). Such an approach is quite easy to implement since different heuristics can use the same event or resource selection rules as their components. We create only 27 pieces of code representing different event/resource selection mechanisms, and only 5 of these pieces of code are substantially different to each other.

4   Hyperheuristics

In this section we describe our hyperheuristics.

Hyperheuristic Greedy selects and applies at each iteration the low level heuristic either providing the greatest improvement to the objective function or leading to the smallest deterioration (or yielding zero improvement) if there are no improving heuristics. Note that improvements upon the current objective value, not upon the best value found so far are considered for all hyperheuristics in this section. Ties are broken randomly.

Hyperheuristics from the "peckish" group combine random and greedy methods [12]. Hyperheuristic Peckish1 (P1) randomly selects a low level heuristic at each iteration from the candidate list of low level heuristics which improve the current solution or from the whole set of low level heuristics if the candidate list is empty. Hyperheuristic Peckish2 (P2) randomly selects a low level heuristic from the candidate list which contains the n best (not necessarily improving) heuristics. Hyperheuristic Peckish3 (P3) attempts to form the candidate list of only improving heuristics and if such a list is not empty, randomly selects a low level heuristic from it. Otherwise, random selection from the candidate list of n best non-improving heuristics is applied. In hyperheuristic Peckish4 (P4) the candidate list size n is dynamically changed. It is initially set to 1. If at some iteration the improving low level heuristics exist and one of them is applied, the candidate list size n is reset to 1. Otherwise, a low level heuristic is randomly selected from n best non-improving heuristics and candidate list size is incremented. The candidate list size n determines how "greedy" and how "random" the peckish hyperheuristic is – increasing n adds randomness and decreasing n makes the hyperheuristic more greedy.

Hyperheuristics from the next group are based on the ideas of tabu search (see [15]). Hyperheuristic TabuHeuristic (TH) employs a tabu list of recently called heuristics. The size of the tabu list is fixed and set to some prespecified value. The algorithm greedily selects the best low level heuristic at each iteration. If such a heuristic leads to an improved objective value, it is always applied and released from the tabu list if there; a non-improving heuristic is chosen only if it is not in the tabu list and immediately becomes tabu after its application. Hyperheuristic TabuEvent (TE) is similar to TabuHeuristic but the tabu list holds recently selected events. In hyperheuristics TabuHeuristicAdaptive (THA) and TabuEventAdaptive (TEA) the tabu list size is changed adaptively as the search progresses (see [13]).

Note that all hyperheuristics above ensure that the solution is *changed* at each iteration by discarding low level heuristics failing to find any alternative resources for the selected event.

5   Learning Techniques

We have tested two approaches to select the subset of promising low level heuristics from a large superset:

1. Warming up approach (WU): the hyperheuristic identifies the specified number of the fittest low level heuristics during some warm up period (given by the number of iterations) and then either continues its run until a stopping condition is met (WU-C) or restarts from a known initial solution with a reduced set of low level heuristics (WU-R).
2. Step-by-step reduction (SSR) approach: the hyperheuristic gradually reduces the set of low level heuristics during its run until some number of the fittest low level heuristics remain in the set and continues with a reduced set until a stopping condition is met.

The following parameters should be specified for the step-by-step reduction:

- $h_{\min}$ – the minimum possible number of low level heuristics remaining in the subset. In other words, when after several reduction steps the number of the fittest low level heuristics in the reduced set reaches $h_{\min}$, no further reductions are allowed.
- s – reduction step, i.e. the number of iterations between two successive reductions.
- $f \in [0.9; 1)$ – reduction factor, i.e. $h_{after} = [h_{before} * f]$, where h_{after} and h_{before} are the number of low level heuristics in the set after and before reduction respectively and $[x]$ denotes the integer nearest to x.

Selecting different values of s and/or f, we can control the speed of reduction. Increasing s or f makes the reduction slower, decreasing one of the parameters speeds up the process.

Note that both WU and SSR approaches preserve the same ratio of low level heuristics for not yet scheduled events and for already scheduled events in the reduced set as in the whole set of 95 low level heuristics. Since this ratio in the superset is approximately $1 : 3$, the algorithm consecutively takes off three low level heuristics for already scheduled events from the set before one low level heuristic for not scheduled events is discarded.

We use *total improvement* as a fitness criterion for learning low level heuristics. For each low level heuristic, the hyperheuristic accumulates the corresponding value of improvement returned at each iteration. When a reduction occurs, the low level heuristics with the smallest total improvement over all previous iterations are discarded.

Table 1. Real-world datasets for the trainer scheduling problem

	Dataset1	Dataset2	Dataset3	Dataset4	Dataset5	Dataset6
Courses	224	147	224	147	83	161
Meetings	0	0	37	71	27	58
Projects	0	0	71	169	31	171
Trainers	53	54	53	54	54	47
Locations	16	16	16	16	19	16
Rooms	37	39	37	39	35	36

6   Experiments and Results

This section describes input data and experimental settings for our methods and contains the comparative analysis of the results of computational experiments.

Input Data. We use real datasets provided by a financial institution. Table 1 presents a summary for 6 datasets used in our experiments. The first two datasets have been used in our previous work (see [13]) and represent simplified versions of datasets 3 and 4 since only training courses have to be scheduled. This paper adds realism to the model of [13] by including project and meeting events and their associated constraints. The meetings are particularly difficult to schedule since the presence of a large number of trainers may be required on the same day and at the same location.

Implementation. The problem model, low level heuristics and hyperheuristics were implemented in Microsoft Visual C++ making good use of object orientation and the experiments were run on a Pentium 4 1600MHz PC with 640Mb RAM running under Windows 2000. A single experiment for hyperheuristic consists of 10 runs with different random seeds and starting from the same initial solution. The stopping condition is 500 iterations for the experiments with the whole set of low level heuristics and 1000 iterations when the reduced sets are used. The set of 95 low level heuristics is reduced down to 20 heuristics when learning strategies are applied and the warming up period length is 200 iterations.

Initial schedules are constructed using a greedy heuristic which takes events one by one and assigns the combination of resources yielding the lowest travel penalty for each event until all events have been tried. The events are pre-sorted in descending order of their priority. If the meetings are present in a dataset, ties in priority are broken by descending order of the number of trainers involved to ensure that the meetings have been considered early in schedule construction. The upper bound of all possible schedules for each problem instance represents the solution for the relaxed version of the problem where all constraints on availability of trainers and rooms, room types and capacities, on starting times for the events and their time windows are ignored (see [13] fore more details).

Table 2. Average performance of hyperheuristics applied to a whole set of 95 low level heuristics (WS) and after learning the fittest low level heuristics using step-by-step reduction approach (SSR) given in distance from the upper bound

Hyper-heuristic	Dataset 1		Dataset 2		Dataset 3		Dataset 4		Dataset 5		Dataset 6	
	WS	SSR	WS	SSR	WS	SSR	WS	SSR	WS	SSR	WS	SSR
Initial	4352		6727		13880		24210		9968		29610	
Greedy	751	541	2382	1507	4931	3941	8197	6911	3188	2197	5741	5064
P1	517	503	2398	2033	4856	4336	8085	**5091**	3687	3690	5944	**3406**
P2(25)	1061	556	2842	2092	5910	4901	9432	6644	4091	2295	4998	4090
P3(25)	567	269	2793	2044	5634	4403	7750	5586	3988	2296	5281	4497
P4	941	465	2584	1782	4675	4114	7533	5468	4285	2194	5870	4195
TH(30)	746	362	1990	1770	4623	3760	7775	7106	3687	2595	6022	5378
THA(45)	749	276	2289	1600	4435	3945	7115	7863	4287	3089	6340	5025
TE(N/2)	553	**256**	1973	**1317**	4237	**3241**	7642	6525	3093	1799	4946	4670
TEA(45)	555	360	2185	1693	4259	3281	7568	7027	3289	**1797**	5526	5047

Results. The average results for hyperheuristics managing a large collection of 95 low level heuristics (denoted by WS) and a reduced set of low level heuristics selected by step-by-step reduction approach (SSR) are compared in Table 2. We choose SSR as the approach producing better results on average than both versions of the WU method (see Figure 1). For SSR method the following values of parameters are choosen based on empirical tests: $h_{min} = 20$, $s = 30$, $f = 0.90$ (fast reduction). These settings provide the most consistent outcomes (although not necessarily the best results for every hyperheuristic and dataset). The figures in Table 2 represent the distances from the corresponding upper bound. The numbers in parentheses after some of hyperheuristics' names are the values of either candidate list size or tabu list size. Hyperheuristics are not particularly sensitive to the values of these parameters and we use moderate values for illustration purposes. The only exception is hyperheuristic TabuEvent which produces the most consistent outcomes when the tabu list may contain up to a half of all the events in the dataset. This is denoted by $N/2$ in Table 2, where N is the number of events in the dataset. It is evident from the table that the learning technique embedded into a hyperheuristic leads to significant improvements in the quality of the schedules. The small deviations in the objective values observed represent practically very significant differences in trainer inconvenience due to additional travel, or additional low priority scheduled events. Analysing the performance of individual hyperheuristics with learning, we can notice from Table 2 that hyperheuristics whose behaviour is primarily greedy (Greedy and tabu list based) outperform their peckish counterparts for Datasets 1,2,3 and 5 with the best results consistently delivered by TabuEvent hyperheuristic. For the most hard-to-schedule Datasets 4 and 6 peckish hyperheuristics are more suc-

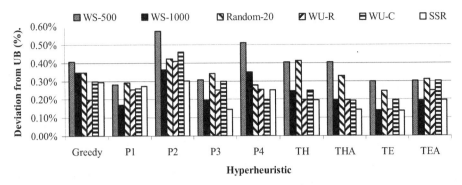

Fig. 1. Comparison of hyperheuristic approaches for Dataset 1 (deviation in % from the upper bound)

cessful, especially Peckish1 where the random component dominates and more frequent calls of low level heuristics which worsen the objective value are possible. This is not surprising since for tightly constrained datasets it is difficult to find alternative resources without increasing travel penalties. Scheduling a new event may require a few rescheduling moves which worsen the current schedule by increasing penalties. Hyperheuristics with a greater degree of randomness accept such moves more often than more greedy hyperheuristics. This provides a smaller number of not scheduled events in the best schedule. On the other hand, the presence of a greedy component in a peckish hyperheuristic guarantees that a worse move will not be accepted too often and therefore the penalties will not be significantly damaged. For "easier" Datasets 1,2,3 and 5 there is more freedom to find resources and greedier hyperheuristics perform better. Figure 1 demonstrates the difference in performance of hyperheuristic approaches with and without learning for Dataset 1. The results for other datasets follow similar patterns. It is clear that SSR is the best method in most occasions. In order to check the quality of low level heuristics selected by SSR algorithm, we have conducted a range of experiments with the reduced sets of 20 randomly selected low level heuristics ("Random-20" columns in Figure 1). The results of these experiments are rather erratic: the high quality solutions comparable to those produced by SSR approach are present along with very poor, unacceptable ones. For example, the worst objective value achieved by hyperheuristic Greedy for Dataset 1 using the Random-20 approach is unacceptable 3227 from the upper bound whereas SSR is never more than 1214 from the bound. These results give strong evidence for the importance of the right selection of low level heuristics in the subset. We can also notice from Figure 1 that WU strategies are less consistent than SSR and their performance varies widely for different hyperheuristics. Comparing the results of WS approach for 500 and 1000 iterations (WS-500 and WS-1000 columns in Figure 1) with those of SSR approach, we can observe that although further improvements have been recorded during additional 500 iterations for the hyperheuristics applied to a large set of low level heuristics, the solutions obtained are generally slightly worse than that of hyperheuristics employing the SSR strategy. The WS-500 and SSR approaches use similar amounts of CPU time, although the SSR approaches carry out 1000 iterations in their time compared to WS-500's 500 iterations.

The low level heuristics in the reduced set are usually those yielding the most frequent and valuable improvements to the current solution since the selection criterion used in the SSR approach is the total improvement achieved by the low level heuristic. The most effective event selection rules for not scheduled events involve random selection of events and selection based on priority. Other event selection rules are also present in the reduced set but less often and depending on the hyperheuristic and dataset. The resource selection rules for not scheduled events are distributed quite evenly among the best low level heuristics. For already scheduled events we observe different preferences in selection rules. All event selection rules are usually represented in the reduced set of low level heuristics although random selection, selection of events with the highest priority and with the highest travel penalty are present in higher ratios. Only four resource selection rules for already scheduled events are contained in low level heuristics from the reduced set. Three of them are well suited to the total improvement criterion selecting appropriate trainers, locations or their combinations which reduce the travel penalty for the selected event. The fourth rule selects alternative timeslots for scheduled events and always yields zero improvement.

Summarising the above analysis, we observe that the reduced sets of low level heuristics obtained by applying our learning approaches are similar for different datasets. It appears that the choice of low level heuristics depends primarily on the method used to select the reduced subset and to a lesser extent on the problem instance and hyperheuristic. Successful individual runs of hyperheuristics with a subsets of randomly selected low level heuristics support the idea that even greater diversity could be useful. Therefore, the development of more advanced and comprehensive learning mechanisms is an interesting research direction, possibly employing weight adaptation schemes similar to those presented by Nareyek in [5] or advanced scoring systems for ranking low level heuristics.

7   Conclusions

Since a hyperheuristic accumulates knowledge about low level heuristic performance rather than directly about the problem, there is growing evidence that the approach is robust for a variety of problem instances and across problem domains. In this paper, we have added to this evidence. We have presented a range of hyperheuristic algorithms managing a large collection of low level heuristics constructed by combining simple selection rules. In order to make the methodology faster and more effective, we have introduced learning techniques into hyperheuristics which allow us to reduce the set of low level heuristics leaving only those most likely providing regular improvements to the current solution for the particular problem instance. The experimental study shows that our hyperheuristics produce high quality results for difficult real-world instances of personnel scheduling problem. Hyperheuristics with learning outperform their counterparts dealing with a large set of low level heuristics both in terms of solution quality and CPU time.

The methodology presented in this paper has the potential to solve other complex real-world scheduling problems. By considering manual schedule generation and repair techniques, it is usually straightforward to design selection rules and therefore to form a set of low level heuristics, even when the problem structure is poorly understood. The choice of which selection rules will work in an automated system is difficult, and we have presented and compared several methods to identify effective low level heuristics from a large set in this paper. Investigating the robustness of the method across other problem domains is our primary goal for the near future.

References

1. Ernst, A.T., Jiang, H., Krishnamoorthy, M., Sier, D.: Staff Scheduling and Rostering: A Review of Applications, Methods and Models. European Journal of Operational Research **153** (2004) 3-27

2. Burke, E., Kendall, G., Newall, J., Hart, E., Ross, P., Schulenburg, S.: Hyperheuristics: An Emerging Direction in Modern Search Technology. In:Glover, F., Kochenberger, G.A. (eds.): Handbook of Metaheuristics. Kluwer Academic Publishers, Boston Dordrecht London (2003) 457-474

3. Cowling, P., Kendall, G., Soubeiga, E.: A Hyperheuristic Approach to Scheduling a Sales Summit. In: Burke, E., Erben, W. (eds.): Practice and Theory of Automated Timetabling III: PATAT2000. Lecture Notes in Computer Science, Vol. 2079. Springer-Verlag, Berlin Heidelberg New York (2000) 176-190

4. Gratch, J., Chien, S.: Adaptive Problem-Solving for Large-Scale Scheduling Problems: A Case Study. Journal of Artificial Intelligence Research **4** (1996) 365-396

5. Nareyek, A.: Choosing Search Heuristics by Non-Stationary Reinforcement Learning. In: Resende, M., de Souza, J. (eds.): Metaheuristics: Computer Decision-Making. Kluwer Academic Publishers, Boston Dordrecht London (2003) 523-544

6. Ross, P., Schulenburg, S., Marín-Blázquez, J. G., Hart, E.: Hyper-Heuristics: Learning to Combine Simple Heuristics in Bin-packing Problems. In: Proceedings of the Genetic and Evolutionary Computation Conference (GECCO 2002). Morgan Kauffmann (2002) 942-948.

7. Hart, E., Ross, P., Nelson, J.: Solving a Real-World Problem Using an Evolving Heuristically Driven Schedule Builder. Evolutionary Computation **6** (1998) 61-80

8. Cowling, P., Kendall, G., Han, L.: An Investigation of a Hyperheuristic Genetic Algorithm Applied to a Trainer Scheduling Problem. In: Proceedings of 2002 Congress on Evolutionary Computation (CEC2002). IEEE Computer Society Press, Honolulu, USA (2002) 1185-1190

9. Han, L., Kendall, G., Cowling., P.: An Adaptive Length Chromosome Hyperheuristic Genetic Algorithm for a Trainer Scheduling Problem. In: Proceedings of the 4th Asia-Pacific Conference on Simulated Evolution and Learning (SEAL'02). Orchid Country Club, Singapore (2002) 267-271

10. Burke, E., Kendall, G., Soubeiga, E.: A Tabu-Search Hyperheuristic for Timetabling and Rostering. Journal of Heuristics **9** (2003) 451-470

11. Cowling, P., Kendall, G., Soubeiga, E.: A Parameter-Free Hyperheuristic for Scheduling a Sales Summit. In: Proceedings of the Third Metaheuristic International Conference (MIC'2001). Porto, Portugal (2001) 127-131

12. Corne, D., Ross, P.: Peckish Initialisation Strategies for Evolutionary Timetabling. In: Burke, E., Ross, P. (eds.): Practice and Theory of Automated Timetabling. Lecture Notes in Computer Science, Vol. 1153, Springer-Verlag, Berlin Heidelberg New York (1995) 227-240

13. Cowling, P., Chakhlevitch, K.: Hyperheuristics for Managing a Large Collection of Low Level Heuristics to Schedule Personnel. In: Proceedings of the 2003 IEEE Congress on Evolutionary Computation (CEC'2003). IEEE Computer Society Press, Canberra, Australia (2003) 1214-1221

14. Chakhlevitch, K.: Hyperheuristics Which Manage Large Collections of Low Level Heuristics. PhD thesis, in preparation

15. Glover, F., Laguna, M.: Tabu Search. Kluwer Academic Publishers, Boston Dordrecht London (1997)

An Attribute Grammar Decoder for the 01 MultiConstrained Knapsack Problem

Robert Cleary and Michael O'Neill

University of Limerick, Ireland
{Robert.Cleary, Michael.ONeill}@ul.ie

Abstract. We describe how the standard genotype-phenotype mapping process of Grammatical Evolution (GE) can be enhanced with an attribute grammar to allow GE to operate as a decoder-based Evolutionary Algorithm (EA). Use of an attribute grammar allows GE to maintain context-sensitive and semantic information pertinent to the capacity constraints of the 01 Multiconstrained Knapsack Problem (MKP). An attribute grammar specification is used to perform decoding similar to a first-fit heuristic. The results presented are encouraging, demonstrating that GE in conjunction with attribute grammars can provide an improvement over the standard context-free mapping process for problems in this domain.

1 Introduction

The NP-Hard 01 Multiconstrained Knapsack Problem (MKP) can be formulated as;

$$maximise \quad \sum_{j=1}^{n} p_j x_j \qquad (1)$$

$$subject\ to \quad \sum_{j=1}^{n} w_{ij} x_j \quad \leq c_i, \qquad (2)$$

$$x_j \in \{0,1\}, \quad j = 1 \ldots n \qquad (3)$$

where, p_j refers to the profit, or worth of item j, x_j refers to the item j, w_{ij} refers to the relative-weight of item j, with respect to knapsack i, and c_i refers to the capacity, or weight-constraint of knapsack i. There exist $j = 1 \ldots n$ items, and $i = 1 \ldots m$ knapsacks.

The objective function (equation 1) tells us to find a subset of the possible items (ie. the vector of items); where the sum of the profits of these items is maximised, according to constraints presented in equation 2. Equation 2 states, that the sum of the relative-weights of the vector of items chosen, is not to be greater than the capacity of any of the m knapsacks. Equation 3 refers to the notion that we wish to generate a vector of items, of size $n(\ j = 1..n$ items), whereby a 0 at the ith index indicates that this item is not in the chosen subset and a 1 indicates that it is.

G.R. Raidl and J. Gottlieb (Eds.): EvoCOP 2005, LNCS 3448, pp. 34–45, 2005.
© Springer-Verlag Berlin Heidelberg 2005

Exact methods such as Branch and Bound have been found as good approximation algorithms to the single-constrained knapsack problem [1], but however Evolutionary Algorithms (EAs) have been found to be most appropriate in solving large instances of the MKP for which exact methods are too slow. As a result EAs, and in particular, decoder-based EAs have been heavily studied in application to the MKP [2, 3, 4, 5, 6, 7, 8]. Their advantage over the more traditional direct representation of the problem, is their ability to always generate and therefore carry out evolution over feasible candidate solutions, and thus focus the search on a smaller more constrained search space [9, 10]. The best EAs for the MKP that we are aware of utilise problem-specific domain knowledge to carry out repair and optimisation to maintain feasible solutions [11, 12, 5].

These EAs have been developed specifically for solving the MKP, and are based heavily on domain knowledge of the problem and efficiency to solution time via locally-optimised initialisation and search techniques. It is not the focus of this paper to attempt to compete with such algorithms, rather, in this instance; we wish to examine the ability of Grammatical Evolution's (GE's) mapping process to be transformed to the role of a decoder for constrained optimisation problems. More specifically, we use constrained optimisation problems as a test-bed to demonstrate how attribute grammars allow the extension of GE to context-sensitive problem domains. As a side effect, we also see the possibility to further our analysis of the internal workings of GE, through merging research in the methods of analysis found within the field of decoder-based EAs. Core to the functioning of such decoder-based EAs is a genotype-phenotype mapping process, and methods have been developed for the effective analysis of the workings of such mapping processes.

The remainder of the paper is structured as follows. An introduction to decoder-based EAs from the literature is presented in Section 2, followed by a short description of Grammatical Evolution in the context of knapsack problems in Section 3. Attribute grammars and their application to knapsack problems are discussed in Section 4 followed by details on the experimental setup in Section 5. Finally the results are presented in Section 6 and conclusion and future work outlined in Section 7.

2   Decoder Approaches from the Literature

The previous section outlined the knapsack problem as that of a constrained optimisation problem. From our literature review we divide the various approaches into two categories; *infeasible*, and *feasible-only*. From this survey we encountered many successful works from both approaches, with Raidl's improved GA [5] being the best infeasible approach, outperforming Chu and Beasley's [11] GA by what is reported to be a non-deterministic local optimisation strategy. Of the feasible-only approaches, the problem space decoder based EA of Raidl [12], marginally outperforms Gottlieb's study of *permutation-based EAs* in [2].

2.1   Infeasible Solutions

The allowance of infeasible solutions within the evolving population of the EA stems from what [9] and [12] refer to as a *direct* representation. That is, chromosomes encode for a set of items to be included in the knapsack. Each gene represents a corresponding item. The most typical use of this approach is the binary bit-string representation of size n where a 1 at the ith index indicates that this item is to be included in the knapsack.

As pointed out in [8], it is important with this kind of approach to ensure that the infeasible solution doesn't end up in the final population, or result in being awarded a fitness better than a feasible solution. Khuri and Olsen [13,14], both report only moderate success rates with such approaches; and as such we focus our attention on the use of decoders in providing feasible-only solutions.

2.2   Feasible-Only Solutions

The simple structure of the MKP is often approached by applying the search algorithm to the space of possible solutions, mapped out by $P \in \{0,1\}^n$. A common alternative to this is the use of an EA which works in some other search space which maps into F the feasible subset of P [9]. The entity which does such a mapping is generally referred to as a decoder, and guarantees the generation of solutions that lie within F, and often further constrain the search to more promising regions of F.

In principle, all decoders work the same way. A decoder can be thought of as a builder of legal knapsacks. It works by attempting to add items to a knapsack, whereby each attempted addition of an item is governed by a capacity-constraints check. That is, items are added as long as no capacity constraint is violated by it's addition.

Decoders differ; in their use of problem specific heuristic information and how they interpret the genome. That is, EAs which choose the decoder approach to constrained optimisation problems, utilise a representation which feeds the decoder's internal knapsack generation algorithm. This representation is generally one of two different classes: *a)* The chromosome representation is a mapping to some permutation of the set of items that implicitly defines the order by which the decoder attempts to build a knapsack. Such EAs are referred to as *order-based*, and have been exhaustively studied in [2,3,4,5,6] and [7,8] where the general conclusion is that this is a very effective approach; *b)* The chromosome uses a symbolic representation, where genes represent actual items; and the decoder works by dropping items which violate capacity constraints. In this case, evolutionary operators must incorporate the decoder heuristic so as to allow only the generation of feasible solutions. Hinterding [7], introduces such an EA as an order-independent mapper and reports it to perform well, although it is outperformed by an order-based approach.

With either approach, the introduction of a many-to-one genotype-phenotype mapping may occur. Raidl et al [3] refers to this as *heuristic bias*, whereby the stronger the restriction of the search space to promising regions of the feasible search space $F \subset P \in \{0,1\}^n$, the stronger the heuristic bias. The building of a legal knapsack is governed by termination at the point of a capacity violation. Thus, many of the same genotypes may decode to the same phenotype as all of their genetic material may not be allowed to be used with such decoder termination.

Although working with the simpler single-constrained knapsack problem, Hinterding reports that a redundant mapping from genotype to phenotype gave better results [8]. This is similarly supported in [3, 4] where it is also observed that although desirable too much redundancy in the decoder may result in degradation of performance. The following section will describe the Grammatical Evolution EA, and demonstrate how it lends itself to act as a decoder for constrained optimisation problems via the introduction of an attribute grammar into the genotype-phenotype mapping process.

3   Grammatical Evolution

Grammatical Evolution (GE) [15], is an evolutionary algorithm that can evolve computer programs in any language, and can be considered as a form of grammar-based genetic programming. Rather than representing the programs as parse trees, as in standard GP [16, 17], a variable-length linear genome representation is used. Fig. 1 provides an illustration of the mapping process over a simple example CFG.

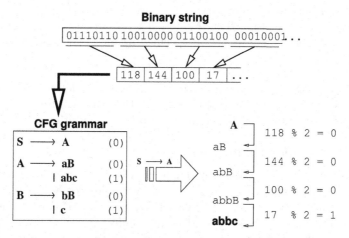

Fig. 1. An illustration of GE's genotype-phenotype mapping process' operation

As illustrated, GE uses the CFG as a *phrase-structure generative grammar*, whereby, rules of the grammar - outline the structure by which syntactically correct sentences of the language can be derived[1]. It can be seen that the grammar

---

[1] Although GE uses a grammar in Backus-Naur-Form (BNF) - for clarity of explanation of the subsequent attribute grammars, we subscribe to the notation of Knuth [18] to do the same; whereby a → denotes a production, as opposed to ::= in BNF.

specification of Fig. 1 defines *a language*. This language, written $L(G)$, determines the set of legal (*or syntactically correct*) sentences, which can be generated by application of the grammar's rules. For example, the grammar within the illustration defines the language $L(G) = ab^+c$: the set of strings starting with the terminal-symbol 'a' - ending in the terminal-symbol c; and having one-or-more of the terminal-symbol 'b' in between (Note: the $^+$ symbol denotes, *one-or-more*).

The non-terminal symbols 'A' and 'B' define the phrase-structure of the language. They define *A-phrases* and *B-phrases*, from which the language is contained. These would be similar to constructs such as *noun-phrases* in spoken language, or for example, a *boolean-expression phrase* from the abstract syntax of a programming language. In terms of the example grammar, a syntactically valid *A-phrase* contains an 'a' followed by a *B-phrase*. A recursive definition of the *B-phrase* thereafter, defines the previously stated language of the grammar. In this way, the structure of the syntax of an entire language can be defined in a concise and effective notation.

Rules of the grammar are referred to as *production-rules* and as such $A \rightarrow aB$, can be read as, "*A*" *produces* "*aB*". Similarly it can be said that "*aB*" is derived from A. Such a production is said to yield a derivation in the *sentential-form*, where by a completed derivation results in a sentential form consisting solely of terminal symbols - a sentence of the language.

The GE mapping process works, by first constructing a map of the grammar, such that left-hand-side non-terminals are used as a key to a corresponding right-hand-side list of rules(*the index of which are specified in parenthesis*). Production-rules are chosen, then, by deriving the production at the index of the current non-terminal's rule-list as specified by the following formula:

$$Rule \;\; = \;\; CodonValue \,\% \, Num. \, Rules$$

where % represents the modulus operator. (So as not to detract from the focus of the paper, we defer the interested reader to the canonical texts of GE for an explanation of the intricacies of the mapping process' algorithm [15, 19, 20, 21].)

3.1   CFG Decoder Limitations

In considering GE as a decoder for knapsack problems then, we wish to use the mapping process to *decode* a genotype, into sentences of the language of knapsacks. That is, we require a CFG definition to represent the language of feasible knapsacks. Let us now consider the viability of such a grammar-based decoder as is afforded by the standard GE mapping process. We can define a grammar for an n item knapsack problem as follows:

$$S \rightarrow K$$
$$K \rightarrow I$$
$$K \rightarrow IK$$
$$I \rightarrow i_1$$
$$\vdots$$
$$I \rightarrow i_n$$

Beginning from the start symbol S a sentence in the language of knapsacks is created by application of productions to S such that only terminal symbols remain; yielding a string from the set of items $\{i_1, \ldots, i_n\}$. Consider however, the problem of generating such a string for a 01 knapsack problem as defined in the previous section. GE essentially carries out a left-most derivation, according to the grammar specified. The following derivation-sequence illustrates the point at which a CFG fails to be able to uphold context-specific information.

$$S \rightarrow K \rightarrow IK \rightarrow i_3 K \rightarrow i_3 IK \rightarrow i_3??$$

What this derivation-sequence provides is a *context*. That is, given the context that i_3 has been derived, the next derivation-step must ensure that i_3 is not produced again. Re-deriving an i_3 violates the semantics of the language of 01 knapsacks. A CFG has no method of encoding this context-sensitive information and hence, cannot be used as a decoder to decode chromosomes to feasible knapsack solutions The answer to these limitations lies in the power of attribute grammars, which allow us to give context to the current derivation step. By employing an attribute grammar as the generative power of the mapping process we can extend GE to become a decoder for feasible-only candidate solutions.

4 Attribute Grammars for Knapsacks

Attribute grammars (AGs) were first introduced by Knuth [18], as a method to extend CFGs by assigning attributes (or pieces of information), to the symbols in a grammar. Attributes can be assigned to any symbol of the CFG, whether terminal or non-terminal, and are defined (given meaning) by functions associated with productions in the grammar. These shall be termed the *semantic functions*. Attributes can take the form of simple data (eg. integers), or more complex data-structures such as lists, which append to each symbol of the grammar. In terms of AGs it's best to think of a derivation by it's tree representation where the root is S and it's children the symbols of the applied production. A portion of a derivation-tree descended from a single non-terminal node comprises the context of a phrase. A sentential-form is the set of nodes *directly* descended from such a non-terminal. Also, the term *terminal-producing production* will be used to refer to a sentential-form which contains one or more terminals.

Attributes are thus pieces of data appended to nodes of the tree, and can be evaluated in one of two ways. In the first, the value of an attribute is determined by the value of the attributes of child nodes. That is, the evaluation of a parent attribute can be *synthesized* or made up of it's child's attribute values. In the second, the value of an attribute is determined by information passed down from parent nodes. That is, a child's attribute is evaluated based on information which is *inherited* down from parent nodes. In either case, attributes of a node *can* be evaluated in terms of other attributes of that same node. Information however, originates either from the root node S or leaf nodes of the tree, which generally provide constant values from which, the value of all other nodes in the tree are synthesized or inherited.

4.1   An Attribute Grammar for 01 Compliance

Consider the following attribute grammar specification to show how attributes can be used to preserve 01 compliance when deriving strings in the language of knapsacks. This attribute grammar is identical to the earlier CFG, with regard to the syntax of the knapsacks it generates. The difference here being the inclusion of attributes associated with both terminal and non-terminal symbols, and their related *semantic functions*. As each symbol in the grammar maintains it's own set of attributes, we use a subscript notation to differentiate between occurrences of like non terminals.

Following the notation of Knuth [18], we have appended the following attributes to the previous CFG grammar:

items(K) : A synthesized attribute that records all the items currently in the knapsack (ie. *items which have been derived thus far*).

item(I) : A string representation, identifying which physical item the current non-terminal will derive. For example $item(I) = "i_1''$ where that I derives or produces i_1 of the problem.

notInKnapsack?(i_n) : A boolean flag, indicating whether the 01 property can be maintained by adding this item (ie. given the current derivation, has this item been previously derived?). This is represented as a string-comparison of *item(I)* over *items(K)*.

The following gives a description of such an attribute grammar, and provides an example to illustrate how it can be used to drive a context-specific derivation

$$S \to K$$
$$K \to I \qquad items(K) = items(K) + item(I)$$

$$K_1 \to IK_2 \qquad items(K_1) = items(K_1) + item(I)$$
$$items(K_2) = items(K_1)$$

$$I \to i_1 \qquad item(I) = "i_1"$$
$$\textbf{Condition}: \quad if(notinknapsack?(i_1))$$

$$\vdots$$

$$I \to i_n \qquad item(I) = "i_n"$$
$$\textbf{Condition}: \quad if(notinknapsack?(i_n))$$

Consider the above attribute grammar, when applied to the following derivation-sequence:

$$S \to K \to IK \to i_1 K \to i_1 IK \to i_1(i_\lambda \in \{i_2...i_n\})K \to ...$$

At the point of mapping I given the above context, it can be seen from the above semantic functions that it's *items(I)* attribute will be evaluated to "i_λ" if the *notinknapsack?()* condition holds. Following this the root node will have it's *items(K_1)* updated to include "i_λ" which can from then on be passed down the tree by the inherited attribute of *items(K_2)*. This in turn allows for the next *notinknapsack()* condition to prevent duplicate items being derived. The next section follows to provide a deeper example, which shows how we can include the evaluation of weight-constraints at the point in a derivation where we carry out a terminal-producing production.

4.2 An Attribute Grammar for Constraints Checking

Further attributes can be added, in order to extend the context-sensitive information captured, during a derivation. The following outlines these attributes and their related semantic-functions in a full AG specification, which maintains both 01 and constraint-violation information.

lim(S) : A global attribute containing each of the m knapsacks' weight-constraints. This can be inherited or passed down to all nodes.

lim(K) : As lim(S) just used to inherit to each K_2 child node.

usage(K) : A usage attribute, records the total weight of the the knapsack to date. That is, the weight of all items which have been derived at this point.

weight(K) : A weight attribute, used as a variable to hold the weight of the item derived by the descendant I to this K.

weight(I) : A synthesized attribute, made-up of the descendant item's physical weight.

weight(i_n) : The physical weight of item i_n(*the weight of item i_n as defined by the problem instance*).

The corresponding attribute grammar is given below with an example showing how it's attributes are evaluated. At the point of deriving a left-hand side production, the corresponding right-hand side semantic functions are evaluated/executed. Conditions govern the firing of the set of semantic functions directly above them at the point of their satisfaction.

$$S \rightarrow K \qquad lim(K) = lim(S)$$

$$K \rightarrow I \qquad weight(K) = weight(K) + weight(I)$$
$$\textbf{Condition}: \quad if(usage(K) + weight(I) <= lim(K))$$
$$items(K) = items(K) + item(I)$$

$$K_1 \rightarrow IK_2 \qquad weight(K_1) = weight(K_1) + weight(I)$$
$$items(K_1) = items(K_1) + item(I)$$
$$usage(K_1) = usage(K_1) + weight(I)$$
$$\textbf{Condition}: \quad if(usage(K_1) < lim(K_1))$$
$$lim(K_2) = lim(K_1)$$
$$usage(K_2) = usage(K_1)$$
$$items(K_2) = items(K_1)$$

$$I \rightarrow i_1 \qquad item(I) = \text{``}i_1\text{''}$$
$$\textbf{Condition}: \quad if(notinknapsack?(i_1))$$
$$weight(I) = weight(i_1)$$

$\vdots$

$$I \rightarrow i_n \qquad item(I) = \text{``}i_n\text{''}$$
$$\textbf{Condition}: \quad if(notinknapsack?(i_n))$$
$$weight(I) = weight(i_n)$$

In terms of the problem being solved, $lim(K)$ is actually a list of constraint-bounds for *each* of the m knapsacks. Similarly, $items(K)$, is a list of the items which have currently been derived by the GE mapping process. For clarity of explanation, the following example will assume a single knapsack weight-constraint, but the more complicated problem can be extracted by altering the below conditions to have $lim(K)$ as an array of constraint-bounds as opposed to a single integer value. Fig. 2 shows the synthesized and inherited message passing

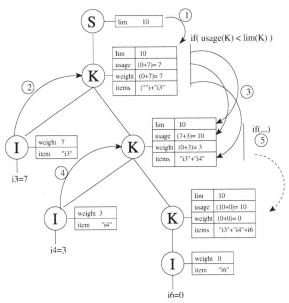

Fig. 2. Diagram showing synthesized and inherited message-passing for evaluating attributes in the derivation tree of an attribute grammar

involved in evaluating derivation trees for the above attribute grammar. We can see, that initially the global limit is passed down to K by the first semantic function. From the grammar, we can see that following this, the first three semantic functions of K are evaluated before a condition checks to see that we haven't violated a weight-constraint[2]. Passing this allows for inheriting values down the tree by the second three semantics functions (otherwise we would have remapped K via another production and repeated the process). The attribute grammar decoder works then by attempting to add items according to I's pro-

---

[2] For clarity, we assume that the $notInKnapsack(i_3)$ condition has passed and the its values have synthesized up the tree.

duction rules, and at the point of constraint-violation or a 01 collision the codon causing the error is skipped (becomes an intron) and the subsequent codon read.

5   Experimental Setup

In this study, we have chosen to apply our analysis to a range of problem instances from the literature [22], which allow us to gauge the effectivity of different grammars to capture the context-sensitive information for the test-bed knapsack problems. Our current attribute grammar decoder as described in the previous section, uses a simple construction heuristic similar to that of the first-fit heuristic described in [8]. From the literature, this set of problems allow us a direct comparison with two different knapsack problem approaches. Primarily, we undergo a direct comparison with the penalty-based GA of Khuri et al. [13], and a secondary comparison with a hybrid GA which uses a problem-space search [23]. It is worth noting however that Khuri et al. use a direct bitstring representation of fixed-length, where a graded penalty function overcomes the problem of infeasiblity. We utilise the standard variable-length binary string representation of GE, with the attribute grammar mapping as a decoder for feasibility. For experiments with the standard GE mapping process, we penalise to zero - all infeasible candidates. Our earlier work in [24] tested a graded penalty term but it provided no improvement for results.

We adopt standard experimental parameters for GE, changing only the population size to that of Khuri et al., whose a population size of $\mu = 50$ running for up to 4000 generations. We adopt a variable length one-point crossover probability of 0.9, bit mutation probability of 0.01, and roulette selection. A steady-state evolutionary process is employed, whereby a generation constitutes the evolution and attempted replacement of $\mu/2$ children into the current population. Replacement occurs if the child is better than the worst individual in the population. The initial population of variable-length individuals were initialised randomly, with an average length of 20 codons, and standard-deviation of 5 codons from average. Standard 8-bit codons are employed, and GE's wrapping operator is turned off. The experimental metric of *percentage of runs yielding an optimum solution* serve to demonstrate GE's ability to solve these problems.

6   Results

A comparison of the standard GE context-free grammar, the 01 attribute grammar, and the full attribute grammar can be seen in Table 1. The benefit of adopting an attribute grammar on these problem instances are clear with the full constraint checking attribute grammar clearly outperforming the two other grammars analysed. On comparison to the results obtained in [13, 23] it can be seen that the results presented show GE with the attribute grammar decoder to clearly outperform the traditional GA of Khuri et al. on some instances, and provide competitive results to the hybrid GA of Cotta which uses local optimisation. The results labeled *DE* show the effect of implementing phenotypic

Table 1. Comparing the three grammars, and results from [13, 23] on the percentage of runs achieving an optimum solution. The effect of phenotypic duplicate elimination (DE) is also presented for the full attribute grammar.

Problem	n	m	GE	AG(01)	AG(Full)	Khuri	Cotta	AG(Full)+DE
knap15	15	10	3.33%	60%	83.33%	83%	100%	96.6%
knap20	20	10	6.66%	33.33%	76.66%	33%	94%	100%
knap28	28	10	0%	3.33%	40%	33%	100%	90%
knap39	39	5	0%	0%	36.66%	4%	60%	43.33%
knap50	50	5	0%	0%	3.33%	1%	46%	16.66%
Sento1	60	30	0%	0%	10%	5%	75%	66.66%
Sento2	60	300	0%	0%	3.33%	2%	39%	30%
Weing7	105	2	0%	0%	0%	0%	40%	0%
Weing8	105	2	0%	0%	6.66%	6%	29%	36.66%

duplicate elimination, as described in [6], where we observe that disallowing duplicates at a phenotypic level has the desired effect in improvement of performance. It should be noted, however, that the best of the attribute grammar results fall short of the number of successful solutions found by the best results in the literature.

7   Conclusions and Future Work

We wished to examine the extension of the standard GE mapping process to handle context-sensitive information via the medium of attribute grammars. The results demonstrated a clear advantage for the attribute grammars over the standard context-free grammar on the problem instances examined. Results have also been provided to support the findings of Raidl and Gottlieb [6], which show that duplicate elimination at a phenotypic level can improve performance. More work is required to improve the performance of this approach, and to analyze the redundancy of the attribute grammar decoder in terms of locality and effect of operators (initial results show that this is preserved).

References

1. Martello, S., Toth, P. (1990). *Knapsack Problems*. J. Wiley & Sons, 1990.
2. Gottlieb, J. (2000). Permutation-Based Evolutionary Algorithms for Multidimensional Knapsack Problem. *Proc. of ACM Symp. on Applied Computing*.
3. Raidl, Günther R., Gottlieb, J. (1999). Characterizing Locality in Decoder-Based EAs for the Multidimensional Knapsack Problem. *4th European Conference on Artificial Evolution*, pp. 38 - 52, Springer-Verlag.
4. Raidl, Günther R., Gottlieb, J. (1999). The Effects of Locality on the Dynamics of Decoder-Based Evolutionary Search. *Proc. of the Genetic and Evolutionary Computation Conference*, pp. 787, Morgan Kaufmann.

5. Raidl, Günther R. (1998). An Improved Genetic Algorithm for the Multicon-strained 0-1 Knapsack Problem. *Proc of 1998 IEEE Congress on Evolutionary Computation*, pp. 207 - 211.
6. Raidl, Günther R., Gottlieb, J. (1999). On the importance of phenotypic duplicate elimination in decoder-based evolutionary algorithms. *Proc. of the Genetic and Evolutionary Computation Conference*, Late-Breaking Papers, pp. 204-211.
7. Hinterding, R. (1994). Mapping, Order-Independant Genes and the Knapsack Problem. *Proc. 1st IEEE Int. Conf. on Evolutionary Computation*, pp. 13-17.
8. Hinterding, R. (1999). Representation, Constraint Satisfaction and the Knapsack Problem. *Proc. of 1999 IEEE Congress on EC*, pp. 1286-1292.
9. Gottlieb J. (1999) Evolutionary Algorithms for Multidimensional Knapsack Prob-lems: the Relevance of the Boundary of the Feasible Region. *Proc. of the Genetic and Evolutionary Computation Conference*, pp. 787, Morgan Kaufman.
10. Gottlieb, J. (1999) On the Effectivity of Evolutionary Algorithms for the Multidi-mensional Knapsack Problems. *Proc. of Artificial Evolution*, Springer LNCS.
11. Chu, P.C. and Beasley, J.E. (1998). A genetic algorithm for the multidimensional knapsack problem. *Journal of Heuristics* 4:63-86.
12. Raidl, Günther R. (1999). Weight-Codings in a Genetic Algorithm for the Multi-constraint Knapsack Problem. *Proc of 1999 IEEE Congress on Evolutionary Com-putation*, pp. 596-603.
13. Khuri, S., Back, T., and Heitkotter, J. (1994). The zero/one multiple knapsack problem and genetic algorithms. In Deaton, E. et al., editors, *Proc. of the 1994 ACM symposium of Applied Computation,* pp. 188-193, ACM Press.
14. Olsen, A. L. (1994): Penalty Functions and the Knapsack Problems. *in Proc. of the 1st Int. Conf. on Evolutionary Computation*, pp. 559-564.
15. O'Neill, M., Ryan, C. (2003). *Grammatical Evolution: Evolutionary Automatic Pro-gramming in an Arbitrary Language*. Kluwer Academic Publishers.
16. Koza, J.R. (1992). *Genetic Programming*. MIT Press.
17. Banzhaf, W., Nordin, P., Keller, R.E., Francone, F.D. (1998). *Genetic Program-ming – An Introduction; On the Automatic Evolution of Computer Programs and its Applications*. Morgan Kaufmann.
18. Knuth, D.E. (1968). Semantics of Context-Free Languages. *Mathematical Systems Theory*, Vol. 2, No. 2. Springer-Verlag.
19. O'Neill, M. (2001). *Automatic Programming in an Arbitrary Language: Evolving Programs in Grammatical Evolution*. PhD thesis, University of Limerick, 2001.
20. O'Neill, M., Ryan, C. (2001). Grammatical Evolution, *IEEE Trans. Evolutionary Computation*, Vol.5, No.4, 2001.
21. Ryan, C., Collins, J.J., O'Neill, M. (1998). Grammatical Evolution: Evolving Pro-grams for an Arbitrary Language. *Proc. of the First European Workshop on GP*, 83-95, Springer-Verlag.
22. Beasley, J.E. (1990). OR-Library: distributing test problems by electronic mail. *Journal of the Operational Research Society* Vol. 41 No. 11, pp. 1069-1072.
23. Cotta, C.,Troya, Jose, M (1998). A Hybrid Genetic Algorithm for the 0-1 Multiple Knapsack Problem. *In Artificial Neural Nets and Genetic Algorithms 3*, pp. 251-255, Springer-Verlag.
24. O'Neill, M., Cleary, R., Nikolov, N. (2004). Solving Knapsack Problems with At-tribute Grammars. *In Proc. of the Grammatical Evolution Workshop 2004.*

EvoGeneS, a New Evolutionary Approach to Graph Generation

Luigi Pietro Cordella[1], Claudio De Stefano[2], Francesco Fontanella[1],
and Angelo Marcelli[3]

[1] Department of Information Engineering and Systems,
University of Naples,
Via Claudio, 21 80125 Naples – Italy
{cordel, frfontan}@unina.it
[2] Department of Automation, Electromagnetism,
Information Engineering and Industrial Mathematics,
University of Cassino,
Via G. Di Biasio, 43 02043 Cassino (FR) – Italy
destefano@unicas.it
[3] Department of Computer Science and Electrical Engineering,
University of Salerno,
84084 Fisciano (SA) – Italy
amarcelli@unisa.it

Abstract. Graphs are powerful and versatile data structures, useful to
represent complex and structured information of interest in various fields
of science and engineering. We present a system, called EvoGeneS, based
on an evolutionary approach, for generating undirected graphs whose
number of nodes is not a priori known. The method is based on a special
data structure, called multilist, which encodes undirected attributed relational graphs. Two novel crossover and mutation operators are defined
in order to evolve such structure. The developed system has been tested
on a wireless network configuration and the results compared with those
obtained by a genetic programming based approach recently proposed in
the literature.

1 Introduction

Graphs are powerful and versatile data structures, useful to represent complex
and structured information. In the last decades, there has been an increasing
interest in studying and using graphs in many applications, also because the
developments of computer technology made high computational cost problems
to be dealt with.

Graphs have been used in various fields of science and engineering. They may
effectively represent physical networks, such as transportation systems, power
systems, and mobile communication infrastructures [1, 2, 3], but have been also
used to model less tangible interactions, as might occur in ecosystems, databases
or in the control flow of a computer program [1]. In fields like pattern recognition and machine vision, the high representational power of graphs make them

G.R. Raidl and J. Gottlieb (Eds.): EvoCOP 2005, LNCS 3448, pp. 46–57, 2005.

very attractive and well-suited to model complex patterns in terms of parts and their relations. Attributes of graph nodes and edges are often added to incorporate further information, leading to a graph representation form generally called Attributed Relational Graph (ARG) [4]. Examples of successful applications include shape analysis and 3-D object recognition [5, 6], character recognition [7], classification of ideograms and symbols in document analysis and technical drawing interpretation [8].

In many cases, a prominent problem is that of generating graphs exhibiting some particular properties. The generation of prototypes in pattern recognition problems, so as the generation of the optimal configuration of a physical network are examples of such problem. Thus, the use of graph representations often requires the definition of effective techniques for generating the graphs representing the desired solutions. To this purpose, two main different approaches can be identified, depending on the nature of the problem: in case of applications in which training samples are available, the graphs may be generated by exploiting the information included in such a training set. In all the other cases the solution is found by defining a function $\mathcal{F}$ able to measure the goodness of tentative solutions in a given space: the graphs representing the solutions are generated by finding all the absolute maxima of the function $\mathcal{F}$. Combinatorial, heuristic and inductive learning approaches have been used, among others, to generate graphs [9]. Several attempts to generate graphs using evolutionary approaches have also been done. Methods have been proposed in the fields of molecular design [10] and electrical circuit design [11], using a direct encoding of the evolving graph. It is worth noting that these methods define evolutionary operators tailored for the considered problem. Indirect encoding of graphs in terms of bit strings [12] or trees [13] has also been used. In the latter approach, for instance, a tree encodes the operations to be applied to a very simple starting graph, in order to transform it into another one arbitrarily complex.

We present a system, called *EvoGeneS* (Evolutionary Graph Generation System), based on an evolutionary approach, for generating graphs whose number of nodes is not a priori known. The proposed method aims at overcoming two major disadvantages of the methods discussed earlier by providing a direct encoding of graphs and two novel, general purpose and problem independent operators. A special structure, called multilist, encodes undirected ARG's, and demonstrated to be particularly convenient for generating new and different graphs under given constraints. For evolving multilists, two basic operators have been devised: the first one, called *crossover* by analogy to genetic algorithms, swaps parts of two multilists, thus swaps subgraphs of two graphs, thus generating graphs of variable length. The second operator, called *mutation*, operates on a multilist in such a way to change a graph into a new one whose node number is unchanged, whereas both node and link attributes can be modified.

In the following, after defining the multilist and the elementary operations defined for it, an application of the proposed evolutionary system will be illustrated. The results obtained by EvoGeneS will be compared with those of

EvoGraph, the approach described in [13], showing that EvoGeneS performance overcomes that of this Genetic Programming based approach.

2   *EvoGeneS*

EvoGeneS is essentially based on two elements: a new data structure encoding undirected relational graphs with attributes and two operators devised for such structure.

2.1   Graph Encoding

Let us consider a graph G of N nodes. Let also denote by A_n and A_a the sets of values for the attributes describing the nodes and the arcs of the graph, respectively. The data structure we have adopted for representing attributed relational graphs has been called *multilist* (ML in the following) since it is based on the list concept and consists of two basic lists. The first one, called *main list*, contains the information on graph node attributes, thus its number of elements is equal to the number N of nodes. Each element of the second list is on its turn a list, called *sublist*. One sublist is associated with each node and includes the attributes of the arcs connected to that node. In order to preserve information about the nodes interconnected by each arc, arc attributes are sorted in each sublist in a suitable order. Namely, the i-th sublist contains information on the arcs connecting the i-th node of the graph to the nodes following it in the main list, in the order they appear in such list. If two nodes are not connected, this information is anyway suitably stored in the proper place of a sublist. In practice, a "null" relation has been defined so that even the absent arcs are encoded in the ML representation of a graph (see Fig. 1). The *length* of a ML is defined as the number of elements of its main list. It is important to notice that, in this paper, we consider simple (i.e., without loops) undirected graphs, so that the relation linking the i-th node to the j-th node coincides with the relation linking the j-th node to the i-th node. For this reason, the length of the sublist associated with a node decreases as the position of the node in the main list increases: the first sublist is made of $N - 1$ elements, the second sublist has $N - 2$ elements and so on. In fact, the information on the link between each node and the previous ones in the main list is already expressed in the previous sublists. As a consequence, the sublist of the last node of the graph is void. Thus a ML has a triangular shape: the base of the triangle is the main list and is long N, while the height is represented by the first sublist and is long $N - 1$. In the following, the operations defined on the ML's will be introduced.

2.2   The Operators

The just described data structure has been devised in such a way to make easier the application of operators able to generate new and different items from previously generated ones. Two basic operators have been defined for the ML: they have been called crossover and mutation by analogy to genetic algorithm

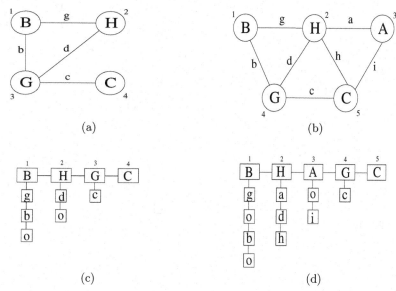

Fig. 1. Two graphs (top) and their encoding multilists (bottom). The horizontal list is the main one and the vertical lists are the sublists. The elements of the set A_n are denoted by capital letters, while those of A_a are denoted by small letters. The null element is denoted by the letter 'o'

operators. The former operator swaps parts of two ML's. In this way, it is possible to generate better solutions by combining solutions that contain only part of a good solution. The mutation operator, instead, generates a new graph whose number of nodes is unchanged, whereas the attributes of both nodes and arcs can be modified.

Crossover Operator. The crossover is applied to two ML's, L' and L'', called in the following *parents*, respectively encoding the graphs G' and G'' and generates two new ML's, M' and M'' called *offspring*, respectively encoding the graphs H' and H''. The crossover operator allows generating ML's of variable length. In fact, if the parents are of length N' and N'' respectively, the length of the offspring may varies in the interval $[2, (N' + N'') - 2]$. The operator is obtained by combining two more elementary operations that can be applied to a ML. The former, called *t-cut*, splits a generic ML L of length N in two ML's, the first one consisting of the first t nodes of L and the second one of the remaining $N - t$ nodes. The latter operation, instead, is called *merge* and, given two ML's L', of length N' and L'' of length N'' yields a new ML of length $(N' + N'')$, encoding a graph including both the nodes of G' and G''. To show how the crossover works, let L' be the ML of Fig. 1(a) ($N' = 4$) and L'' that of Fig. 1(b) ($N'' = 5$). Then, to apply the crossover to these ML's, two integers $t_1 \in [1, N' - 1]$ and $t_2 \in [1, N'' - 1]$ have to be randomly chosen. Let $t_1 = 2$ and $t_2 = 1$, then the 2-cut operation is applied to L' and two ML's are obtained: L'_1 and L'_2, both

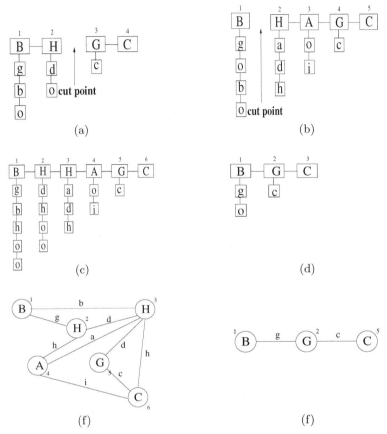

Fig. 2. The crossover operator. (a) and (b) the application of the t-cut to the multilists of Fig. 1. (c) and (d) The offspring obtained after the merge. (e) and (f) The resulting graphs

of length 2 (see Fig. 2(a)). Afterwards, the 1-cut operation is applied to L'', which yields two ML's: L_1'' of length 1 and L_2'' of length 4 (see Fig. 2(b)). At this point, the merge operation is applied to L_1'' and L_2'': it yields a ML of length 6, which represents our first offspring (see Fig. 2(c)). The merge operation has to be applied also to the remaining ML's, L_2' and L_1''. In this case a ML of length 3 is obtained, which represents the second offspring (Fig. 2(d)). The obtained graphs are shown in Figures 2(e) and 2(f). Note that the length of the offspring depends on the values chosen for t_1 and t_2.

Mutation Operator. The mutation operator defined here, actually gives place to a sort of micro-mutation, because it does not modify the structure of the ML to which it is applied, but only the values of the elements of the main list and of the sublists. Such an operation is based on a probability value, called *mutation probability* (p_m in the following). For each element in the main list, p_m represents the probability to replace its value with another one randomly chosen from the

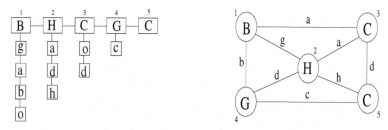

Fig. 3. The multilist (left) and the corresponding graph (right), derived from the application of the mutation operator, with a probability equal to 0.1, to the multilist of Fig. 1(b). The mutation modifies the attribute of node 3 and that associated with the arc which links nodes 3 and 5. Moreover, the mutation added a new arc, which links nodes 1 and 3, absent in the graph before of the application of the mutation operator

set A_n. The same occurs for the elements of the sublists, but in this case the value can be replaced either by one belonging to the set A_a or by the null value. Let L be a generic ML and L' be the ML produced by the application of the mutation operator to L. Let us examine the possible differences between the graphs encoded by L and L'. Both graphs contain the same number of nodes, while the number of arcs of the two graphs may be different. In fact, when the mutation is applied to an element of the main list it changes only the attributes of that node leaving the number of nodes unchanged. Instead, when the mutation is applied to an element of a sublist, and either a null element is changed to a not null one or vice versa, the corresponding arc is added to or removed from the original graph, respectively. Finally, if a not null element is replaced by a different not null element, then both the graphs G and G' will contain the arc represented by that element, but the relation associated with the arc in G is different from that in G' (see Fig. 3). We should note that, generally, the differences between the graphs G and G' are directly proportional to the mutation probability p_m.

2.3   The Algorithm

The evolutionary algorithm implemented in EvoGeneS starts by generating at random a population of P individuals. Each individual is a ML encoding a graph representing a solution of the problem to be solved. The length of these initial individuals range from 2 to N_{max} nodes. Afterwards, the fitness of the individuals generated is evaluated. To generate a new population, first the best E individuals are selected and copied in the new population in order to implement an elitist strategy. Then $(P - E)/2$ couples of individuals are selected using the tournament method, to control loss of diversity and selection intensity. The crossover operator is applied to each of the selected couples, according to a chosen probability factor p_c. Afterwards, the mutation is applied to the individuals according to a probability factor p_m. Finally these individuals are added to the new population. The process just described is repeated for N_g generations.

3   Testing the Approach

In order to ascertain the effectiveness of the proposed approach, we have chosen a planning and optimization problem. To cope with this kind of problems, several approaches have been proposed in the literature, including genetic programming [13], simplex method [14], simulated annealing [15], Tabu search [16] and genetic algorithms [17]. In particular in [13], the wireless access point configuration problem, a hard non-linear optimization problem, has been considered. We have chosen the same problem, in order to compare our results with those presented in [13].

The scenario of the problem is the following: a community is planning to provide wireless Internet service to its citizens (clients) who are scattered around a given area. A certain number of access points need to be placed to cover all clients, because each access point has a limited service radius. All access points are wired and one of them is connected to an Internet gateway. The design problem consists in determining the optimal configuration of the AP's in the area to cover. To reduce the cost, a configuration with minimal number of AP's and minimum length of the wires connecting them is considered optimal. According to the constraints imposed, the wireless access point configuration problem can be formulated in different ways. E.g., [18] assumes that the AP's are located at a specified set of possible points. We assume that the AP's can be located at any place. More precisely, the problem to solve is defined as follows:

GIVEN a set of N_C clients located at (x_i^c, y_i^c) $i = 1 \ldots N_C$ in an area of size $W \times H$ where $x_i^c \in [0, W]$ and $y_i^c \in [0, H]$, and the gateway G located at (x^g, y^g), let us assume that all AP's are equal and that the service radius of an AP is r_s;
FIND a configuration of wired access points located at (x_i^{AP}, y_i^{AP}) with $i = 1 \ldots N_{AP}$, connected to the gateway port G in such a way that each client is covered by at least one AP and the total cost of the AP's and the wires is minimal. Thus, let C_{AP} be the cost of each AP and C_w the cost of a unit length wire, the aim is:
minimize $f = C_{AP} * N_{AP} + C_w * \sum |L_i|$
where the L_i are the lengths of the connections among AP's.

A more precise solution of the problem would require considering some parameters of an AP like transmission power, channel allocation and antenna directionality. In this paper, such parameters are not considered. Nevertheless, the proposed formulation keeps the essentials of the wireless configuration problem, avoiding a time-consuming simulation for evaluating the fitness of a configuration.

The Fitness Function. To solve the problem, a configuration of AP's is represented with a graph whose nodes are the AP's and whose arcs are the wire segments connecting the AP's. The set of node attributes is made up of the AP coordinates in the area to cover (see Fig. 4). In the problem at hand, it is necessary to know only which nodes are linked to a given node. Hence, in the ML representation encoding the graph, the value 1 is used to indicate the presence of an arc, while the 0 indicates the absence of an arc.

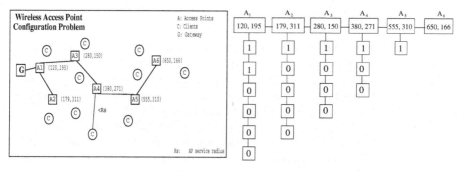

Fig. 4. An instance of the wireless access point configuration problem (left) and the multilist encoding it (right). The citizens (clients) are labeled as circled C, while the access point are represented by squares

As mentioned in the previous section the fitness function has to consider three aspects of the problem: the percentage of covered clients, the number of AP's employed and the total length of the wires connecting them. For this reason, the fitness function is the weighted sum of three terms. The first term F_{cover} should measure how well the clients are covered by the configuration of AP's: the more clients are covered, the better. The second term F_{wires} should measure how good the connection topology is: the shorter the wires used, the better. Finally, the term F_{AP} should estimate the goodness of a configuration as regards the number of wireless AP's employed: now, the fewer AP's are used, the better. It may be convenient that the fitness terms are normalized and suitably weighted, so as to reflect their different importance for evaluating the goodness of a configuration. Since the aim of this paper is that of presenting a general purpose method for graph generation, for the specific problem considered we have adopted the same fitness function as proposed in [13], in order to ascribe any difference in the performance of the methods to the way the solutions are generated, not to the way they are evaluated. Namely, the fitness terms are:

$$F_{\text{cover}} = \frac{4.0 * C_c}{C_c + C_T}; \qquad F_{\text{wires}} = \frac{10000}{10000 + L_w}; \qquad F_{\text{AP}} = \frac{C_T}{C_T + 1.5 * N_{AP}} \qquad (1)$$

where C_c is the number of clients covered, C_T the total number of clients, N_{AP} the number of AP's and L_w is the total length of wire segments connecting the AP's. Then, the fitness function F_{tot} is the weighted sum of the above three terms:

$$F_{\text{tot}} = 0.7 * F_{\text{cover}} + 0.1 * F_{\text{wires}} + 0.2 * F_{\text{AP}} \qquad (2)$$

Moreover, solutions (i.e., configurations) containing isolated AP's are penalized by multiplying their fitness by 0.5.

4   Experimental Results

By analogy with the experimental framework presented in [13], we have considered a search space whose length and width are both equal to 1000, and we have assumed, for the sake of simplicity, that the clients are represented in this space by points having integer coordinates. The values of the evolutionary parameters have been experimentally determined and are summarized in Table 1.

In the experiments reported below, the number of clients has been varied starting from 25 up to 50 with increments equal to 5. For each considered value, a distribution of clients has been randomly generated, and 50 runs have been performed with different initialization of the population, so as to reduce the effects of randomness embedded in the evolutionary algorithms. At the end of each run, the best solution found by the algorithm is stored. The corresponding length of the wires connecting the AP's is computed and stored as well. For each distribution of clients, we have computed the mean $\overline{N}_{AP}$ and the standard deviation $\sigma_{N_{AP}}$ of the number of AP's found by our method while performing 50 runs (see Fig. 5(a)). The mean $\overline{L}$ and the standard deviation σ_L of the lengths of the wires have been computed as well (see Fig. 5(b)).

In order to highlight the effectiveness of the obtained results, for each solution provided by our method, we have separately computed the Minimum Spanning Tree (MST) [1] of the corresponding graph. In fact, the MST represents the connection topology with minimal wire cost, thus we have compared the length of the wires relative to the MST with that of our solution. In practice, for the sake of comparison, we have computed the mean $\overline{L}'$ of the wire lengths for each distribution of clients. Each length refers to one of the 50 runs and is obtained by using the MST over the set of AP's provided by our method for that run. The plot of $\overline{L}'$ as a function of the number of clients, is shown in Fig. 5(b). The results are very encouraging because in every run a complete coverage of the clients has been obtained and the wire costs are very close to those computed on the MST. Moreover, the number of AP's needed to solve the problem, as well as the lengths of their connections, slightly increase with the number of clients. Finally, the standard deviations of both the number of AP's and the wire lengths assume small values, thus indicating that the solutions are widely independent of the initial conditions. Note that, for each distribution of clients, the evolutionary

Table 1. Values of the basic evolutionary parameters used in the experiments

Parameter	symbol	value
Population size	P	1000
Tournament size	$\mathcal{T}$	60
elitism size	E	40
Crossover probability	p_c	0.3
Mutation probability	p_m	0.04
Number of Generations	N_g	500
Maximum number of nodes	N_{max}	50

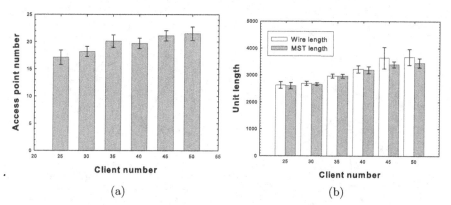

Fig. 5. (a) The mean number of access points and its standard deviation as a function of the number of clients are respectively represented by bars and segments on top of the bars. (b) The mean of the wire lengths and its standard deviation as a function of the number of clients, computed by our method and by using the MST for finding the connection topology

algorithm converges to solutions having almost the same number of AP's and the same wire cost.

Fig. 6 illustrates some results of one of the experiments performed by using a randomly generated distribution of 40 clients. In particular, the best solutions obtained at generation 10, 100, 300 and 500 are shown. During the initial generations, the evolutionary process tends to improve the client covering by adding more and more AP's, without optimizing the connection topology. Only after an almost complete coverage has been obtained, the system tries to reduce the number of AP's and focuses the search on optimizing the connection topology. This behavior can be explained considering that the term F_{cover} in the fitness function has the highest weight, while the term F_{wires} the lowest. For instance, at generation 10, all clients but one are covered using 23 AP's, but the connection topology is messy. At generation 100, all the clients are covered with 21 AP's and the connection topology is significantly improved. At generation 300 the total coverage is obtained with 20 clients and the connection topology is nearly optimal. At generation 500, finally, 19 AP's are used and the topology connection is optimal (i.e., it coincides with the MST of the related graph).

In comparing the results obtained by EvoGeneS with those reported in [13] it is worth noting that our system perform a global optimization in that both the number of AP's and their connections are simultaneously exploited for computing the fitness function. On the contrary, the genetic programming based method, is able to find a solution only when a sequential approach is adopted: first, solving the coverage problem by GP and then using a MST algorithm for determining the connection topology. Thus, it succeeds only when problem specific knowledge can be exploited to reformulate the original global optimization problem as a sequence of partial optimization problems.

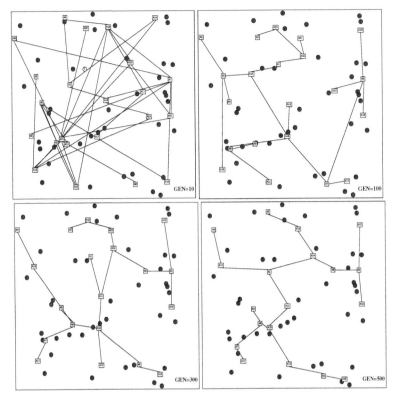

Fig. 6. The best solutions obtained at generation 10, 100, 300, 500, relative to an experiment with 40 clients. Black circles represent the clients covered by at least one access point, while white circles represent uncovered clients. The access points are represented by white squares

5   Conclusion

We have presented a system, called EvoGeneS, based on an evolutionary approach, for generating undirected graphs whose number of nodes is not a priori known. The proposed approach solves two key problems encountered when using genetic algorithms for evolving graphs: it provides a suitable data structure for the direct representation of graphs and defines crossover and mutation operators for manipulating graphs with a variable number of nodes, i.e. representations of variable length.

The results obtained by EvoGeneS on a wireless access point network configuration problem show a significant improvement with respect to those reported in the literature. Moreover, the approach does not rely on any problem specific knowledge and therefore it is suitable to deal with any problem whose solutions can be represented by graphs.

References

1. Gross, J., Yellen, J.: Graph Theory and Its Application. McGrawHill (2001)
2. Cascetta, E.: Transportation systems engineering: theory and methods. Kluwer Academic (2001)
3. Crow, M.: Computational Methods for Electric Power Systems. CRC Press (2003)
4. Eshera, M.A., Fu, K.S.: A graph distance measure between attributed relational graphs for image analysis. In: Proceedings of 7th Int. Conf. on Pattern Recognition, IEEE Press (1984) 75–77
5. Pelillo, M., Siddiqi, K., Zucker, S.W.: Matching hierarchical structures using association graphs. Lecture Notes in Computer Science **1407** (1998) 3–13
6. Arcelli, C., Cordella, L., di Baja, G.S., eds.: Visual Form 2001, LNCS 2059. Springer-Verlag (2001)
7. Filatov, A., Gitis, A., Kil, I.: Graph-based handwritten digit string recognition. In: Proceedings of the Third International Conference on Document Analysis and Recognition (Volume 2), IEEE Computer Society (1995) 845
8. Cordella, L.P., Vento, M.: Symbol recognition in documents: A collection of techniques. International Journal on Document Analysis and Recognition (IJDAR) **3** (2000) 73–78
9. Cordella, L.P., Foggia, P., Sansone, C., Vento, M.: Learning structural shape descriptions from examples. Pattern Recognition Letters **23** (2002) 1427–1437
10. Globus, A., Lawtonb, J., Wipkeb, T.: Automatic molecular design using evolutionary techniques. In Globus, A., Srivastava, D., eds.: The Sixth Foresight Conference on Molecular Nanotechnology, Westin Hotel in Santa Clara, CA, USA (1998)
11. Naofumi Homma, T.A., Higuchi, T.: Multiplier block synthesis using evolutionary graph generation. In: Proceedings of the 2004 NASA/DoD Conference on Evolvable Hardware. (2004) 79–82
12. Lohn, J.D., Colombano, S.P.: Automated analog circuit synthesis using a linear representation. Lecture Notes in Computer Science **1478** (1998) 125+
13. Hu, J., Goodman, E.: Wireless access point configuration by genetic programming. In: Proceedings of the 2004 IEEE Congress on Evolutionary Computation, Portland, Oregon, IEEE Press (2004) 1178–1184
14. Wright, M.H.: Optimization method for base station placement in wireless applications. In: Proceedings of the 1998 IEEE Conference on Vehicular Technology, IEEE Press (1998) 287–291
15. Hurley, S.: Planning effective cellular mobile radio networks. IEEE Transactions on Vehicular Technology **51** (2002) 243–253
16. Lee, C., Kang, H.: Cell planning with capacity expansion in mobile communications a tabu search approach. In: IEEE VTC2000. (2000) 1678–1691
17. K.Lieska, Laitinen, E., Lahteenmaki, J.: Radio coverage optimization with genetic algorithms. In: Proceedings of PIMRC. Volume 1. (1998) 318–322
18. Koichi, E., Yoishinori, W.: Automatic cell design for wide area wireless lan systems. Special Issue on Devices and Systems for Mobile Communications **44(4)** (2003)

On the Application of Evolutionary Algorithms to the Consensus Tree Problem

Carlos Cotta

Dept. Lenguajes y Ciencias de la Computación, University of Málaga,
ETSI Informática, Campus de Teatinos, 29071 - Málaga, Spain
ccottap@lcc.uma.es

Abstract. Computing consensus trees amounts to finding a single tree that summarizes a collection of trees. Three evolutionary algorithms are defined for this problem, featuring characteristics of genetic programming (GP), evolution strategies (ES) and evolutionary programming (EP) respectively. These algorithms are evaluated on a benchmark composed of phylogenetic trees computed from genomic data. The GP-like algorithm is shown to provide better results than the other evolutionary algorithms, and than two greedy heuristics defined *ad hoc* for this problem.

1 Introduction

Trees are ubiquitous data structures. They appear in diverse domains such as information retrieval [1], scheduling [2], computer graphics [3], and bioinformatics [4] among others. In all these cases, there exists the need of representing data in a hierarchical fashion, and hence the use of trees. Unfortunately, it is generally the case that finding or constructing the optimal tree for one of these applications is a very hard problem. Consider for example the inference of phylogenetic trees, a problem from the bioinformatics domain. This problem seeks a tree representing the evolutionary history of a collection of species. This is typically done on the basis of molecular information –e.g., DNA sequences– from these species, and can be approached in a number of ways: maximum likelihood, parsimony, distance matrices, etc. [5]. The problem is NP−hard under most models [6, 7, 8].

Hardness results motivate heuristic approaches for finding near-optimal trees. Sticking with the phylogeny problem, both greedy heuristics [9] and metaheuristics [10, 11, 12, 13] have been used. A number of different high-quality trees can be found, each possibly telling something about the *true* solution. Furthermore, the fact that data come from biological experiments, which are not exact, makes near-optimal solutions (even near-optimal with respect to different criteria) be almost as relevant as the actual optimum. It is in this situation where the consensus tree problem comes into play. Essentially, a consensus method tries to summarize a collection of trees provided as input, returning a single tree [14]. This implies identifying common substructures in the input trees and representing these in the output tree.

Consensus trees are extremely important in many domains. For example, in phylogenetic inference, it has been observed that independently derived trees are

G.R. Raidl and J. Gottlieb (Eds.): EvoCOP 2005, LNCS 3448, pp. 58–67, 2005.

unlikely to have spurious clades (or clusters) in common [15]. Thus, the clades appearing in most or all the input trees can be considered reliable. Unfortunately again, finding consensus trees is also a hard problem in general (see e.g. [16].) The use of heuristic techniques is thus in order.

We consider the use of several evolutionary algorithms (EAs) for constructing consensus trees. These evolutionary algorithms differ in the operator set and in the evolution model, and will be compared on a benchmark from the phylogeny domain. The comparison will also include two greedy heuristics defined *ad hoc* for this problem.

2   Background on Consensus Methods

Let T be a strictly binary rooted tree; a LISP-like notation will be used to denote the structure of the tree. Thus, (sLR) is the tree with root s, and with L and R as subtrees, and $()$ is an empty tree. The notation (a) is a shortcut for $(a()())$. Let $\mathcal{L}(T)$ be the set of leaves of T. Each edge e in T defines a bipartition $\pi_T(e) = \langle S_1, S_2 \rangle$, where S_1 are the leaves in $\mathcal{L}(T)$ that can be reached from the root passing through e, and S_2 are the remaining leaves. We define $\Pi(T) = \{\Pi_T(e) \mid e \in T\}$.

There is a variety of consensus methods defined in the literature, differing on the characteristics of the input, and on the output sought. As mentioned before, we concentrate here on trying to represent a collection of trees $\{T_1, \cdots, T_m\}$ as a single tree over $\cup_{i=1}^{m}\mathcal{L}(T_i)$. This can be approached in several ways, such as the tree compatibility problem [17], strict consensus [18], and the median tree [19], among others (see also [20].) While the two first models focus on finding a tree such that $\Pi(T) = \cup_{i=1}^{m}\Pi(T_i)$ (resp. $\Pi(T) = \cap_{i=1}^{m}\Pi(T_i)$), the median tree tries to minimize the sum of differences between T and the input trees, i.e., $\min \sum_{i=1}^{m} d(T, T_i)$. Typically, the distance $d(T, T')$ between trees is defined as the number of non-common bipartitions in $\Pi(T)$ and $\Pi(T')$. This is also termed the partition metric.

The partition metric has several drawbacks. For example, two trees differing solely in the position of one leaf can be maximally different [21]. Alternative metrics can be considered. For example, one can cite the nearest neighbor interchange (NNI), subtree prune and regraft (SPR), and tree bisection and reconnection (TBR) (see [22].) These metrics have their own drawbacks though: computing the NNI metric or the TBR metric is $NP-$hard [22, 23]; the complexity of computing SPR is unknown, but it is conjectured to be $NP-$hard as well.

We employ, then, an alternative metric called TreeRank [24]. Given trees T and T', the TreeRank score provides a measure of the topological relationships in T that are found to be the same or similar in T'. This is done with the help of an auxiliary data structure termed the *UpDown matrix*. This structure consists of a pair of matrices for each tree T, the *Up matrix* U_T, and the *Down matrix* D_T. These matrices are intended to capture information on the hierarchical structure

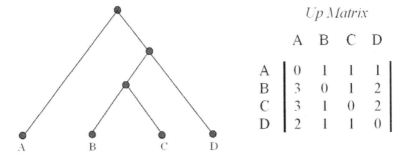

Fig. 1. A tree and its Up matrix

of a tree. More precisely, they store the number of edges that must be traversed upwards or downwards in order to go from a certain leaf to another one (see Fig. 1.) Formally,

$$\forall u \in L \; \forall v \in R \begin{cases} U_{(sLR)}[u,v] = 1 + depth(u,L) \\ U_{(sLR)}[v,u] = 1 + depth(v,R) \end{cases} \tag{1}$$

where $depth(x,T)$ is the depth of node x in tree T (i.e., the length of the path from the root of T to x.) Notice that $U[u,u] = 0$, and $U[u,v] = D[v,u]$ (hence it suffices using just one of the matrices to have complete information on the hierarchical structure.) Now, the TreeRank score is computed as[1]:

$$TreeRank(T,T') = \left(1 - \frac{UpDownDist(T,T')}{\sum_{u,v \in \mathcal{L}(T)} U_T[u,v]}\right) \times 100\% \tag{2}$$

where

$$UpDownDist(T,T') = \sum_{u,v \in \mathcal{L}(T)} |U_T[u,v] - U_{T'}[u,v]| \tag{3}$$

The UpDown matrix can be calculated in $O(|\mathcal{L}|^2)$ as shown in [24]. Notice that if T and T' are identical, then the TreeRank score is 100%. If no common substructure exists in T and T', then the TreeRank score will be near 0%, or even negative if the hierarchical structure is "reversed" in both trees. This TreeRank measure is currently being used in TreeBASE[2] –one of the most widely used phylogenetic databases– for the purposes of handling queries for similar trees.

---

[1] We have adapted this definition of TreeRank to the fact that $\mathcal{L}(T) = \mathcal{L}(T')$. See [24] for a general formulation in the case that $\mathcal{L}(T) \neq \mathcal{L}(T')$.

[2] http://www.treebase.org

3   Heuristics for the Consensus Tree Problem

We describe two greedy and three evolutionary heuristics for the consensus tree problem.

3.1   Greedy Heuristics

The most typical heuristics for tree construction are essentially greedy (see e.g. [25].) In this sense, we firstly consider a greedy heuristic that constructs a consensus tree incrementally, adding one leaf at a time. This is done by testing all possible insertion points within the partially-constructed tree, and retaining the position that yields the best value of the TreeRank distance. This algorithm is denoted by H1, and its pseudocode is shown below:

> **Heuristic H1** $(\{T_1 \cdots, T_m\}, \langle o_1, \cdots, o_n \rangle)$
> 1. Let $A \leftarrow (h(o_1)(o_2))$
> 2. **for each** $j \in \mathbb{N}_{n-2}$ **do**
>   (a) Let the branches of A be numbered from 1 to $2j$.
>   (b) **for each** insertion point $i \in \mathbb{N}_{2j}$ **do**
>   i. Let $A_i \leftarrow A$
>   ii. Insert leaf o_{j+2} in branch i of A_i.
>   (c) Let $A \leftarrow \arg\min \{\sum_{i=1}^{m} TreeRank(A', T_i) \mid A' \in \{A_1, \cdots, A_{2j}\}\}$.
> 3. **return** A.

In this definition, $\mathbb{N}_k$ stands for the natural numbers in $[1..k]$. As it can be seen, heuristic H1 is fed with both the collection of trees whose consensus is sought, and with a permutation of the leaves that determines the order in which they will be inserted in the tree. The two first leaves in the permutation are used to construct the initial subtree.

The second greedy heuristic we consider is a variant of the previous one. It is also close in spirit to an approximation algorithm that was devised by Phillips and Tarnow [16] for the asymmetric median tree problem. The main idea of this heuristic –that we denote by H2– is to consider all possible pairs of trees in the target collection, constructing a greedy consensus tree for each of these pairs. This can be done on the basis of the H1 heuristic described before, as shown in the following pseudocode:

> **Heuristic H2** $(\{T_1 \cdots, T_m\}, \langle o_1, \cdots, o_n \rangle)$
> 1. Let $best \leftarrow -\infty$.
> 2. **for each** $i, j \in \mathbb{N}_m, i < j$ **do**
>   (a) Let $A \leftarrow$ H1$(\{T_i, T_j\}, \langle o_1, \cdots, o_n \rangle)$.
>   (b) Let $d \leftarrow \sum_{k=1}^{m} TreeRank(A, T_k)$.
>   (c) **if** $d > best$ **then**
>   i. Let $best \leftarrow d$.
>   ii. Let $T \leftarrow A$.
> 3. **return** T.

This algorithm admits a minor variant in which the leaf sequence is not the same for all pairs of trees, but separately computed for each of these pairs. This is precisely the version of H2 that we have considered in the experimentation, as it will be described in Sect. 4.

3.2   Evolutionary Heuristics

When using an EA for evolving a non-trivial data structure such as a tree, there exist two main alternatives [26, 27]: either perform the search directly on the space of all n−leaf trees, or conducting the search in an auxiliary space, using a decoder in order to construct the actual trees. In this work, the evolutionary heuristics considered are all defined in the line of the former approach, i.e., each individual in the EA population directly represents a tentative consensus tree. Given this choice of representation, appropriate operators for recombination and mutation must be defined.

First of all, consider the recombination operator. This operator must take information chunks from two parents, and combine them to create a descendant. In this case, these information chunks can be naturally expressed as subtrees. Hence, the recombination process can be approached much like it is typically done in genetic programming (GP), in terms of pruning and grafting subtrees. An important consideration must be nevertheless made: all trees must have all n different leaves, with neither repetitions nor omissions. This means that when attempting to transfer a subtree T' from a parent to another, all leaves in $\mathcal{L}(T')$ must be deleted in advance from the latter parent. Let A_1 and A_2 be the trees being recombined; the whole process would be then as follows (cf. [10, 12]):

Operator Prune-Delete-Graft Recombination (T_1, T_2)
1. Select a subtree T from T_2.
2. **for each** leaf $o \in \mathcal{L}(T)$ **do**
   (a) Find subtree U in T_1 such that $U = (h(o)U')$ or $U = (hU'(o))$.
   (b) Replace U by U' in T_1.
3. Select a random subtree V from T_1.
4. Replace V by $V' = (h'TV)$ in T_1, where h' is a new internal node.

Notice that the fact that the set of leaves is the same for all trees during all the run, makes in principle the mutation operator be dispensable: topological diversity can be produced by recombination as well (for instance, notice that recombining a tree with itself can yield a different tree.) We thus consider a first EA in which reproduction is done exclusively by means of recombination, and denote it by GP.

An alternative approach to that described above is possible, namely using no recombination but just mutation. In this sense, there are numerous possibilities for performing mutation on trees, see e.g. [11]. In this work, we have used two mutation operators:

– SCRAMBLE: Let T be the tree to be mutated; firstly, a subtree T' in T is selected at random and pruned from T. Then, a new random subtree T'' is generated, with $\mathcal{L}(T') = \mathcal{L}(T'')$, and grafted at the original location of T'. In other words, the topology of a certain subtree of T is rearranged at random.

- SWAP: it consists of selecting two leaves of T at random, subsequently swapping their places.

We have considered two mutation-based evolutionary heuristics. The first one is denoted by EP, and consists of applying either SCRAMBLE or SWAP to an individual with 50% probability. The second one is denoted by ES, and just utilizes SCRAMBLE. However, an internal parameter is used to control the size of the subtrees to be rearranged. This parameter can be regarded as a step size, and evolves with each individual. To be precise, whenever a tree is to be mutated, the step size is firstly mutated using a gaussian distribution, and the mutated parameter value is used to select a subtree of the appropriate size to be fed to SCRAMBLE.

The acronyms –GP, EP, ES– used to denote each of the algorithms are intended to reflect their similarity with the corresponding EA family, namely genetic programming, evolutionary programming, and evolution strategies. However, it must be noted that the algorithms have been kept simple, and do not fully exploit the potential of the corresponding paradigm. This has been done so in order to obtain a first assessment on the usefulness of the different operators and evolution models for this problem.

4   Computational Results

The experiments have been performed using two test suites. Both of them comprise three different instances obtained from the biological domain. To be precise, three datasets for phylogenetic inference have been downloaded from TreeBASE. The size of these datasets ranges from 134 up to 178 leaves, as shown in Table 1. In the first test suite –termed AGGLOM– each dataset has been fed to three classical agglomerative clustering techniques: *single-link* [28], *complete-link* [9], and *average link* [29]. Thus, a collection of three trees is obtained in each case. In the second test suite –termed SCATTER– we consider for each dataset the different trees obtained in ten runs of a scatter search metaheuristic [30].

Table 1. The test suites considered in the experimentation. Mean distance refers to the average of distances of each tree in the collection to the whole tree set

	M877		M971		M808	
	AGGLOM	SCATTER	AGGLOM	SCATTER	AGGLOM	SCATTER
mean distance	44.78	94.56	56.71	89.79	31.59	88.80
#leaves	134		158		178	
source	[31]		[32]		[33]	

The parameterization of the algorithms is the following: GP is a steady-state EA, with a population of 100 individuals, using binary tournament for selection; EP has also a population size of 100 individuals, but uses flat selection, i.e., each individual is mutated once, thus yielding a population of 100 descendants.

Subsequently, the best 100 out of the 100 existing individuals and the 100 newly created descendants constitute the population for the next generation (i.e., a *plus* replacement strategy); as to the ES algorithm, it follows a (16,100) evolution model, and uses $n/10$ (where n is the number of leaves) as the hyperparameter for the gaussian mutation of step sizes. In all cases, the algorithms are run for a total number of 250,000 fitness evaluations. Also, trees in the target collection are injected in the initial population.

Regarding the greedy heuristics, H1 is run using leaf permutations *compatible* with the topologies of trees in the collection (one leaf permutation for each tree)[3]. To be precise, let $T = (hLR)$; the sequence $\langle T \rangle = \langle L \rangle :: \langle R \rangle$, where :: represents sequence concatenation, and $\langle L \rangle$ and $\langle R \rangle$ are computed recursively ($\langle \langle a \rangle \rangle = \langle a \rangle$), is compatible with T. H1 has then been run using sequences $\langle T_i \rangle, \cdots, \langle T_m \rangle$. The same is done in the internal invocations of H1 within H2.

Table 2. Results (averaged for 20 runs) of the EAs and the greedy heuristics on the two test suites considered. *sdv.* and *med.* stand for standard deviation and median respectively

AGGLOM test suite								
	M877			M971			M808	
	best	mean ± sdv. med.		best	mean ± sdv. med.		best	mean ± sdv. med.
GP	51.80	51.68 ± 0.06 51.68		63.32	63.26 ± 0.04 63.27		48.45	48.41 ± 0.02 48.42
EP	50.91	50.87 ± 0.03 50.87		62.55	62.49 ± 0.02 62.48		48.16	48.16 ± 0.00 48.16
ES	50.55	50.34 ± 0.06 50.33		62.48	62.48 ± 0.01 62.48		48.15	48.15 ± 0.00 48.15
H1	43.30	36.81 ± 5.80 37.92		61.21	57.80 ± 4.22 60.33		38.01	35.29 ± 1.94 34.23
H2	49.53	40.08 ± 6.73 39.40		61.06	55.37 ± 5.39 57.74		47.92	32.22 ± 12.98 32.30

SCATTER test suite								
	M877			M971			M808	
	best	mean ± sdv. med.		best	mean ± sdv. med.		best	mean ± sdv. med.
GP	96.03	96.03 ± 0.00 96.03		91.90	91.90 ± 0.00 91.90		91.04	90.93 ± 0.09 90.97
EP	95.81	95.81 ± 0.00 95.81		91.90	91.90 ± 0.00 91.90		89.96	89.94 ± 0.03 89.94
ES	95.81	95.81 ± 0.00 95.81		91.90	91.90 ± 0.00 91.90		89.89	89.87 ± 0.02 89.87
H1	83.81	81.33 ± 2.22 82.31		86.62	73.06 ± 8.94 74.07		84.58	78.27 ± 3.22 77.93
H2	90.65	83.69 ± 3.43 84.25		86.49	74.48 ± 7.53 75.29		81.96	76.47 ± 3.70 77.52

The results are shown in Table 2. Notice firstly that all evolutionary heuristics perform clearly better than the greedy heuristics. The latter can hardly produce consensus trees of score similar to those already in the corresponding collection. The evolutionary algorithms can however produce consensus trees of high quality, achieving overall scores superior to the original trees. This implies that the topological information of the collection is being effectively summarized

---

[3] A leaf sequence is said to be compatible with a tree topology if there exists a layout of the tree in which its leaves are ordered from left to right as in the sequence, and no two branches cross.

Table 3. Results (averaged for 20 runs) of the GP algorithm without seeding the initial population on AGGLOM (#1) and SCATTER (#2)

	M877			M971			M808		
	best	mean ± sdv.	med.	best	mean ± sdv.	med.	best	mean ± sdv.	med.
#1	46.80	45.19 ± 1.16	45.32	57.51	56.63 ± 0.58	56.67	40.78	39.21 ± 0.86	39.18
#2	81.99	79.86 ± 1.17	79.93	79.88	78.05 ± 0.78	78.01	79.12	77.05 ± 0.85	77.03

in the consensus tree. The relative performance of the EAs indicates that EP is better than ES, and that GP provides the best outcome. A non-parametric statistical test (a Wilcoxon ranksum test [34]), has been used to corroborate the significance of these results. Fitness differences are always statistically significant (at the standard 5% significance level), except for the EP vs ES comparison on the m877 instance of the SCATTER test suite, and for all EAs on the m971 instance of the same test suite.

A final experiment has been done to confirm the usefulness of injecting the original tree collection in the initial population. The results are shown in Table 3 just for the GP algorithm. As it can be seen, the performance drop is dramatic. Without seeding, the algorithm would require much longer execution times in order to achieve the performance level of its seeded counterpart.

5   Conclusions

We have presented five heuristics for summarizing a collection of trees into a consensus tree, using the TreeRank measure as the scoring metric. From these five, the three evolutionary heuristics have been shown to be effective in summarizing within a single tree topological information contained in the target tree collection. Furthermore, a recombination-based EA has been shown to provide the better results. An important part of the performance of these algorithms is due to the seeding of the initial population.

Regarding the EAs, there is much room for improvement in the underlying evolution model, as anticipated in Sect. 3.2. For example, the EP algorithm could incorporate self-adaptation as well [35], evolving the number of times each mutation operator is used, or a hyperparameter controlling a probability distribution –e.g., a Poisson distribution– over this number of mutations. This would bring closer the ES and EP approaches, and constitutes a line for future developments.

With respect to the greedy heuristics, their performance is not satisfactory. Nevertheless, if not as stand-alone techniques, they can still be useful embedded within an EA. As mentioned in Sect. 3.1, both H1 and H2 must be fed with a particular leaf sequence. It is then conceivable to have an EA evolving leaf permutations to be fed to these heuristics. This way, they would act as decoders, and the EA could benefit from their greedy functioning. This is another line for future developments.

Acknowledgements

Thanks are due to the anonymous reviewers for their useful suggestions. The author is partially supported by Spanish MCyT, and FEDER under contract TIC2002-04498-C05-02.

References

1. Foster, C.: Information retrieval: information storage and retrieval using AVL trees. In: Proceedings of the 1965 20th ACM National Conference, New York NY, ACM Press (1965) 192–205
2. Garofalakis, M., Özden, B., Silberschatz, A.: Resource scheduling in enhanced pay-per-view continuous media databases. In Jarke, M., et al., eds.: Proceedings of the 1997 International Conference on Very Large Databases, Athens, Greece, Morgan Kaufmann (1997) 516–525
3. Naylor, B.: Partitioning tree image representation and generation from 3D geometric models. In Booth, K., Fournier, A., eds.: Proceedings of the 1992 Conference on Graphics Interface, San Francisco CA, Morgan Kaufmann (1992) 201–212
4. Holmes, S.: Phylogenies: An overview. In Halloran, M., Geisser, S., eds.: Statistics and Genetics. Springer-Verlag, New York NY (1999) 81–119
5. Kim, J., Warnow, T.: Tutorial on phylogenetic tree estimation. In Lengauer, T., et al., eds.: Proceedings of the 7th International Conference on Intelligent Systems for Molecular Biology, Heidelberg, AAAI Press (1999) 196–205
6. Day, W., Johnson, D., Sankoff, D.: The computational complexity of inferring rooted phylogenies by parsimony. Mathematical Biosciences **81** (1986) 33–42
7. Foulds, L., Graham, R.: The Steiner problem in phylogeny is NP−complete. Advances in Applied Mathematics **3** (1982) 439–49
8. Wu, B., Chao, K.M., Tang, C.: Approximation and exact algorithms for constructing minimum ultrametric trees from distance matrices. Journal of Combinatorial Optimization **3** (1999) 199–211
9. King, B.: Step-wise clustering procedures. Journal of the American Statistical Association **69** (1967) 86–101
10. Moilanen, A.: Searching for the most parsimonious trees with simulated evolution. Cladistics **15** (1999) 39–50
11. Andreatta, A., Ribeiro, C.: Heuristics for the phylogeny problem. Journal of Heuristics **8** (2002) 429–447
12. Cotta, C., Moscato, P.: Inferring phylogenetic trees using evolutionary algorithms. In Merelo, J., et al., eds.: Parallel Problem Solving From Nature VII. Volume 2439 of Lecture Notes in Computer Science. Springer-Verlag, Berlin (2002) 720–729
13. Barker, D.: LVB: parsimony and simulated annealing in the search for phylogenetic trees. Bioinformatics **20** (2004) 274–275
14. Bryant, D.: A classification of consensus methods for phylogenetics. In Janowitz, M., et al., eds.: Bioconsensus. DIMACS-AMS (2003) 163–184
15. Swofford, D.: When are phylogeny estimates from molecular and morphological data incongruent? In Miyamoto, M., Cracraft, J., eds.: Phylogenetic analysis of DNA sequences. Oxford University Press (1991) 295–333
16. Phillips, C., Warnow, T.: The asymmetric median tree a new model for building consensus trees. Discrete Applied Mathematics **71** (1996) 311–335

17. Gusfield, D.: Efficient algorithms for inferring evolutionary trees. Networks **21** (1991) 19–28
18. Day, W.: Optimal algorithms for comparing trees with labeled leaves. Journal of Classiffication **2** (1985) 7–28
19. Barthélemy, J.P., McMorris, F.: The median procedure for n-trees. Journal of Classiffication **3** (1986) 329–334
20. Östlin, A.: Constructing evolutionary trees. Algorithms and complexity. PhD thesis, Department of Computer Science, Lund University (2001)
21. Penny, D., Hendy, M.: The use of tree comparison metrics. Systematic Zoology **34** (1985)
22. Allen, B., Steel, M.: Subtree transfer operations and their induced metrics on evolutionary trees. Annals of Combinatorics **5** (2001) 1–15
23. DasGupta, B., He, X., Jiang, T., Li, M., Tromp, J., Zhang, L.: On computing the nearest neighbor interchange distance. In Du, D., Pardalos, P., Wang, J., eds.: Proceedings of the DIMACS Workshop on Discrete Problems with Medical Applications. Volume 55 of DIMACS Series in Discrete Mathematics and Theoretical Computer Science., American Mathematical Society (2000) 125–143
24. Wang, J., Shan, H., Shasha, D., Piel, W.: Treerank: A similarity measure for nearest neighbor searching in phylogenetic databases. In: Proceedings of the 15th International Conference on Scientific and Statistical Database Management, Cambridge MA, IEEE Press (2003) 171–180
25. Jain, A., Murty, N., Flynn, P.: Data clustering: A review. ACM Computing Surveys **31** (1999) 264–323
26. Raidl, G., Julstrom, B.: Edge sets: an effective evolutionary coding of spanning trees. IEEE Transactions on Evolutionary Computation **7** (2003) 225–239
27. Rothlauf, F.: Representations for Genetic and Evolutionary Algorithms. Studies in Fuzziness and Soft Computing. Physica, Heidelberg (2002)
28. Sneath, P., Sokal, R.: Numerical Taxonomy. Freeman, London, UK (1973)
29. Ward, J.: Hierarchical grouping to optimize an objective function. Journal of the American Statistical Association **58** (1963) 236–244
30. Cotta, C.: Scatter search with path relinking for phylogenetic inference. European Journal of Operational Research (2005) (to appear).
31. Hibbett, D., Donoghue, M.: Analysis of character correlations among wood decay mechanisms, mating systems, and substrate ranges in homobasidiomycetes. Systematic Biology **50** (2001) 1–27
32. Binder, M., Hibbett, D., Molitoris, H.: Phylogenetic relationships of the marine gasteromycete Nia vibrissa. Mycologia **93** (2001) 679–688
33. Hibbett, D., Gilbert, L.B., Donoghue, M.: Evolutionary instability of ectomycorrhizal symbioses in basidiomycetes. Nature **407** (2000) 506–508
34. Lehmann, E., D'Abrera, H.: Nonparametrics: Statistical Methods Based on Ranks. Prentice-Hall, Englewood Cliffs, NJ (1998)
35. Fogel, D.: Evolutionary Computation: Toward a New Philosophy of Machine Intelligence. IEEE Press, Piscataway, NJ (2000)

Analyzing Fitness Landscapes for the Optimal Golomb Ruler Problem

Carlos Cotta and Antonio J. Fernández

Dept. Lenguajes y Ciencias de la Computación, ETSI Informática,
University of Málaga, Campus de Teatinos, 29071 - Málaga, Spain
{ccottap,afdez}@lcc.uma.es

Abstract. We focus on the Golomb ruler problem, a hard constrained combinatorial optimization problem. Two alternative encodings are considered, one based on the direct representation of solutions, and one based on the use of an auxiliary decoder. The properties of the corresponding fitness landscapes are analyzed. It turns out that the landscape for the direct encoding is highly irregular, causing drift to low-fitness regions. On the contrary, the landscape for the indirect representation is regular, and exhibits comparable fitness-distance correlation to that of the former landscape. These findings are validated in the context of variable neighborhood search.

1 Introduction

Golomb rulers are a class of undirected graphs that, unlike usual rulers, measure more discrete lengths than the number of marks they carry. This is due to the fact that on any given ruler, all differences between pairs of marks are unique. This feature makes Golomb rulers really interesting in many practical applications, such as carrier frequency assignment [1], radio communication [2], X-ray crystallography [3], pulse phase modulation [4], and design of orthogonal codes [5, 6], among others [7, 8, 9]. Needless to say, it also introduces numerous constraints that hinder the search of short feasible rulers, let alone *optimal* Golomb rulers (OGR, i.e., the shortest Golomb ruler for a number of marks).

To date, no efficient algorithm is known for finding the shortest Golomb ruler for a certain number of marks: massive parallelism projects have been undertaken for several months in order to find the optimum instances of up to 23 marks [10]. Being such an extremely difficult combinatorial task, the Golomb ruler problem represents an ideal scenario for deploying the arsenal of evolutionary optimization.

In Sect. 2.2 we discuss some of the non-evolutionary techniques employed so far to solve OGRs. With respect to evolutionary ones, to the best of our knowledge, there have been four attempts to apply evolutionary algorithms (EAs) to the search for OGRs (see Sect. 2.3). These works are essentially empirical, and little has been so far done on the analysis of the properties of the underlying combinatorial landscapes. In this paper, we tackle this issue by analyzing two major problem representations under which evolutionary search can be conducted on this problem. To be precise, we consider the direct representation of

G.R. Raidl and J. Gottlieb (Eds.): EvoCOP 2005, LNCS 3448, pp. 68–79, 2005.

solutions, and an indirect, decoder-based representation that uses a GRASP-like mechanism to perform the genotype-to-phenotype mapping. These landscapes are examined in Sect. 3, paying special attention to landscape regularity, and correlation measures. The variable neighborhood search metaheuristic is used to corroborate the outcome of this analysis in Sect. 4.

2   Background

The OGR problem can be classified as a fixed-size subset selection problem, such as e.g., the p–median problem [11]. It exhibits some very distinctive features though. A brief overview of the problem, and how it has been tackled in the literature is provided below.

2.1   Golomb Rulers

A n-mark Golomb ruler is an ordered set of n distinct non-negative integers, called marks, $a_1 < ... < a_n$, such that all the differences $a_i - a_j$ $(i > j)$ are distinct. Clearly we may assume $a_1 = 0$. By convention, a_n is the length of the Golomb ruler. A Golomb ruler with n marks is an optimal Golomb ruler if, and only if, (i) there exists no other n-mark Golomb rulers having smaller length, and (ii) the ruler is canonically "smaller" with respect to the the equivalent rulers. This means that the first differing entry is less than the corresponding entry in the other ruler. Fig. 1 shows an OGR with 4-marks. Observe that all distances between any two marks are different.

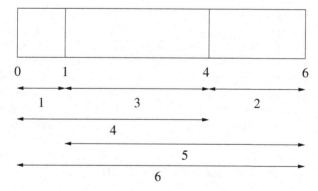

Fig. 1. A Golomb ruler with 4 marks

Typically, Golomb rulers are represented by the values of the marks on the ruler, i.e., in a n-mark Golomb ruler, $a_i = x$ $(1 \leqslant i \leqslant n)$ means that x is the mark value in position i. The sequence $(0, 1, 4, 6)$ would then represent the ruler in Fig. 1. An alternative representation consists of representing the Golomb ruler via the lengths of its segments, where the length of a segment of a ruler is defined as the distance between two consecutive marks. Therefore, a Golomb ruler can be represented with $n-1$ marks specifying the lengths of the $n-1$ segments that compose it. In the previous example, the sequence $(1, 3, 2)$ would encode the ruler depicted in Fig. 1.

2.2   Finding OGRs

The OGR problem has been solved using very different techniques. The evolutionary techniques found in the literature to obtain OGRs are described in Sect. 2.3. We provide here a brief overview of some of the most popular non-evolutionary techniques used for this problem.

Firstly, it is worth mentioning some classical algorithms used to generate and verify OGRs such as the *Scientific American* algorithm [12], the *Token Passing* algorithm (created by Professor Dollas at Duke University) and the *Shift* algorithm [13], all of them compared and described in [8].

In general, both non-systematic and systematic methods have been applied to find OGRs. Regarding the former, we can cite for example the use of geometry tools (e.g., projective plane construction and affine plane construction). With these approaches, one can compute very good approximate values for OGR with up to 158 marks [14]. As to systematic (exact) methods, we can mention the utilization of branch-and-bound algorithms combined with a depth first search strategy (i.e., backtracking algorithms), making use of upper-bounds set equal to the minimum length in the experiments. In this sense there exist several proposals: for example, Shearer [15] computed OGRs up to 16 marks. This approach has been also followed in massive parallelism initiatives such as the OGR project mentioned before. This project has been able to find the OGRs with a number of marks between 20 and 23, although it took several months to find optimum for each of those instances [8, 9, 16, 10].

Constraint programming techniques have also been used, although with limited success. For example, Smith and Walsh [17] obtained interesting results in terms of nodes in the branching schema. However, computation times are far from the results obtained by previous approaches. More recently, Galinier *et al.* [18] proposed a combination of constraint programming and sophisticated lower bounds for finding OGRs. They showed that using the same bound on different ways affects not only to the number of branches in the search tree but also to the computation time.

2.3   Evolutionary Approaches to the OGR

In this section will restrict here just to the evolutionary approaches to solve OGRs considered so far in the literature. In essence, two main approaches can be considered for tackling this problem. The first one is the *direct* approach, in which the EA conducts the search in the space $\mathcal{S}_G$ of all possible Golomb rulers. The second one is the *indirect* approach, in which an auxiliary $\mathcal{S}_{aux}$ space is used by the EA. In this latter case, a decoder [19] must be utilized in order to perform the $\mathcal{S}_{aux} \longrightarrow \mathcal{S}_G$ mapping. Both approaches will be discussed below.

Direct Approaches. In 1995, Soliday, Homaifar and Lebby [20] used a genetic algorithm on different instances of the Golomb ruler problem. They chosen the alternative formulation already mentioned where each chromosome is composed by a permutation of $n-1$ integers that represents the sequence of the $n-1$ lengths of its segments. Two evaluation criteria were followed: the overall length of the ruler, and the number of repeated measurements. This latter quantity was used in order to penalize infeasible solutions. The mutation operator consisted of either a permutation in the segment order, or a change in the segment lengths. As to crossover, it was designed to guarantee that descendants are valid permutations.

Later, Feeney studied the effect of hybridizing genetic algorithms with local improvement techniques to solve Golomb rulers [7]. The representation used consisted of an array of integers corresponding to the marks of the ruler. The crossover operator was similar to that used in Soliday *et al.*'s approach although a sort procedure was added at the end. The mutation operator consisted in adding a random amount in the range $[-x, x]$ –where x is the maximum difference between any pair of marks in any ruler of the initial population– to the segment mark selected for mutation. As it will be shown later, we can use a similar concept in order to define a distance measure on the fitness landscape.

Indirect Approaches. Pereira *et al.* presented in [21] a new EA approach using the notion of random keys [22] to codify the information contained in each chromosome. The basic idea consists of generating n random numbers (i.e., the keys) sampled from the interval $[0, 1]$ and ordered by its position in the sequence $1, \ldots, n$; then the keys are sorted in decreasing order. The indices of the keys thus result in a feasible permutation of $\{1, \cdots, n\}$. A similar evaluation criteria as described in [20] was followed. They also presented an alternative algorithm that adds a heuristic, favoring the insertion of small segments.

A related approach has been presented in [23]. This proposal incorporates ideas from greedy randomized adaptive search procedures (GRASP) [24] in order to perform the genotype-to-phenotype mapping. More precisely, the mapping procedure proceeds by placing each of the $n-1$ marks (the first mark is assumed to be $a_1 = 0$) one at a time; the $(i + 1)^{\text{th}}$ mark can be obtained as $a_{i+1} = a_i + l_i$, where $l_i \geqslant 1$ is the i–th segment length. Feasible segment lengths (i.e., those not leading to duplicate measurements) can be sorted in increasing order. Now, the EA needs only specifying at each step the index of a certain segment within this list (obviously, the contents of the list are different in each of these steps). This implies that each individual would be a sequence $\langle r_1, \cdots, r_{n-1} \rangle$, where r_i would be the index of the segment used in the i–th iteration of the construction algorithm. Notice that in this last placement step it does not make sense to pick any other segment length than the smallest one. For this reason, $r_{n-1} = 1$; hence, solutions need only specify the sequence $\langle r_1, \cdots, r_{n-2} \rangle$. This representation of solutions is orthogonal [25], i.e., any sequence represents a feasible solution, and hence, standard operators for crossover and mutation can be used to manipulate them. This GRASP-based approach is reported to perform better than the previous indirect approach, and hence we use it in our further analysis.

3   Fitness Landscapes for the Golomb Ruler Problem

The notion of fitness landscapes was firstly introduced in [26] to model the dynamics of evolutionary adaptation in Nature. The fitness landscape analysis of a problem can help to identify its structure in order to improve the performance of search algorithms (e.g., to predict the behavior of a heuristic search algorithm, or to exploit some of its specific properties). For this reason, this kind of analysis has become a valuable tool for evolutionary-computation researchers.

In this section, we will analyze the fitness landscapes resulting from the two problem representations described before, the direct encoding of rulers, and the use of a GRASP-based decoder. We will assume below that n is the number of marks for a specific Golomb ruler $\mathbb{G}_n$, and that $a = \langle a_1, \ldots, a_n \rangle$ and $b = \langle b_1, \ldots, b_n \rangle$ are arbitrary solutions from $\mathbb{G}_n$. Analogously, $r = \langle r_1, \cdots, r_{n-2} \rangle$ and $r' = \langle r'_1, \cdots, r'_{n-2} \rangle$ are arbitrary vectors from $\mathbb{N}^{n-2}$, representing the vector of indices for selecting segment lengths. We denote by ψ the bijective function performing the genotype-to-phenotype mapping $\mathbb{N}^{n-2} \to \mathbb{G}_n$.

3.1   Distance Measures and Neighborhood Structure

We define a fitness landscape for the OGR as a triple $\langle S, f, d \rangle_n$ where $S = \mathbb{G}_n$ is the set of all the n-mark Golomb rulers (i.e., the solution set), f is a fitness function that attaches a fitness value to each of the points in S (i.e., $f(a)$ is equal to a_n, the length of a), and $d : S \times S \to \mathbb{N}$ is a function that measures a distance between any two points in S. We have defined one distance function for each of the Golomb ruler representations already commented. Specifically for the direct formulation (i.e., that based on lists of marks) we have defined the distance function d as follows:

$$d(a, b) = \max\{| b_i - a_i |, 1 \leq i \leq n\} . \tag{1}$$

In other words, $d(a, b)$ returns the maximum difference between any two corresponding marks in a and b. Also, for our indirect formulation (i.e., the GRASP-based formulation) we have defined the distance function d as the L1 norm (the Manhattan distance) on the vector of indices, i.e.,

$$d(a, b) = d(\psi(r), \psi(r')) = \sum_{i=1}^{n-2} | r_i - r'_i | . \tag{2}$$

A first issue to be analyzed regards the neighborhood structure induced by these distance measures. More precisely, consider the number of solutions reachable from a certain point in the search space, by a search algorithm capable of making jumps of a given distance. In the direct formulation, this number of solutions turns out to be variable for each point of S, as shown in Fig. 2. We have implemented and used a logic-programming based constraint solver to solve the Golomb ruler constraint satisfaction problem for an arbitrary number of marks. Our solver, implemented in GNU Prolog [27], is based on the model proposed in [28]. In particular, the solver generates a list of all possible distances between

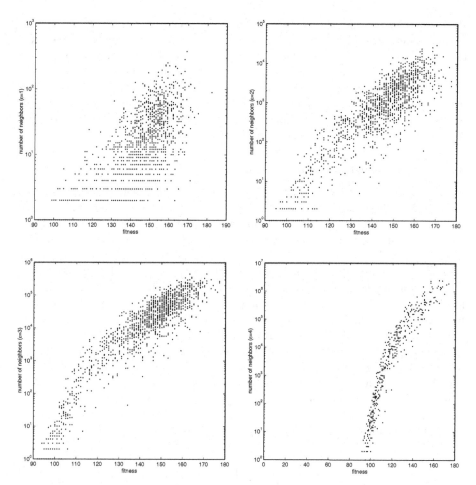

Fig. 2. Number of neighbors for different values of the local radius ϵ in a 12-mark Golomb ruler problem. From top to bottom and left to right, $\epsilon = 1, 2, 3$, and 4. Notice the log-scale in the Y-axis.

any pair of marks i, j ($i < j$ and $i, j \in \{1, \ldots, n\}$) in the ruler and then imposes a global constraint *all-different* on this list instead of imposing the set of binary inequalities between any two marks i, j. The efficiency is further improved by adding some redundant constraints leading to an improvement of the domain pruning. This solver calculates the number of possible neighbors that are located within a given distance ϵ (called the *local radius*) of certain solution a (i.e., it obtains the cardinality of the set $\{\langle c_1, \ldots, c_n \rangle \in S \mid a_i - \epsilon \leq c_i \leq a_i + \epsilon, 1 \leq i \leq n\}$). The solver is then applied to a large sample of solutions covering a wide range of fitness values.

The outcome of this experiment indicates that the connectivity of the fitness landscape increases with worse fitness values. Furthermore, this effect is stronger as we increase the neighborhood radius (see Fig. 2). This kind of irregularity is detrimental for search algorithm navigating this landscape [29], since the neighborhood structure tends to guide the search towards low-fitness regions. This means that a search algorithm on this landscape would have to be continuously fighting against this drifting force. On the contrary, notice that the fitness landscape of the indirect formulation is perfectly regular, since its topology is isomorphic to $\mathbb{N}^{n-2}$. In principle, this regularity makes this landscape more navigable since no underlying drift effect exists.

3.2   Fitness-Distance Correlation

Fitness-distance correlation (FDC) [30] is one of the most widely used measures for assessing the structure of the landscape. It also constitutes a very informative measure to evaluate the problem difficulty for evolutionary algorithms [31]. FDC allows quantifying the correlation between fitness values, and the distance to the nearest optimum in the search space. Landscapes with a high FDC typically exhibit a *big valley structure* [32] (this is not always the case though [30, 33]).

It is typically assumed that low FDC is associated with problem difficulty for local search. Nevertheless, the interplay of this property with other landscape features is not yet well understood. Indeed, it will be later shown how landscape ruggedness and neighborhood irregularity can counteract high FDC values.

Focusing on the problem under consideration, the optimum value opt_n for n-mark Golomb rulers is known (up to $n = 24$, enough for our analysis). We can then obtain a sample of m locally-optimal solutions $A = \{a_1, \ldots, a_m\} \subset S$ and easily calculate the sets $F = \{f_i \mid f_i = f(a_i), 1 \le i \le m, a_i \in A\}$ and $D = \{d_i \mid d_i = d(a_i, opt_n), 1 \le i \le m, a_i \in A\}$. Then we can compute the correlation coefficient as $FDC = C_{FD}/(\sigma_F \sigma_D)$, where

$$C_{FD} = \frac{1}{m} \sum_{i=1}^{m} (f_i - \overline{f})(d_i - \overline{d}) \tag{3}$$

is the covariance of F and D, and $\sigma_F, \sigma_D, \overline{f}$ and $\overline{d}$ are, respectively, the standard deviations and means of F and D. Observe that this definition depends on the definition of the distance function, and as shown in Section 3.1, we consider two different definitions for the two problem representations.

The FDC values computed for the two representations are shown in Fig. 3. In all cases, locally optimal solutions are computed by using hill climbing from a fixed sample of seed feasible solutions. Notice firstly the high correlation for the direct formulation, specially for low values of the local radius ϵ. This can be explained by the fact that the fitness of a solution is actually the value of the last mark, and this value will not change above the given ϵ within the neighborhood. FDC starts to degrade for increasing values of this local radius. To be precise, FDC values for $\epsilon = 1$ up to $\epsilon = 4$ are 0.9803, 0.9453, 0.8769, and 0.8221 respectively. In the case of the indirect formulation ($\epsilon = 1$), the FDC value is 0.8478, intermediate between $\epsilon = 3$, and $\epsilon = 4$. These results indicate that the indirect formulation can attain FDC values comparable to those of the

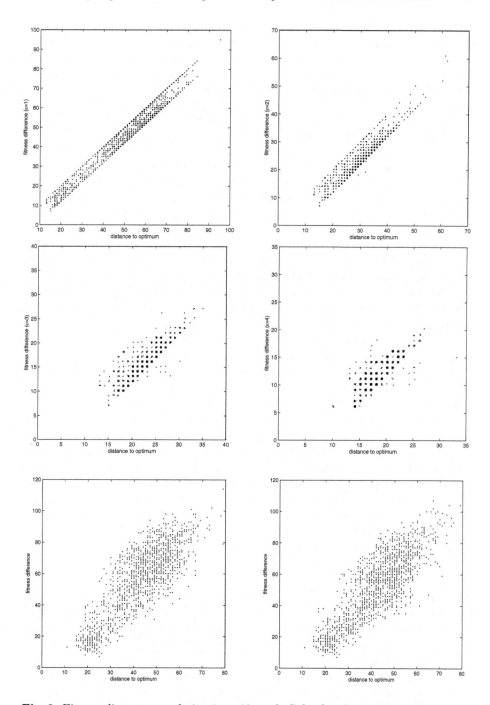

Fig. 3. Fitness distance correlation in a 12-mark Golomb ruler problem. The upper four figures correspond to the direct formulation (from top to bottom and left to right, $\epsilon = 1, 2, 3$, and 4), and those at the bottom to the indirect formulation ($\epsilon = 1$, and 2).

direct formulation, but without suffering from some of the problems of the latter. Actually, the high FDC values for the direct formulation are compensated by two related facts, namely that there is a drift force towards low-fitness regions as mentioned in Sect. 3.1, and that the number of local optima is higher for low values of the local radius, specially in the high-fitness region.

4   Experimental Validation

In order to confirm our findings from the landscape analysis, we have performed some experiments using a variable neighborhood search algorithm (VNS) [34]. This is a generalization of the conspicuous hill climbing algorithm in which different neighborhoods are used during the search. More precisely, a collection of neighborhoods $\mathcal{N}_1, \cdots, \mathcal{N}_k$ is considered. The search starts from the first neighborhood in the collection, and proceeds to the next one when no improvement can be found. Whenever an improvement is found, the search continues from the first neighborhood again. The underlying idea here is the fact that locally optima solutions for one neighborhood are not necessarily locally optimal in the next one. Hence, the algorithm can escape from such non-common local optima, and progress further towards the global optimum. The search finishes when a solution is found that is locally optimal for all neighborhoods.

The VNS algorithm has been deployed on the two problem representations considered. In both cases, neighborhoods $\mathcal{N}_1$ up to $\mathcal{N}_4$ have been considered, where $\mathcal{N}_i(x)$ refers to the set of solutions within distance i (in the corresponding fitness landscape) of solution x. Neighborhoods are explored by sampling 100 solutions, and retaining the best one. If no improving solution is found, the neighborhood is considered exhausted.

The results for $n = 12$ marks are shown in Table 1. VNS^i indicates that VNS is restricted to neighborhoods $\mathcal{N}_1$ to $\mathcal{N}_i$. As it can be seen, the results of the indirect representation are better than those of direct representation for VNS^1 and VNS^2. The difference between both representations tends to decrease for increasing radius: very similar results (no statistical difference according to a Mann-Whitney U test) are obtained in both cases for VNS^3, and the direct representation turns out to be better for VNS^4.

Table 1. Results (averaged for 30 runs) of variable neighborhood search on the two representations. As a reference, starting solutions have a mean value of 127.57 ± 7.64

	direct		indirect		indirect (exhaustive)	
	mean ± std.dev.	median	mean ± std.dev.	median	mean ± std.dev.	median
VNS^1	127.43 ± 7.73	125	114.10 ± 4.11	115	112.80 ± 3.91	113
VNS^2	120.63 ± 6.09	120.5	107.43 ± 3.86	107	108.93 ± 3.71	109
VNS^3	104.70 ± 5.19	105	105.83 ± 2.61	105	101.77 ± 3.03	101.5
VNS^4	98.87 ± 2.49	100	105.17 ± 2.55	105	97.33 ± 1.97	97

Two facts must be noted here. First of all, the magnitude of the radius has not the same meaning in the different representation, and hence, the data in Table 1 should not be interpreted as paired columns. Secondly, the computational cost (not shown in Table 1) of exploring each neighborhood is quite different (around three orders of magnitude larger in the case of the direct representation, as measured in a P4–3GHz 1GB PC under Windows XP). This is so, even allowing an exhaustive exploration of the neighborhood for the indirect representation. The results in this latter case are shown in the two rightmost columns of Table 1. Notice the improvement with respect to the direct representation.

5 Conclusions

This work has tried to shed some light on the question of what makes a problem hard for a certain search algorithm. We have focused on the Golomb ruler problem, an extremely interesting problem due to its simple definition yet tremendous hardness. It is also a problem for which several representations had been tried, but that lacked an analysis of the combinatorial properties of the associated fitness landscapes.

Our analysis indicates that the high irregularity of the neighborhood structure for the direct formulation introduces a drift force towards low-fitness regions of the search space. This contrasts with other problems in which the drift force is beneficial, since it guides the search to high-fitness regions (see [29]). The indirect formulation that we have considered does not have this drawback, and hence would be in principle more amenable for conducting local search in it. The fact that fitness-distance correlation is very similar in both cases also support this hypothesis.

The empirical validation provides consistent results: a VNS algorithm using the indirect formulation can outperform a VNS counterpart working on the direct representation, in a low computational cost scenario. It is also very interesting to note that these results are also consistent with the performance of evolutionary algorithm on this problem, despite the fact that even when the representation may be the same, they do not explore exactly the same landscape.

Future work will be directed to confirm these conclusions in the context of other constrained problems. We will also try to identify new problems in which the irregularity of the neighborhood structure play such a central role, and study alternative formulations for these.

Acknowledgements

This work is partially supported by Spanish MCyT and FEDER under contracts TIC2002-04498-C05-02 and TIN2004-7943-C04-01.

References

1. Fang, R., Sandrin, W.: Carrier frequency assignment for non-linear repeaters. Comsat Technical Review **7** (1977) 227–245
2. Babcock, W.: Intermodulation interference in radio systems. Bell Systems Technical Journal (1953) 63–73
3. Bloom, G., Golomb, S.: Aplications of numbered undirected graphs. Proceedings of the IEEE **65** (1977) 562–570
4. Robbins, J., Gagliardi, R., Taylor, H.: Acquisition sequences in PPM communications. IEEE Transactions on Information Theory **33** (1987) 738–744
5. Klove, T.: Bounds and construction for difference triangle sets. IEEE Transactions on Information Theory **35** (1989) 879–886
6. Robinson, J., Bernstein, A.: A class of binary recurrent codes with limited error propagation. IEEE Transactions on Information Theory **13** (1967) 106–113
7. Feeney, B.: Determining optimum and near-optimum Golomb rulers using genetic algorithms. Master thesis, Computer Science, University College Cork (2003)
8. Rankin, W.: Optimal Golomb rulers: An exhaustive parallel search implementation. Master thesis, Duke University Electrical Engineering Dept., Durham, NC (1993)
9. Dollas, A., Rankin, W.T., McCracken, D.: A new algorithm for Golomb ruler derivation and proof of the 19 mark ruler. IEEE Transactions on Information Theory **44** (1998) 379–382
10. OGR project: `http://www.distributed.net/ogr/` (on-going since September 14, 1998)
11. Mirchandani, P., Francis, R.: Discrete Location Theory. Wiley-Interscience (1990)
12. Dewdney, A.: Computer recreations. Scientific American (1986) 14–21
13. McCracken, D.: Minimum redundancy linear arrays. Senior thesis, Duke University, Durham, NC (1991)
14. Shearer, J.B.: Golomb ruler table. Mathematics Department, IBM Research, `http://www.research.ibm.com/people/s/shearer/grtab.html` (2001)
15. Shearer, J.: Some new optimum Golomb rulers. IEEE Transactions on Information Theory **36** (1990) 183–184
16. Garry, M., Vanderschel, D., et al.: In search of the optimal 20, 21 & 22 mark Golomb rulers. GVANT project, `http://members.aol.com/golomb20/index.html` (1999)
17. Smith, B., Walsh, T.: Modelling the Golomb ruler problem. In: Workshop on non-binary constraints (IJCAI'99), Stockholm (1999)
18. Galinier, P., Jaumard, B., Morales, R., Pesant, G.: A constraint-based approach to the Golomb ruler problem. In: 3rd International Workshop on Integration of AI and OR Techniques (CP-AI-OR'2001). (2001)
19. Koziel, S., Michalewicz, Z.: A decoder-based evolutionary algorithm for constrained parameter optimization problems. In Bäeck, T., Eiben, A., Schoenauer, M., Schwefel, H.P., eds.: Parallel Problem Solving from Nature V. Volume 1498 of Lecture Notes in Computer Science. Springer-Verlag, Berlin Heidelberg (1998) 231–240
20. Soliday, S., Homaifar, A., Lebby, G.: Genetic algorithm approach to the search for Golomb rulers. In Eshelman, L., ed.: 6th International Conference on Genetic Algorithms (ICGA'95), Pittsburgh, PA, USA, Morgan Kaufmann (1995) 528–535
21. Pereira, F., Tavares, J., Costa, E.: Golomb rulers: The advantage of evolution. In Moura-Pires, F., Abreu, S., eds.: Progress in Artificial Intelligence, 11th Portuguese Conference on Artificial Intelligence. Number 2902 in Lecture Notes in Computer Science, Berlin Heidelberg, Springer-Verlag (2003) 29–42

22. Bean, J.: Genetic algorithms and random keys for sequencing and optimization. ORSA Journal on Computing **6** (1994) 154–160
23. Cotta, C., Fernández, A.: A hybrid GRASP - evolutionary algorithm approach to Golomb ruler search. In Yao, X., et al., eds.: Parallel Problem Solving From Nature VIII. Number 3242 in Lecture Notes in Computer Science, Birmingham, UK, Springer (2004) 481–490
24. Resende, M., Ribeiro, C.: Greedy randomized adaptive search procedures. In Glover, F., Kochenberger, G., eds.: Handbook of Metaheuristics. Kluwer Academic Publishers, Boston MA (2003) 219–249
25. Radcliffe, N.: Equivalence class analysis of genetic algorithms. Complex Systems **5** (1991) 183–205
26. Wright, S.: The roles of mutation, inbreeding, crossbreeding and selection in evolution. In Jones, D., ed.: 6th Intenational Congress on Genetics. Volume 1. (1932) 356–366
27. Diaz, D., Codognet, P.: GNU Prolog: beyond compiling Prolog to C. In Pontelli, E., Costa, V., eds.: 2nd International Workshop on Practical Aspects of Declarative Languages (PADL'2000). Number 1753 in LNCS, Boston, USA, Springer-Verlag (2000) 81–92
28. Barták, R.: Practical constraints: A tutorial on modelling with constraints. In Figwer, J., ed.: 5th Workshop on Constraint Programming for Decision and Control, Gliwice, Poland (2003) 7–17
29. Bierwirth, C., Mattfeld, D., Watson, J.P.: Landscape regularity and random walks for the job shop scheduling problem. In Gottlieb, J., Raidl, G., eds.: Evolutionary Computation in Combinatorial Optimization. Volume 3004 of Lecture Notes in Computer Science., Berlin, Springer-Verlag (2004) 21–30
30. Jones, T.: Evolutionary Algorithms, Fitness Landscapes and Search. Phd thesis, Santa Fe Institute, University of New Mexico, Alburquerque (1995)
31. Jones, T., Forrest, S.: Fitness distance correlation as a measure of problem difficulty for genetic algorithms. In Eshelman, L., ed.: Proceedings of the 6th International Conference on Genetic Algorithms, San Francisco, CA, Morgan Kaufmann Publishers (1995) 184–192
32. Reeves, C.: Landscapes, operators and heuristic search. Annals of Operational Research **86** (1999) 473–490
33. Boese, K., Kahng, A., S., M.: A new adaptive multi-start technique for combinatorial global optimizations. Operations Research Letters **16** (1994) 101–113
34. Mladenović, N., Hansen, P.: Variable neighborhood search. Computers and Operations Research **24** (1997) 1097–1100

Immune Algorithms with Aging Operators for the String Folding Problem and the Protein Folding Problem

V. Cutello, G. Morelli, G. Nicosia, and M. Pavone

Department of Mathematics and Computer Science,
University of Catania,
V.le A. Doria 6, 95125 Catania, Italy
{cutello, morelli, nicosia, mpavone}@dmi.unict.it

Abstract. We present an Immune Algorithm (IA) based on clonal selection principle and which uses memory B cells, to face the protein structure prediction problem (PSP) a particular example of the String Folding Problem in 2D and 3D lattice. Memory B cells with a longer life span are used to partition the funnel landscape of PSP, so to properly explore the search space. The designed IA shows its ability to tackle standard benchmarks instances substantially better than other IA's. In particular, for the 3D HP model the IA allowed us to find energy minima not found by other evolutionary algorithms described in literature.

Keywords: Clonal selection algorithms, aging operator, memory B cells, protein structure prediction, HP model, functional model proteins.

1 The String and Protein Folding Problems

A *d-dimensional lattice* is a graph $\mathcal{L} = (V, E)$, where $V \subseteq Z^d$, i.e. the vertices are points in the Euclidean space with integer coordinates, and the set of edges $E \subseteq \{(\boldsymbol{x}, \boldsymbol{y}) : \boldsymbol{x} = (x_1, \ldots, x_d), \boldsymbol{y} = (y_1, \ldots, y_d)) : \sum_{i=1}^{d} \mid x_i - y_i \mid = 1\}$. Let now $s = <s_1, \ldots, s_j, \ldots, s_n>$ be a string of length n in the alphabet $\{0,1\}^*$. By *folding* of the string s we mean its embedding into the lattice $\mathcal{L}$, i.e., a one-to-one mapping f from the set $\{1 \leq j \leq n\}$ to V such that for all $1 \leq j \leq n-1$ we have $(f(j), f(j+1)) \in E$. The points $f(j)$ and $f(j+1)$ are called $f-neighbors$. A folding can therefore be seen as a walk in the lattice. A folding of a string s is a self-avoiding walk iff two characters s_i, s_j with $i \neq j$ do not occupy the same node of the lattice. Given a folding f of a string s, we can also introduce a measure to "assess" the quality of f. We say that an edge $(\boldsymbol{x}, \boldsymbol{y}) \in L$ is a *loss* if the two vertices are not $f - neighbors$, and exactly one of them is the image under f of a symbol $s_j = 1$ Given a *d-dimensional lattice* $\mathcal{L} = (V, E)$, a string $s \in \{0,1\}^*$, and an integer k, the STRING FOLDING PROBLEM (SFP) is be defined as the problem of checking whether there exists a self-avoiding folding of s into the lattice $\mathcal{L}$, with k or fewer losses. Let now $d = 2$, and let $L = \{p = (x_p, y_p) : 0 \leq x_p, y_p \leq n-1\}$. The neighborhood of a point p is defined

G.R. Raidl and J. Gottlieb (Eds.): EvoCOP 2005, LNCS 3448, pp. 80–90, 2005.

as the set of points in the lattice L connected by a single edge to p, i.e. as the set $\mathcal{N}(p) = \{q \in L : |x_p - x_q| + |y_p - y_q| = 1\}$. One of the main characteristics of a lattice "geometry" is the number of nodes directly connected to each node. Such a number is usually the same for all nodes, and it is known as the *coordination number* (C_n) of the lattice. For instance, in two dimensions we have honeycomb lattice $(C_n = 3)$, square lattice $(C_n = 4)$, and triangular lattice $(C_n = 6)$. In three dimensions possible lattice geometries are the following: diamond lattice $(C_n = 4)$, cubic lattice $(C_n = 6)$, and tetrahedral lattice $(C_n = 12)$.

1.1   The Protein Folding Problem

The special case of the STRING FOLDING PROBLEM with $d = 2$, and square lattice $(C_n = 4,)$ captures the protein folding problem in the 2D HP model [6]. Analogously for $d = 3$ and cubic lattice $(C_n = 6)$ we have the 3D HP model [6]. The HP model was shown to be NP complete problem for 2D lattice [12] (the NP-hardness is shown by reduction from an interesting variation of the planar Hamilton cycle problem), and for 3D lattice [13] (the NP-hardness is shown by reduction from a variation of the Bin Packing problem).

The HP model is a well-known approach to face the protein folding problem. It models proteins as 2D or 3D *self-avoiding walks* (i.e. two residues cannot occupy the same side of the lattice) of ℓ monomers on the square lattice. There are only two monomer types: the H and the P monomers, respectively for hydrophobic and polar monomers. Then, the HP model reduces the alphabet from 20 characters to 2, where our protein sequences take the form of strings belonging to the alphabet $\{H, P\}^+$. Any feasible conformation in the HP model is assigned a free energy level: each H–H topological contact, that is, each lattice nearest-neighbor H–H contact interaction, has energy value $\epsilon \leq 0$, while all other contact interaction types (H–P, P–P) contribute with $\delta \geq 0$ value to the total free energy. In general, in the HP model the residues interactions can be defined as follows: $e_{HH} = \epsilon$ and $e_{HP} = e_{PH} = e_{PP} = \delta$. When $\epsilon = -1$ and $\delta = 0$ we have the typical interaction energy matrix for the standard HP model [6], whereas when $\epsilon = -2$ and $\delta = 1$ we have the energy matrix for the shifted HP model [7, 17]. The *native state* of a protein is a conformation that minimizes the free energy function and, hence, the conformation that maximizes the number of contacts H–H.

In this paper we will present experimental results on the PFP, i.e. the SFP using square lattice $(C_n = 4,)$ and cubic lattice $C_n = 6$. In particular, we design and test an Immune Algorithm (IA) using classical PFP instances of the *Tortilla 2D HP Benchmarks*[1], 3D cubic lattice HP instances (taken from [8, 9]), and the classical benchmarks for the Functional Model Proteins[2], into 2D square lattice. The 3D HP benchmark uses the same protein sequences of Tortilla 2D HP Benchmarks using a 3D cubic lattice. We also note that each instance of the functional model proteins benchmarks has a unique native fold with minimal energy value, E^*, and an energy gap between E^* and the first excited state (best suboptimal).

---

[1] http://www.cs.sandia.gov/tech_reports/compbio/tortilla-hp-benchmarks.html
[2] http://www.cs.nott.ac.uk/~nxk/HP-PDB/2dfmp.html

Finally we note that in the HP model, a protein is represented as a sequence in a lattice in either two or three dimensions. The sequence is the set of coordinates that give the position in the lattice of each ammino-acid of the protein. Given the position in a lattice for the first ammino-acid of the protein, the sequence can be also identified by a set of moves that allow to find the position in the lattice of an ammino-acid using the position of the previous one. In this scenario the folding is represented by a sequence of moves (directions) that allow to find a sequence with the maximum number of topological contact.

2   The Clonal Selection Algorithm for the PFP

In this article we describe an improved version of a previously proposed immune algorithm [14], that uses only two entity types: antigens (Ag) and B cells. The Ag is the given input string $s \in \{0,1\}^*$ of the SFP and it models the hydrophobic-pattern of the given protein, that is a sequence $s \in \{H, P\}^\ell$, where ℓ is the protein length. The B cells population, $P^{(t)}$ (of size k), represents a set of candidate solution in the current fitness landscape at each generation t. The B cell, or B cell receptor, for the 2D HP model is a sequence of directions $r \in \{F, L, R\}^{\ell-1}$, with $F = Forward$, $L = Left$, and $R = Right$, where each r_i, with $i = 2, \ldots, \ell-1$, is a *relative direction* [11] with respect to the previous direction r_{i-1} (i.e., there are $\ell-2$ relative directions) and r_1 is the non-relative direction. Analogously, for the 3D HP model, the B cell receptor is a sequence $r \in \{F, L, R, B, U, D\}^{\ell-1}$, where $B = Backward$, $U = Up$, and $D = Down$. The sequence r detects a conformation suitable to compute the energy value of the hydrophobic-pattern of the given protein. For the 2D protein instances we use the *relative directions* because their performance are better with respect to *absolute directions* in the square lattice [11], while for the 3D protein instances we use both coding: relative and absolute directions to assess the effectiveness of the IA. The initial population, at time $t = 0$, is randomly generated in such a way that each B cell in $P^{(0)}$, represents *self-avoiding* conformations. The function $Evaluate(P^{(t)})$ computes the fitness value F of each B cell $\boldsymbol{x} \in P^{(t)}$. Then, $F(\boldsymbol{x}) = e$ is the energy of conformation coded in the B cell receptor $\boldsymbol{x}$, with $-e$ being the number of topological contacts $H - H$ in the lattice (2D or 3D). The function $Termination\_Condition()$ returns true if a solution is found, or a maximum number of fitness function evaluations (T_{max}) is reached. The cloning operator, simply, clones each B cell dup times, producing an intermediate population P^{clo} of size $Nc = k \times dup$. We tested our IA using the combination of inversely proportional hypermutation and hypermacromutation operators. In the Inversely Proportional Hypermutation operator the number of mutations is inversely proportional to the fitness value. In particular, at each generation t, the operator will perform at most the following mutations:

$$M_i(F(\boldsymbol{x})) = \begin{cases} ((1 - \frac{E^*}{-1}) \times \beta) + \beta), & \text{if } F(\boldsymbol{x}) = 0 \\ ((1 - \frac{E^*}{F(\boldsymbol{x})}) \times \beta)), & \text{if } F(\boldsymbol{x}) > 0 \end{cases} \quad (1)$$

with $\beta = c \times \ell$. In this case, $M_i(F(\boldsymbol{x}))$ has the shape of an hyperbola branch. In the Hypermacromutation operator the number of mutations is determined by a simple random process. It tries to mutate each B cell receptor M times, maintaining the self-avoiding property. The number of mutations M is at most $M_m(x) = j - i + 1$, in the range $[i, j]$, with i and j being two random integers such that $(i + 1) \leq j \leq \ell$. The number of mutations is independent from the fitness function F and any other parameter. The hypermacromutation operator, for each B cell receptor, randomly selects a perturbation direction, either from left to right ($k = i, \ldots, j$) or from right to left ($k = j, \ldots, i$).

Table 1. Pseudo–code of the Immune Algorithm for the PFP

$$
\begin{array}{l}
\textbf{Immune Algorithm}(\ell, k, dup, \tau_B, c) \\
t := 0; \ Nc := k * dup; \\
P^{(t)} := \text{Initial\_Pop}(); \\
\text{Evaluate}(P^{(t)}); \\
\textbf{while } (\neg \text{ Termination\_Condition}()) \textbf{ do} \\
\quad P^{(clo)} := \text{Cloning } (P^{(t)}, Nc); \\
\quad P^{(hyp)} := \text{Hypermutation } (P^{(clo)}, c, \ell); \\
\quad \text{Evaluate}(P^{(hyp)}); \\
\quad P^{(macro)} := \text{Hypermacromutation } (P^{clo}); \\
\quad \text{Evaluate } (P^{(macro)}); \\
\quad (P_a^{(t)}, P_a^{(hyp)}, P_a^{(macro)}) := \text{Aging}(P^{(t)}, P^{(hyp)}, P^{(macro)}, \tau_B); \\
\quad P^{(t+1)} := (\mu + \lambda)\text{-Selection } (P_a^{(t)}, P_a^{(hyp)}, P_a^{(macro)}); \\
\quad t := t + 1; \\
\textbf{end\_while}
\end{array}
$$

In the hypermutation phase, we use the *stop at the First Constructive Mutation* (*FCM*) strategy: if a constructive mutation occurs, the mutation procedure will move on to the next B cell. We adopted such a mechanism to slow down (premature) convergence, exploring more accurately the search space. Formally, the mutation operator acts on the population $P^{(clo)}$, where each B cell is a feasible candidate solution, i.e. it is a self-avoiding walk, generating the new populations $P^{(hyp)}$ and $P^{(macro)}$. In 2D lattice, given a protein conformation sequence R, the mutation operator randomly selects a direction r_j, with $1 \leq j \leq \ell - 1$, or a subsequence $R_{ij} =< r_i, r_{i+1}, \ldots, r_{j-1}, r_j >$, with $i > 1$ and $j \leq \ell - 1$. For each relative direction $D = r_j$, a new direction $D' \neq D \in \{F, L, R\}$ is randomly selected. If the new conformation is again self-avoiding then the operator accepts it, otherwise the procedure repeats the process with a new and last direction $D'' \neq< D, D' >\in \{F, L, R\}$.

Aging. The aging process reflects the attempt to benefit from modelling the limited life spans of B cells and longer life spans of Memory B cells. Starting from this basic observation, the aging operator eliminates old B cells from the populations $P^{(t)}$, $P^{(hyp)}$ and $P^{(macro)}$, so to avoid premature convergence. The parameter τ_B (and τ_{B_m} for the memory B cells) sets the maximum number of generations allowed to B cells to remain in the population. When a B cell is τ_B+1 old (or $\tau_{B_m} + 1$ old) it is erased from the current population, no matter what its fitness value is. We call this strategy, *static pure aging*. During the cloning expansion, a cloned B cell takes the age of its parent. After the hypermutation phase, a cloned B cell which successfully mutates, i.e. with a better fitness value, will be considered to have age equal to 0. Thus, an equal opportunity is given to each "new genotype" to effectively explore the fitness landscape. We note that for τ_B greater than the maximum number of allowed generations, the IA works essentially without aging operator. In such a limit case the algorithm uses a strong elitist selection strategy. Aging is a new operator that causes a turn-over in the populations of the IA. Its goal is to generate diversity and to avoid getting trapped into local minima. It is an operator inspired by the biological immune system where there is an expected mean life for the B cell [15], and it is, in general, problem- and algorithm-independent. After clonal expansion and aging phase, a new population $P^{(t+1)}$, of k B cells, for the next generation $t+1$, is obtained by selecting the best B cells which "survived" the aging operator, from the populations $P^{(t)}$, $P^{(hyp)}$ and $P^{(macro)}$. No redundancy is allowed: each B cell receptor is unique, i.e. each genotype is different from all other genotypes. If only $k' < k$ B cells survived, new randomly created B cells (with $age = 0$) are added by the *Elitist_Merge* function into the population (the *Birth phase*). In general, the selection operator chooses the k best elements from both parent and offspring B cells sets, thus guaranteeing monotonicity in the evolution dynamic. In table 1 we show the pseudo-code of the proposed Immune Algorithm.

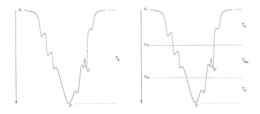

Fig. 1. Typical funnel landscape for the PSP problem (left plot); partitioning of the funnel landscape in three region using memory B cells with two aging parameter values, τ_B, and τ_{B_m}

2.1   Partitioning the Funnel Landscape Using Memory B Cells

The proposed IA uses memory B cells to better handle the space of solutions and improve the performance, for the protein structure prediction problem. We have not tested its performance on general instances of the String Folding Problem. One characteristic feature of the PSP problem is its rugged *funnel landscape* (see fig.1 on the left), where the number of feasible conformations decreases with low free energy values, i.e. many conformations have high energy and few have low energy. Thus, very likely, we could get trapped in a local minimum.

Starting from this simple topological observation we used memory B cells to partition the funnel landscape into three regions. Each partition is obtained using two threshold energy values: if the protein native fold has energy value E^* we have $E_{level} = -E^* + 1$ energy levels, thus the boundary of the first partition and secondary partition are respectively $E_{FP} = -(E_{level} \times 0.67)$ and $E_{SP} = -(E_{level} \times 0.85)$. Theoretical findings [3] and experimental results (not reported in this paper), show that the hardest region to search is the middle one. It is typically rugged with many local minima. So we apply memory B cells only to such a region. Conformations whose energy value is in the middle region, are allowed to maturate.

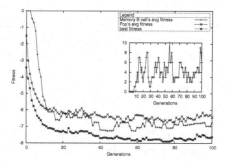

Fig. 2. The best fitness value, average fitness function values of memory B cells and $P^{(t)}$ versus generations on protein sequence *Seq2*, with parameter $d = 10, dup = 2, \tau_B = 5$ and $\tau_{B_m} = 10$. In the inset plot we report the number of memory B cells versus generations

In figure 2 we show memory B cells dynamic. We set the minimal population size value $d = 10$, with $dup = 2, \tau_B = 5$ and $\tau_{B_m} = 10$. All curves are averaged on 30 independent runs. We plot the best fitness value, the average fitness of $P^{(t)}$ and memory B cell populations, whereas in the inset plot we show as change the number of memory B cells with respect to generations.

3   Experimental Results

To assess the overall performance of the new version of the IA using memory B cells we tested it using the well-known tortilla benchmarks in the standard 2D and 3D HP model and the classical protein instances for the Functional Model Proteins.

3.1   HP Model in 2D Square Lattice

In table 2 we compare our improved IA with the previous versions, respectively IA with Hypermacromutation Operator only (with and without elitism) [4], and IA with Inversely Proportional Hypermutation and Hypermacromutation

Table 2. IA with memory B cells compared to other types of IA. Results averaged on 30 independent runs

No.	ℓ	E^*	Macro with Elit.		Macro without Elit.		Inv.Prop.+Macro		IA with mem.	
			SR	AES	SR	AES	SR	AES	SR	AES
1	20	-9	96.67	20508.9	100	25418.8	100	**14443.7**	100	15439.6
2	24	-9	100	**37659.7**	100	39410.9	100	39644.1	100	46034.9
3	25	-8	96.67	58905.3	100	**79592.1**	100	95147	100	99865.7
4	36	-14	36.67	310291.4	16.67	466176.4	23,33	388323,4	**100**	2032504
5	48	-23	3.33	277454	6.67	483651.5	*(b.f. -22)*	//	**56.67**	2403985.3
6	50	-21	53.33	459868	16.67	469941.2	53.33	538936.4	**100**	1011377.4
7	60	-36	*(b.f. -35)*	//	*(b.f. -34)*	//	*(b.f. -34)*	//	**(b.f. -35)**	//
8	64	-42	*(b.f. -39)*	//	*(b.f. -36)*	//	*(b.f. -39)*	//	**(b.f. -39)**	//
9	20	-10	96.67	27719.14	100	27852.1	100	**17293.9**	100	20135.4

operators [14]. Comparisons were done in terms of Success Rate (SR) and Average number of Evaluations to Solution (AES). The IA used standard parameter values: $k = 10, dup = 2, c = 0.4$, as described in [14] while the standard B cells have aging parameter $\tau_B = 5$ and the memory B cells $\tau_{B_m} = 10$.

Table 2 shows that the improved IA is comparable on the simplest protein instances to the previous IA versions, and outperforms them on the hard instances. Indeed, the new IA obtained $SR = 100$ on the *Seq4* and *Seq6*, and $SR = 56.67$ on the *Seq5*, where the other versions failed. For the hardest instances the new IA obtained the lowest energy values. These results show that partitioning the landscape in three groups, is an effective approach for the PSP in the standard 2D HP model.

3.2   HP Model in 3D Cubic Lattice

In the 3D cubic lattice, each point has 6 different neighbors and 5 available locations. We use two different schemes of moves (absolute and relative directions) to represent and embed a protein in the lattice. The relative encoding has been described in section 2: the residues direction are relative to direction of the previous move, while in the *absolute directions* encoding the residues direction are relative to the axes defined by the lattice. Both for the absolute and relative coding not all moves give a feasible conformation. In our work we force the self-avoidance constraint so each set of moves will correspond to a feasible sequence (feasible conformation). Concerning the experimental results, for all considered instances the IA (working with feasible solutions) has found the known minimum value and for all instances the found mean value is lower than the results obtained in [9] using Evolutionary Algorithms (EAs) working on Feasible-Space. For several sequence presented in [9], as shown in table 3, we have found new best lowest energy values for 3D protein sequences 5, 7, and 8. The IA used the standard parameter values: $k = 10, dup = 2, c = 0.4$, as described in [14], B cells have aging parameter $\tau_B = 5$ and the memory B cells $\tau_{B_m} = 10$. For the experimental protocol we adopt the same values used in [9]: 50 independent runs and a maximum number of evaluations equal to 10^5. In [9] the author does not use the SR and AES values as quality metrics, but the following parameters: Best found solution (Best), mean and standard deviation (σ). We designed an

Table 3. Results of the IA for the 3D HP model

		Absolute Encoding						Relative Encoding					
		F-EA			IA			F-EA			IA		
Seq.	ℓ	Best	Mean	σ	Best	Mean	σ	Best	Mean	σ	Best	Mean	σ
1	20	-11	-10.32	0.61	-11	**-11**	0	-11	-9.84	0.86	-11	-10.90	0.32
2	24	-13	-10.90	0.98	-13	**-13**	0	-11	-10.00	0.87	-13	-12.22	0.65
3	25	-9	-7.98	0.71	-9	**-9**	0	-9	-8.64	0.69	-9	-8.88	0.48
4	36	-18	-14.38	1.26	-18	**-16.76**	1.02	-18	-13.72	1.41	-18	-16.08	1.02
5	48	-25	-20.80	1.61	**-29**	**-25.16**	0.45	-28	-18.90	2.08	**-28**	-24.82	0.71
6	50	-23	-20.20	1.50	-23	**-22.60**	0.40	-22	-19.06	1.46	-23	-22.08	1.43
7	60	-39	-34.18	2.31	**-41**	**-39.28**	0.24	-38	-32.28	3.09	**-41**	-39.02	0.50
8	64	-39	-33.01	2.49	**-42**	**-39.08**	0.95	-36	-30.84	2.55	**-42**	-39.07	1.20

IA which uses a Penalty strategy and a Repair-based approach as reported in [9] which obtained similar experimental results to [9] (not shown due to space limitation). The used penalty strategy is based on the fact that not all moves of a conformation maintain the self-avoiding property; when a move does not satisfy the self-avoiding constraint, the energy value assigned to the conformation is increased of a penalty coefficient and, also, the aminoacid involved in a collision will be not considered in H-H contact. The repair strategy is applied, after the *HyperMutation* and *HyperMacroMutation* phases, on each conformation of the populations that contains a collision, and starting the process from the last collision found. Let i be the collision position, the repair process determines a free position L in the lattice such that it is possible to reach the position $i + 1$ with only one move. Moreover, either L is reachable directly from the position $i - 1$ (in this case i will move to position L) or there exists another free position C, reachable from $i - 2$ with only one move, and from which is possible to reach L, with one move (in this case i will move to position L and $i - 1$ to position C). In both cases, the moves of the conformation will be modified according to the new coordinates. Such an IA proved to be very efficient for both absolute and relative encoding, and allowed us to find energy minima not found by other EAs working in feasible spaces and described in literature [9].

3.3   Functional Model Proteins in 2D Square Lattice

We show here the experimental results on the classical benchmarks for the Functional Model Proteins using memory B cells. In table 4 we report the experimental results obtained by our IA using different life span values for the memory B cells: $(\tau_B = 3, \tau_{B_m} = 5)$, $(\tau_B = 4, \tau_{B_m} = 8)$, and $(\tau_B = 5, \tau_{B_m} = 10)$. In the last two columns we report the performance of an IA without memory B cells using the standard parameter values: $k = 10, dup = 2, c = 0.4, \tau_B = 5$. All the experimental results reported are averaged on 30 independent runs. Like in the standard HP Model, the proposed IA obtained the best results using the pair $(\tau_B = 5, \tau_{B_m} = 10)$ values. However, for the functional model proteins the IA without memory B cells is more effective. The Table shows how the IA without memory B cells outperforms the IA with memory B cells in term of SR and AES, obtaining $SR = 100$ values on all functional model protein instances, except for the $SeqC$, where the algorithm reaches $SR = 56.67$ with $mean = -15.13$, and

Table 4. Improved IA performances using memory B cells (τ_B, τ_{B_m}) in the Functional Model Proteins. Each protein instance has $\ell = 23$ monomers

No.	E^*	$\tau_B = 3, \tau_{B_m} = 5$		$\tau_B = 4, \tau_{B_m} = 8$		$\tau_B = 5, \tau_{B_m} = 10$		$\tau_B = 5$	
		SR	AES	SR	AES	SR	AES	SR	AES
A	-20	100	253393	100	60664.8	100	38586.63	100	32847.7
B	-17	100	41189.7	100	258387	100	28434.9	100	17526.7
C	-16	16.67	31311300	36.67	2448820	43.33	2583300.8	56.67	2667430
D	-20	100	568485	100	261439.1	100	130849	100	128015.1
E	-17	100	16726.3	100	17586.13	100	20834.46	100	12095.3
F	-13	96.67	1083238.6	100	711828.33	100	483126.76	100	332938.5
G	-26	100	1171346.4	100	936008.9	100	588057.5	100	584179.8
H	-16	100	107131	100	54432.7	100	42562.53	100	38262.6
I	-15	100	506368	100	75273.8	93.33	907962.4	100	281720.8
J	-14	100	226564.87	100	141515.23	100	100085.43	100	104155.4
K	-15	100	43327.2	100	82361.1	100	71903.1	100	27743.7

$\sigma = 0.65$. This confirm the optimal searching ability and diversity generation of the pure aging operator. Finally, in table 5 we show the number of energy evaluations required by the best run (in [10] the authors use only this metric to assess the effectiveness of their algorithm) to reach the optimum or a sub-optimum energy value. We compare the performances of the IA with and without memory B cells with the state of art algorithm for the functional model proteins, the Multimeme Algorithm [10]. Comparing the results both versions of the IA, with or without memory B cells outperforms the Multimeme Algorithm on all the protein instances, in particular the IA without memory B cells obtains the best results (9 instances over 11).

Table 5. Comparison of best runs for MultiMeme Algorithm (MMA) [10] and IA, with and without memory B cells for the Functional Model Proteins. Each protein instance has $\ell = 23$ monomers

No.	E^*	MMA	IA without memory B cells	IA with memory B cells
1	-20	15170	**3372**	3643
2	-17	61940	**578**	1488
3	-16	132898	319007	**100234**
4	-20	66774	**4955**	20372
5	-17	53600	**1047**	1956
6	-13	32619	**2828**	7482
7	-26	114930	**10061**	37841
8	-16	28425	**1818**	1937
9	-15	25545	**3845**	10399
10	-14	111046	**2847**	3462
11	-15	52005	3176	**1007**

4   Conclusions

In this paper we propose an improved version of an IA for the protein structure prediction problem, in the standard 2D and 3D HP model and the Functional Model Proteins. In [4, 14] the results obtained for the 2D HP model suggested that the previous IA version was comparable to and, in many protein instances, outperformed folding algorithms which are present in literature. The results obtained in this research work established the new IA for the 2D HP model as the state-of-art algorithm for this discrete lattice model. Moreover, for the 3D HP model the IA allowed us to find energy minima not found by other EAs described in literature. Finally, for the functional model protein, the IA, with or without memory B cells, outperforms the Multimeme Algorithm on all the protein instances. Our algorithm proved to be very effective and very competitive, compared to the existing state-of-art EAs.

We intend to analyze now the impact, on the efficiency and efficacy of the Immune Algorithm, of the parameters τ_B and τ_{B_m}. We also believe that it could be worthwhile to implement a mutation rate dependent upon the B cells age.

Acknowledgements. we are grateful to the anonymous referees for their valuable comments.

References

1. Cutello V., Nicosia G.: The clonal selection principle for in silico and in vitro computing. In L. N. De Castro and F. J. Von Zuben editors, Recent Developments in Biologically Inspired Computing. Idea Group Publishing, Hershey, PA (2004).
2. De Castro L. N., Von Zuben F. J.: Learning and optimization using the clonal selection principle. IEEE Trans. Evol. Comput., 6(3), pp. 239–251 (2002).
3. Plotkin S. S., Onuchic J. N.: Understanding protein folding with energy landscape theory. Quarterly Reviews of Biophysics, 35(2), pp. 111-167 (2002).
4. Cutello V., Nicosia G., Pavone M: An immune algorithm with hyper-macromutations for the 2D hydrophilic-hydrophobic model. CEC'04, 1, pp. 1074–1080, IEEE Press (2004).
5. Cutello V., Nicosia G., Pavone M.: A hybrid immune algorithm with information gain for the graph coloring problem. GECCO'03, Lectures Notes in Computer Science, 2723, pp. 171–182 (2003).
6. Dill K. A.: Theory for the folding and stability of globular proteins. Biochemistry, 24(6), pp. 1501–1509 (1985).
7. Hirst J. D.: The evolutionary landscape of functional model proteins. Protein Engineering, 12(9), pp. 721–726 (1999).
8. Unger R., Moult J.: Genetic algorithms for protein folding simulations. J. Molecular Biology, 231(1), pp. 75–81 (1993).
9. Cotta C.: Protein Structure Prediction using Evolutionary Algorithms Hybridized with Backtracking. IWANN '03, Lecture Notes in Computer Science, 2687, pp. 321–328, (2003).
10. Krasnogor N., Blackburne B. P., Burke E. K., Hirst J. D.: Multimeme algorithms for protein structure prediction. PPSN VII, Lectures Notes in Computer Science, 2439, pp. 769–778 (2002).
11. Krasnogor N, Hart W. E., Smith J., Pelta D. A.: Protein structure prediction with evolutionary algorithms. GECCO'99, pp. 1596–1601 (1999).
12. Crescenzi P., Goldman D., Papadimitriou C., Piccolboni A., Yannakakis M.: On the complexity of protein folding. Journal of Computational Biology, 5(3), pp. 423–466 (1998).
13. B. Berger and T. Leighton, "Protein folding in the hydrophobic-hydrophilic model is np complete," *J. Comput. Biol.*, vol. 5, pp. 27–40, 1998.

14. Cutello V., Nicosia G., Pavone M.: Exploring the capability of immune algorithms: A characterization of hypermutation operators. ICARIS'04, Lectures Notes in Computer Science, 3239, pp. 263–276 (2004).
15. Seiden P. E., Celada F.: A model for simulating cognate recognition and response in the immune system. J. Theor. Biology, 158, pp. 329–357 (1992).
16. Shmygelska A., Hoos H. H.: An Improved Ant Colony Optimization Algorithm for the 2D HP Protein Folding Problem. Proc. Conf. on Artificial Intelligence, Lectures Notes in Computer Science, 2671, pp. 400–417 (2003).
17. Blackburne B. P., Hirst J. D.: Evolution of functional model proteins. J. Chemical Physics, 115(4), pp. 1935–1942 (2001).
18. Chan H. S., Dill K. A.: Comparing folding codes for proteins and polymers. Proteins: Struct., Funct., Genet., 24, pp. 335–344 (1996).

Multiobjective Quadratic Assignment Problem Solved by an Explicit Building Block Search Algorithm – MOMGA-IIa

Richard O. Day and Gary B. Lamont

Department of Electrical Engineering,
Graduate School of Engineering & Management,
Air Force Institute of Technology,[**]
WPAFB (Dayton) OH, 45433, USA
{Richard.Day, Gary.Lamont}@afit.edu

Abstract. The multi-objective quadratic assignment problem (mQAP) is an non-deterministic polynomial-time complete (NPC) problem with many real-world applications. The application addressed in this paper is the minimization of communication flows in a heterogenous mix of Organic Air Vehicles (OAV). A multi-objective approach to solving the general mQAP for this OAV application is developed. The combinatoric nature of this problem calls for a stochastic search algorithm; moreover, two linkage learning algorithms, the multi-objective fast messy genetic algorithm (MOMGA-II) and MOMGA-IIa, are compared. Twenty-three different problem instances having three different sizes (10, 20, and 30) plus two and three objectives are solved. Results indicate that the MOMGA-IIa resolves all pareto optimal points for problem instances < 20.

1 Introduction

The scalar quadratic assignment problem (QAP) was introduced in 1957 by Koopmans and Beckmann. In 2002, Knowles and Corne extended the QAP to be multi-objective and it became the multi-objective quadratic assignment problem (mQAP) [11]. Explicit Building Block (BB) search Algorithms are good at solving a multitude of NPC problems [3, 10, 21], including the mQAP. This investigation illustrates our latest achievement in finding a better building block builder by way of a good competitive template selection mechanism added into the multi-objective fast messy GA (MOMGA-II) [22]. The new MOMGA-II is called the MOMGA-IIa. MOMGA-IIa originated as a single objective messy GA

[**] The views expressed in this article are those of the authors and do not reflect the official policy of the United States Air Force, Department of Defense, or the U.S. Government. The authors also wish to acknowledge the following individuals: Jesse Zydallis for the use of his MOMGA-II code; Mark Kleeman, Todd Hack and Justin Kautz for their persistent help and discussions.

G.R. Raidl and J. Gottlieb (Eds.): EvoCOP 2005, LNCS 3448, pp. 91–100, 2005.

(mGA) and evolved into a multi-objective mGA called the MOMGA [6]. Many different Multi-objective Evolutionary Algorithms (MOEAs) were produced during this time period; however, the MOMGA is the only MOEA explicitly using good BBs to solve problems – even the Bayesian optimization algorithm (BOA) uses a probabilistic model to find good building blocks. The MOMGA has a population size limitation: as the BB size increases so does the population size during the Partially Enumerative Initialization (PEI) phase. This renders the MOMGA less useful on large problems. To overcome this problem, the MOMGA-II, based on the single objective fmGA, was designed. The fmGA is similar to the mGA in that it specifically uses BBs to find solutions; however, it requires smaller population sizes and has a lower run time complexity when compared to the mGA. MOMGA-II includes many different repair, selection, and crowding mechanisms. Unfortunately, the MOMGA-II is found to be limited when solving large problems [5]. This called for the development of basis function diversity measures in the MOMGA-IIa which are designed for smart BB searching in both the geno- and pheno-type domains. The problem under investigation is the mQAP. Test instances used in this study for the mQAP were designed by Knowles and Corne [12]. Results are compared with deterministic results (where available) and previously published attempts at solving these test instances. This paper begins with this introduction and then is followed by a description of the Organic Air Vehicles (OAV) problem mapped to the mQAP problem domain. Next the algorithm domain is presented. A short discussion of the deterministic approach is also included in the algorithm domain discussion. This is followed by the design of experiments section which includes the resources and parameter settings. Finally, results are presented and conclusions are drawn.

2   Problem Domain

Today, OAVs are operated in an independent role where they each have their own mission and a single controller. Future operation of OAVs must include collaboration and autonomous operation of a package (heterogeneous mix) of OAVs. During flight operations of an autonomous package of OAVs, vehicles must communicate in an efficient manner. Flight vehicle patterns play an important role in communication effectiveness (power consumption) during long range missions. In this investigation, the communication and flight pattern of a heterogeneous set of OAVs is mapped to the mQAP.

The QAP was originally designed to model a plant location problem [2]. Mapping the OAV problem into a QAP is accomplished with replacement. By inserting OAVs for plants, flight formation positions for plant locations, and communication traffic for supply flow, the OAVs problem is mapped directly onto the QAP. The mQAP is similar to the scalar QAP[1], with the exception of having multiple types of flows coming from each object.

---

[1] See http://www.seas.upenn.edu/qaplib/ for more info about the QAP.

For example, the OAVs may use one communication channel for passing reconnaissance information, another channel for target information, and yet another channel for OAV operational messages. The end goal is to minimize all the communication flows between OAVs. The mQAP[2] is defined mathematically in Equations 1 and 2.

$$minimize\{C(\pi)\} = \{C^1(\pi), C^2(\pi), \ldots, C^m(\pi)\} \tag{1}$$

$$C^k(\pi) = \underset{\pi \in P(n)}{min} \sum_{i=1}^{n} \sum_{j=1}^{n} a_{ij} b^k_{\pi_i \pi_j}, k \in 1..m \tag{2}$$

where n is the number of objects/locations, a_{ij} is the distance between location i and location j, b^k_{ij} is the kth flow from object i to object j, π_i gives the location of object i in permutation $\pi \in P(n)$, and 'minimize' means to obtain the Pareto front [12].

Many algorithm approaches have been used on the QAP. QAP researchers can only optimally solve for problems that are of size < 20. Furthermore, problem sizes of 15 are extremely difficult [2]. When feasible, optimal solutions are found using branch and bound methods [8, 2]. However, since many real-world problems are larger than 20 instances, other methods need to be employed in order to find a good solution in a reasonable amount of time. The use of Stocastic Local Searches and Ant Colonies has been explored. These have been found to do well when compared to some of the best heuristics available for the QAP and mQAP [7, 15, 18]. Evolutionary algorithms have also been applied [17, 9]. Additionally, several researchers have compared the performance of different search methods [20, 16].

3   Algorithm Domain

While many different algorithms have been used to solve the QAP [1, 13], only a few have been applied to mQAP [10]. This investigation compares results found by all attempts at solving the mQAP test instances developed in [12]. Table 1 list all multi-objective problem (MOP) instances solved in this investigation. Unfortunately, some researchers do not have access to solutions found by their algorithm, so a direct pareto front comparison cannot be made for a more accurate differentiation between solution quality.

Knowles and Corne [11] collected results by running 1000 local searches from each of 100 (for 2-objective instances) or 105 (3-objective instances) different λ vectors, thus giving them ≈ 200000 records. This technique is an interesting one; however, they do not include the actual data points found on their pareto front. Also, we previously used a multi-objective evolutionary algorithm (MOEA), MOMGA-II, to solve the problem [4]. The next section describes both the MOMGA-II and MOMGA-IIa.

---

[2] See http://dbk.ch.umist.ac.uk/knowles/mQAP/ for more info about the mQAP.

Table 1. Multi-objective problems numbered and listed according to size and number of objectives. There are *real like* (#rl) and *uniform* (#uni) instances. The size of each problem is indicated by the two digit number following the *KC* (KC##). The number of objectives for each problem is indicated by the number preceeding the *fl* (#fl). Each column list the sizes of MOPs used: 10, 20, and 30. The shaded area of the table is identifying the MOPs with 3 objectives - others have 2 objectives

(MOP #)	Name (size 10)	(#)	Name (size 20)	(#)	Name (size 30)
1	KC10-2fl-1rl	9	KC20-2fl-1rl	17	KC30-2fl-1rl
2	KC10-2fl-1uni	10	KC20-2fl-1uni	18	KC30-3fl-1rl
3	KC10-2fl-2rl	11	KC20-2fl-2rl	19	KC30-3fl-1uni
4	KC10-2fl-2uni	12	KC20-2fl-2uni	20	KC30-3fl-2rl
5	KC10-2fl-3rl	13	KC20-2fl-3rl	21	KC30-3fl-2uni
6	KC10-2fl-3uni	14	KC20-2fl-3uni	22	KC30-3fl-3rl
7	KC10-2fl-4rl	15	KC20-2fl-4rl	23	KC30-3fl-3uni
8	KC10-2fl-5rl	16	KC20-2fl-5rl		

3.1   Extended Multi-objective fmGA (MOMGA-IIa)

The MOMGA-IIa is a multi-objective version of the fmGA that has the ability to achieve a semi-partitioned search in both the genotype and phenotype domains during execution. It is an algorithm that exploits "good" building blocks (BBs) in solving optimization problems. These explicit BBs represent "good" information in the form of partial strings that can be combined to obtain even better solutions. The MOMGA-IIa algorithm executes in three phases: Initialization, Building Block Filtering, and Juxtapositional Phase. See Figure 1 for diagram of the program flow.

The algorithm begins with the Probabilistically Complete Initialization (PCI) Phase where it randomly generates a user specified number of population members. These population members are a specified chromosome length and each is evaluated to determine its respective fitness values. Our implementation utilizes a binary scheme in which each bit is represented with either a 0 or 1.

The Building Block Filtering (BBF) Phase follows by randomly deleting loci and their corresponding allele values in each of the population member's chromosomes. This process completes once the length of the population member's chromosomes have been reduced to a predetermined BB size. These reduced chromosomes are referred to as underspecified[3] population members. In order to evaluate population members that have become underspecified, competitive templates (CTs) are utilized to fill in the missing allele values. Evaluation consists of the partial string being overlayed onto a CT just prior to evaluation. CTs are fully specified chromosomes that evolve as the algorithm executes. CT replacement is done after each BB generation. In the MOMGA-II, future CTs are updated with the best individuals found with respect to each objective function. However, the MOMGA-IIa selects a *competent* CTs that partitions both the phenotype and genotype. This innovative balance is achieved through two mechanisms: Orthogonal CT generation and Target Vector (TV) guidance. Orthogonal CT generation is used to partition the genotype space, while keeping a

---

[3] An underspecified chromosome is chromosome where some, but not all locus positions have an associated allele value.

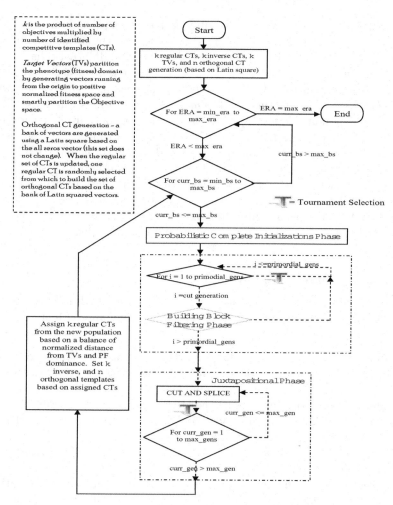

Fig. 1. This figure illustrates the MOMGA-IIa program flow. Note the placement of each phase and where tournament selection is performed. Additionally, the MOMGA-IIa exploits and partitions in both the phenotype and genotype domains by updating and generating regular, inverse, and orthogonal competitive templates. See Section 3.1 for a detailed description of the algorithm

good partition in the phenotype space requires TV guidance. TVs are normalized fitness markers that capture one solution per vector for future CT replacement. In the MOMGA-IIa, target vectors are used in a manner to divide the normalized fitness space of pareto-front members and select a distribution of CTs that fall nearest to each TV. Also, an orthogonal bank of chromosomes is used to filter a randomly selected CT through for creation of a set of orthogonal CTs.

The BBF process is alternated with a selection mechanism to keep only the strings with the "best" BBs found, or those with the best number of fitness values. In the case of a tie, where two strings each have an equal number of better fitness values (i.e. each have $\frac{m}{2}$ "best" fitness values), the string is randomly selected between the two. It should be noted that the MOMGA-IIa has a more complex selection mechanism than MOMGA-II because it maintains more fitness values per solution. In the MOMGA-II each string has m fitness values, while in MOMGA-IIa each string has $f = (c * m + i + o) * m$ fitness values associated with it – corresponding to the m objective functions to optimize, c competitive templates, i inverse templates (equal to $c * m$), and o orthogonal templates.

Finally, the juxtapositional phase uses the BBs found through the BBF phase and recombination operators to create chromosomes that are fully specified. A chromosome is referred to as fully specified if it is not missing any locus positions, or in other words, does not need to use the CT for evaluation.

The MOMGA-IIa has an outer and inner loop that must be completely iterated through using each BB size and epoch before terminating.

3.2   Non-stochastic Approaches

Two different approaches are discussed in this section. The first is the type of approach used by Knowles and Corne in [12]. The second approach is simply our exhaustive search algorithm.

Local Search Approach: The local search (LS) method employed for the mQAP problem is where positions of facilities (or objects) are switched [11]. The new positioning is kept if the new configuration yields a lower fitness value. This search method works for solving the QAP [19]; however, the mQAP makes employing a strict LS approach difficult for the deceptive hyperplanes that accompany multi-objective problems. Specifically, a researcher is faced with how to initialize the LS method. Knowles and Corne concluded that the starting points would be randomly selected out of a basin of attraction [12,19]. After a new point is picked, the LS method is applied for a specified number of generations. This is the technique used by [12] finding most, if not all, pareto front (PF) solutions. To our knowledge, the solutions for the larger problems have not been published, making comparisons difficult.

Complete Iterative Approach: The complete iterative approach is an exhaustively deterministic approach that can be accomplished on MOPs 1∼8 (See Table 1). The number of solutions that must be evaluated is calculated by Equation 3 where n is the number of facilities and k is the number of locations. Consequently, for the mQAP, $k = n$.

$$x \approx \frac{n!}{(n-k)!}, \; where \; k = n \; this \; reduces \; to \; \mathbf{n!} \tag{3}$$

Function calculations for each MOP are $m * 10!$, $m * 20!$, and $m * 30!$ or $m * 3628800$, $\approx m * 2.43e18$ and $\approx m * 2.65e32$. These numbers are not to be confused with the search space size. For each search space solution, m calculations must occur.

Table 2. System Configuration

Cluster 1 (TAHOE)	Cluster 2 (ASPEN)	Cluster 3 (Polywells)
Fedora Core 2/Raid 5	Redhat Linux 9.0/Raid 5	Redhat Linux 7.3/Raid 5
Dual Opteron 2.2 ghz	Ath XP 3000+ 2.1ghz	Ath XP 2800+ 2.0ghz
RAM 4 GByte/Cache(L1 I 64,D 64/L2 1024)KB	1 GByte/(64,64/512)KB	1 GByte/(64,64/512)KB
Crossbar Switch/Gb Ethernet	Crossbar Switch/Fast Ethernet	Crossbar Switch/Gb Ethernet
65 node,2 CPUS/node	48 node,2 CPUS/node	16 node,1 CPU/node

Table 3. Summary of Results for all experiments. Included in this table are the number of optimal pareto front points (when known), and the number of PF points found by each algorithm {MOMGA-IIa (M-IIa), MOMGA-II (M-II), and LS}. u indicates that it is unknown how many dominated solutions this particular algorithm found when compared to the best PF solutions set found by all the algorithms considered. In addition, diameter (dia) and entropy (ent) is calculated for M-II's and M-IIa's solutions

Algorithm	**mQAP Number**, Size 10, (Deterministic PF True Points) True PF Pts Found/Total PF pts Found							
	1(58)	**2**(13)	**3**(15)	**4**(1)	**5**(55)	**6**(130)	**7**(53)	**8**(49)
LS	58/58	13/13	15/15	1/1	55/55	130/130	53/53	49/49
M-II	57/59	13/13	11/17	0/3	50/53	122/122	25/34	36/45
M-IIa[†]	58/58	11/12	15/15	1/1	55/55	130/130	53/53	49/49
†Time(mins)	21.5	62.3	29.8	10.9	45.5	68.1	45.4	15.8
mQAP Number, Size 20								
	9	**10**	**11**	**12**	**13**	**14**	**15**	**16**
LS	u/541	u/80	u/842	u/19	u/1587	u/178	u/1217	u/966
M-II	0/17	0/24	0/12	0/5	0/29	0/51	0/28	0/17
(dia/ent)	11.6/0.43	11.4/0.48	11.01/0.45	7.2/0.25	12.1/0.54	12.3/0.55	10.39/0.43	11.76/0.46
M-IIa[†]	36/36	33/33	31/31	7/7	63/63	139/139	48/48	44/44
(dia/ent)	13.0/0.58	13.7/0.69	11.17/0.47	3.67/0.16	14.1/0.73	15.5/0.88	12.76/0.60	11.37/0.50
†Time(days)	9.8	8.3	8.3	8.3	8.8	8.3	8.3	1.7
mQAP Number, Size 30								
	17	**18**	**19**	**20**	**21**	**22**	**23**	
LS	-	u/1329	u/705	u/1924	u/168	u/1909	u/1257	
M-II	n/a	0/507	0/552	10/552	0/104	0/795	0/755	
(dia/ent)	-	24.1/0.79	20.1/0.50	24.3/0.78	22.3/0.64	21.2/0.57	20.4/0.56	
M-IIa[†]	40/40	507/507	552/552	542/552	104/104	795/795	755/755	
(dia/ent)	17.2/0.42	23.9/0.80	23.2/0.76	23.1/0.74	21.9/0.59	24.0/8.11	25.1/0.90	
†Time(days)	8	∼8	8	8	8	8	8	

4   Design of Experiments

Experiments for the MOMGA-II were conducted on Clusters 2 and 3 listed in Table 2. Experiments for the MOMGA-IIa were done on Cluster 1 in the same table. The MOMGA-II was run 10 times in parallel and the data was then processed incrementally so as to show solutions gradually being found. The MOMGA-IIa ran 10 experiments in serial and kept one pool of PF solutions at all times. The MOMGA-II was run using BB sizes 1 through (10, 10, and 10) while the MOMGA-IIa was run using BB sizes 1 through (10, 15, and 20) for each MOP sized (10, 20, and 30). These experiments are run to determine how well each algorithm can solve the MOPs in this study. This research group's hypothesis is that the MOMGA-II's CT generation and evolution mechanism limited the exploration and building block finding ability of the algorithm, while the MOMGA-IIa now has the enhancement required to overcome this limitation.

5   Results and Analysis

Overall we are pleased with results of the MOMGA-IIa. In the MOP of size less than 20, the MOMGA-IIa found all true PF points available in a short amount of time (under 16 minutes in some cases) - the MOMGA-II did not. Additionally, in MOPs of size 20, the MOMGA-IIa solutions dominated the MOMGA-II's in every case (Illustrated by Figure 2). Finally, in MOPs of size 30 the MOMGA-IIa found more solutions than the MOMGA-II and were found to be dominate in all except for the MOP 20 case where 10 solutions found by the MOMGA-II are non-dominated. The reader should note also that MOPs of size 20 and 30 all took several days to solve. As far as the results for the LS method, these results are good in quantity, but without the actual data to compare we cannot claim that either algorithm LS or MOMGA-IIa is better at solving these MOPs. PF points found for each MOP will soon be posted on our web site,

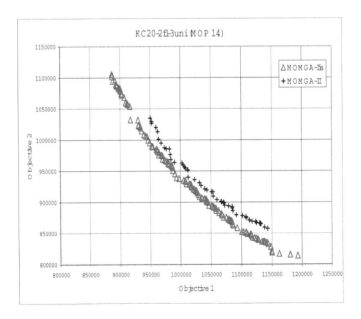

Fig. 2. Results for MOP 14 illustrating that the MOMGA-IIa's CT generator is superior at fining good BBs

`http://en.afit.edu/agct`. We conclude that the reason for the dominance of the MOMGA-IIa over the MOMGA-II is due to the CT generation and selection mechanism. In addition, MOMGA-II's limited number of CTs might be causing it to destroy some good building blocks. Lastly, the CT selection mechanism for the MOMGA-IIa allows for better multi-objective building blocks to be found - thus the MOMGA-IIa is a better building block builder. This phenomena is reflected in the data for each MOP.

Future Analysis: Further analysis of the MOMGA-IIa in solving the mQAP is required including a comparison to recent algorithm designs to solve the Biobjective QAP using Ant Colony Optimization (ACO) by Luis Paquete [14]. In addition, a memetic adjustment to MOMGA-IIa by adding a local search onto the end of the Juxtapositional Phase should be evaluated.

References

1. R.E. Burkard, S.E. Karisch, and F. Rendl. A quadratic assignment problem library. *Journal of Global Optimization*, pages 391–403, 1997.
2. Eranda Çela. *The Quadratic Assignment Problem - Theory and Algorithms*. Kluwer Academic Publishers, Boston, MA, 1998.
3. Richard O. Day. A multiobjective approach applied to the protein structure prediction problem. Ms thesis, Air Force Institute of Technology, March 2002. Sponsor: AFRL/Material Directorate.
4. Richard O. Day, Mark P. Kleeman, and Gary B. Lamont. Solving the Multiobjective Quadratic Assignment Problem Using a fast messy Genetic Algorithm. In *Congress on Evolutionary Computation (CEC'2003)*, volume 1, pages 2277–2283, Piscataway, New Jersey, December 2003. IEEE Service Center.
5. Richard O. Day and Gary B. Lamont. Multi-objective fast messy genetic algorithm solving deception problems. *Congress on Evolutionary Computation; Portland, Oregon*, 4:1502–1509, June 19 - 23 2004.
6. Carlos M. Fonseca and Peter J. Fleming. Genetic Algorithms for Multiobjective Optimization: Formulation, Discussion and Generalization. In Stephanie Forrest, editor, *Proceedings of the Fifth International Conference on Genetic Algorithms*, pages 416–423, San Mateo, California, 1993. University of Illinois at Urbana-Champaign, Morgan Kauffman Publishers.
7. L. M. Gambardella, E. D. Taillard, and M. Dorigo. Ant colonies for the quadratic assignment problems. *Journal of the Operational Research Society*, 50:167–176, 1999.
8. Peter Hahn, Nat Hall, and Thomas Grant. A branch-and bound algorithm for the quadratic assignment problem based on the hungarian method. *European Journal of Operational Research*, August 1998.
9. Jorng-Tzong Horng, Chien-Chin Chen, Baw-Jhiune Liu, and Cheng-Yen Kao. Resolution of quadratic assignment problems using an evolutionary algorithm. In *Proceedings of the 2000 Congress on Evolutionary Computation*, volume 2, pages 902–909. IEEE, IEEE, 2000.
10. Mark P. Kleeman. Optimization of heterogeneous uav communications using the multiobjective quadratic assignment problem. Ms thesis, Air Force Institute of Technology, March 2004. Sponsor AFRL.

11. Joshua Knowles and David Corne. Towards Landscape Analyses to Inform the Design of Hybrid Local Search for the Multiobjective Quadratic Assignment Problem. In A. Abraham, J. Ruiz del Solar, and M. Koppen, editors, *Soft Computing Systems: Design, Management and Applications*, pages 271–279, Amsterdam, 2002. IOS Press. ISBN 1-58603-297-6.

12. Joshua Knowles and David Corne. Instance generators and test suites for the multiobjective quadratic assignment problem. In Carlos Fonseca, Peter Fleming, Eckart Zitzler, Kalyanmoy Deb, and Lothar Thiele, editors, *Evolutionary Multi-Criterion Optimization, Second International Conference, EMO 2003, Faro, Portugal, April 2003, Proceedings*, number 2632 in LNCS, pages 295–310. Springer, 2003.

13. Eliane Maria Loiola, Nair Maria Maia de Abreu, Paulo Oswaldo Boaventura-Netto, Peter Hahn, and Tania Querido. An analytical survey for the quadratic assignment problem. Technical report, Council for the Scientific and Technological Development, of the Brazilian gov, 2004.

14. M. López-Ibáñez, L. Paquete, and T. Stützle. On the design of ACO for the biobjective quadratic assignment problem. In M. Dorigo, M. Birattari, C. Blum, L. Gambardella, F. Montada, and T. Stützle, editors, *Proceedings of the Fourth International Workshop on Ant Colony Optimization (ANTS 2004)*, volume 3172 of *Lecture Notes in Computer Sience*. Springer Verlag, 2004. (©Springer Verlag).

15. Vittorio Maniezzo and Alberto Colorni. The ant system applied to the quadratic assignment problem. *IEEE Transactions on Knowledge and Data Engineering*, 11:769–778, 1999.

16. Peter Merz and Bernd Freisleben. A comparison of memetic algorithms, tabu search, and ant colonies for the quadratic assignment problem. In *Proceedings of the 1999 Congress on Evolutionary Computation, 1999. CEC 99*, volume 3, pages 1999–2070. IEEE, IEEE, 1999.

17. Peter Merz and Bernd Freisleben. Fitness landscape analysis and memetic algorithms for the quadratic assignment problem. *IEEE Transactions on Evolutionary Computation*, 4:337–352, 2000.

18. L. Paquete, M. Chiarandini, and T. Stützle. Pareto local optimum sets in the biobjective traveling salesman problem: An experimental study. In X. Gandibleux, M. Sevaux, K. Sörensen, and V. T'kindt, editors, *Metaheuristics for Multiobjective Optimisation*, volume 535 of *Lecture Notes in Economics and Mathematical Systems*. Springer Verlag, 2004. (©Springer Verlag).

19. Thomas Stntzle. Iterated local search for the quadratic assignment problem. *Technical Report AIDA-99-03*, 1999.

20. Eric D. Taillard. Comparison of iterative searches for the quadratic assignment problem. *Location science*, 3:87–105, 1995.

21. Jesse B. Zydallis. *Explicit Building-Block Multiobjective Genetic Algorithms: Theory, Analysis, and Development*. Dissertation, Air Force Institute of Technology, AFIT/ENG, BLDG 642, 2950 HOBSON WAY, WPAFB (Dayton) OH 45433-7765, Feb 2002.

22. Jesse B. Zydallis, David A. Van Veldhuizen, and Gary B. Lamont. A Statistical Comparison of Multiobjective Evolutionary Algorithms Including the MOMGA–II. In Eckart Zitzler, Kalyanmoy Deb, Lothar Thiele, Carlos A. Coello Coello, and David Corne, editors, *First International Conference on Evolutionary Multi-Criterion Optimization*, pages 226–240. Springer-Verlag. Lecture Notes in Computer Science No. 1993, 2001.

Lot-Sizing in a Foundry Using Genetic Algorithm and Repair Functions

Jerzy Duda

AGH University of Science and Technology,
Faculty of Management, Dept. of Applied Computer Science,
ul. Gramatyka 10, 30-067 Kraków, Poland
jduda@zarz.AGH.edu.pl

Abstract. The paper presents a study of genetic algorithms applied to a lot-sizing problem, which has been formulated for an operational production planning in a foundry. Three variants of genetic algorithm are considered, each of them using special crossover and mutation operators as well as repair functions. The real size test problems, based on the data taken from the production control system, are presented for assessment of the proposed algorithms. The obtained results show that the genetic algorithm with two repair functions can generate good suboptimal solutions in the time, which can be acceptable from the decision maker point of view.

1 Introduction

The lot-sizing models allow to determine the production quantities at all production planning levels. The reviews of them can be found in [1], [3] or in [5]. The problem presented in this paper comes from a real production environment in a foundry and focuses on a short-term production planning.

The considered foundry is a typical foundry, which produces iron castings and uses hand-operated moulding machines. Among the shops existing in such a foundry two are the most important regarding operational production planning: a melting shop in which hot iron is prepared and a moulding shop where the moulds are made. Pouring and moulding operations must be coordinated, as melted iron cannot wait too long to be poured into the moulds and the space for the moulds waiting for pouring is limited.

Thus the main weekly task for the planners is to prepare a moulding plan together with a pouring schedule for the furnaces. While building those plans many technological and organizational constraints must be taken into consideration. The most significant are:

- capacities of furnaces and moulding machines,
- the number, desired delivery date and cast iron grade of ordered castings,
- the number of different castings, which can be produced during one shift (setup times are included in moulding times),
- the number of flasks of various size available during a working shift.

G.R. Raidl and J. Gottlieb (Eds.): EvoCOP 2005, LNCS 3448, pp. 101–111, 2005.

2 Optimisation Model

A mathematical model is built around the classical discrete capacitated lot-sizing problem with single level and multi item production. Presented model can be classified as a small bucket model, because only limited number of different items can be produced during one period of time.

Only few models dedicated strictly to planning in iron foundries can be found in literature. Van Voorhis et al. [7] provide a description of they work for Steel Foundries Society of America to develop software for generating pouring schedules. The objective function proposed by them is the sum of the non-utilization costs of heats and moulding lines, the costs of putting production for a given order into a particular lot, inventory costs and the penalty value for lateness. The constraints reflect all the capacity limitations as well as metallurgical ones. They used two stage heuristic, which solves an LP problem in the first stage and an IP problem in its second stage.

A model which is closer to the classical lot-sizing model can be found in dos Santos-Meza et al. paper [6]. The authors present a lot-sizing problem in a foundry with automated moulding machines. They use a minimization of item production costs as the only objective function and apply a relaxation method for the problems, which are then solved using CPLEX 4.0 library.

The objective function used in the model presented herein is similar to the objective function proposed by Van Voorhis et al. Instead of the non-utilization costs of furnaces and moulding lines, which may not always be estimated precisely, the combined utilization value is used directly. Also the inventory costs are omitted, as they are more or less fixed for the considered foundry (to some, but high enough limit).

The following symbols are used:

Decision variables:

   x_{ijtz} – number of castings planned for order i to be manufactured on machine j during day t and shift z,

   v_{htz} – number of heats of grade h during day t and shift z,

Data:

   τ – week for which the plan is created,

   k – number of working days in a week,

   l – number of machine type,

   m_j – number of working shifts for machines type j,

   n_j – number of active orders for machines type j,

   C_P – daily furnaces melting capacity [kg],

   W – weight of single heat [kg],

   C_{Fj} – capacity of moulding machines type j during a working shift [minutes],

   w_{ij} – total iron weight needed to produce single i casting [kg],

   a_{ij} – time of making a mould for casting i on machine j [minutes],

   d_{ij} – ordered number of castings of type i to be produced on machine j,

   γ – number of iron grades,

   g_{ij} – iron grade for casting i, $g_{ij} \in \{1,...,\gamma\}$,

   ω – number of flask types,

   S_o – flask number of type o available during a working shift,

q_{ij} – flask type in which a mould for casting i is prepared, $q_{ij} \in \{1,...,\omega\}$,

κ_j – number of different castings which can be produced on machine type j during one working shift,

δ_{ij} – due week for castings of type i to be produced on machine j,

Maximize:

$$\sum_{j=1}^{l}\sum_{i=1}^{n_j}\sum_{t=1}^{k}\sum_{z=1}^{m_j}(\frac{x_{ijtz}w_{ij}}{kC_P} + \frac{x_{ijtz}a_{ij}}{km_jC_{Fj}}) - \sum_{j=1}^{l}\sum_{i=1}^{n_j}((d_{ij} - \sum_{t=1}^{k}\sum_{z=1}^{m_j}x_{ijtz})(\tau - \delta_{ij})(\tau < \delta_{ij}))/1000 \qquad (1)$$

Subject to:

$$\sum_{h=1}^{\gamma}v_{hzt}W \leq C_P, \quad t=1,...,k, \quad z=1,...,m_j \qquad (2)$$

$$\sum_{i=1}^{n_j}x_{ijtz}a_{ij} \leq C_{Fj}, \quad j=1,...,l, \quad t=1,...,k, \quad z=1,...,m_j \qquad (3)$$

$$\sum_{t=1}^{k}\sum_{z=1}^{m_j}x_{ijtz} \leq d_{ij}, \quad j=1,...,l, \quad i=1,...,n_j \qquad (4)$$

$$\sum_{j=1}^{l}\sum_{i=1}^{n_j}(x_{ijtz}w_{ij}(g_{ij}=h)) \leq v_{htz}W, \quad h=1,...,\gamma, t=1,...,k, z=1,...,m_j \qquad (5)$$

$$\sum_{i=1}^{n_j}(x_{ijtz}>0) \leq \kappa_j, \quad j=1,...,l, \quad t=1,...,k, \quad z=1,...,m_j \qquad (6)$$

$$\sum_{j=1}^{l}\sum_{i=1}^{n_j}(x_{ijtz}(q_{ij}=o)) \leq S_o, \quad o=1,...,\omega, \quad t=1,...,k, \quad z=1,...,m_j \qquad (7)$$

The objective function (1) maximizes two elements. The first is the utilization level of furnaces and moulding machines, which are the main bottlenecks in the production system. Both utilization values are treated equally, however in reality the decision maker may use a weighted sum of them. The second element of the sum maximizes the penalty for the backlogging. The penalty function is proportional to the backlogged quantity and the number of overdue weeks. Those two criteria have been indicated directly by the planners in the considered foundry.

Constraints (2) and (3) are the capacity constraints for the furnaces and the moulding machines, respectively. Constraint (4) limits the production of a given casting to the quantity ordered by the customer. Constraint (5) limits the weight of the planned castings of a particular cast iron grade to the weight of the metal which is to be melted. Constraint (6) limits the number of different items which may be produced during one working shift. The last constraint (7) limits the flask availability.

The model is formulated as a discrete nonlinear problem. It was changed into an integer programming formulation by entering additional binary variables for the sake of the comparison between the results obtained by genetic algorithm and CPLEX solver.

3 Test Problems

Two test problems have been chosen from the data existing in the production control system, which is used in the described foundry.

The first test problem (*fixed1*) consists of 84 orders while the second one (*fixed2*) has 100 orders. There are four moulding lines in the considered factory, each consisting of two moulding machines, one for making a cope and one for making a drag (top and bottom part of a flask). However, there are only three types of moulding machines (denoted here as *A*, *B* and *C*). The type of a machine, which has to be used for making a mould for a particular casting is stated in a casting operation sheet.

Detailed orders specification for problems *fixed1* and *fixed2* are shown in Table 1 and Table 2, respectively.

Table 1. Detailed specification of *fixed1* problem

machine	order no.	flasks left to make	weight [kg]	moulding time [min]	iron grade	due week	machine	order no.	flasks left to make	weight [kg]	moulding time [min]	iron grade	due week	machine	order no.	flasks left to make	weight [kg]	moulding time [min]	iron grade	due week
A	1	282	61.2	30.5	4	-3	B	13	91	25.0	13.9	4	0	B	41	184	23.8	12.1	4	5
A	2	37	82.0	32.1	4	0	B	14	212	10.4	12.1	2	0	B	42	52	8.3	13.0	5	5
A	3	26	61.6	29.0	4	0	B	15	159	12.2	12.1	2	0	B	43	59	4.2	11.5	5	5
A	4	3	54.0	31.8	4	0	B	16	4	9.0	12.6	5	0	B	44	545	28.9	13.2	4	5
A	5	125	43.0	27.3	4	0	B	17	47	13.8	12.0	5	0	C	1	3	15.6	17.2	4	-3
A	6	226	65.0	32.6	4	1	B	18	16	12.4	13.9	5	0	C	2	257	18.2	15.1	4	-3
A	7	102	48.0	25.6	4	2	B	19	16	11.6	13.9	5	0	C	3	26	10.7	16.4	4	-2
A	8	16	30.4	25.4	4	3	B	20	16	11.0	13.9	5	0	C	4	25	10.8	18.1	4	-2
A	9	16	37.3	25.4	4	3	B	21	16	12.0	13.9	5	0	C	5	58	52.2	19.0	4	-2
A	10	22	34.7	30.2	5	3	B	22	133	12.0	12.3	5	1	C	6	196	29.6	17.7	4	-1
A	11	14	51.0	27.3	4	3	B	23	16	12.9	13.2	5	1	C	7	4	70.0	19.2	5	0
A	12	249	62.8	29.3	4	3	B	24	37	12.8	13.2	5	1	C	8	26	18.6	17.3	4	0
A	13	30	43.0	26.1	4	4	B	25	26	15.9	13.2	5	1	C	9	37	62.0	16.5	5	0
A	14	6	54.6	31.8	4	5	B	26	24	21.4	15.1	5	1	C	10	43	29.5	19.0	5	0
A	15	44	80.0	35.0	4	5	B	27	229	13.5	11.5	4	2	C	11	265	23.0	18.0	4	0
A	16	548	79.0	37.4	4	5	B	28	8	6.8	14.3	5	3	C	12	67	18.3	15.9	4	1
B	1	32	24.0	14.2	5	-3	B	29	16	6.0	14.3	5	3	C	13	36	6.9	2.9	5	2
B	2	35	24.0	14.2	5	-3	B	30	31	1.8	3.6	5	3	C	14	36	3.4	1.4	5	2
B	3	231	18.0	11.8	2	-3	B	31	6	10.4	13.9	5	3	C	15	83	5.8	2.4	5	2
B	4	424	9.3	11.6	4	-3	B	32	11	9.1	13.9	5	3	C	16	122	9.0	6.8	5	3
B	5	8	3.4	5.5	4	-2	B	33	16	10.8	13.9	5	3	C	17	96	23.6	16.7	5	3
B	6	31	15.6	13.9	2	-2	B	34	5	10.9	14.0	5	3	C	18	249	13.6	10.2	5	3
B	7	404	15.1	14.2	4	-2	B	35	5	13.1	14.0	5	3	C	19	22	21.7	18.6	5	3
B	8	538	15.1	14.2	4	-1	B	36	19	15.2	13.1	2	3	C	20	62	26.8	18.4	4	5
B	9	432	16.2	15.3	5	0	B	37	10	13.9	12.3	2	3	C	21	108	30.2	14.1	4	5
B	10	44	14.3	12.7	4	0	B	38	112	9.6	13.3	5	3	C	22	27	36.6	14.7	4	5
B	11	28	18.1	14.3	4	0	B	39	458	12.2	12.7	5	3	C	23	401	30.4	17.4	4	5
B	12	83	25.0	13.9	4	0	B	40	32	12.6	13.0	5	3	C	24	53	39.2	18.6	4	5

Table 2. Detailed specification of *fixed2* problem

machine	order no.	flasks left to make	weight [kg]	moulding time [min]	iron grade	due week	machine	order no.	flasks left to make	weight [kg]	moulding time [min]	iron grade	due week	machine	order no.	flasks left to make	weight [kg]	moulding time [min]	iron grade	due week
A	1	19	143	38.6	1	-2	B	3	45	9.9	13.0	4	-3	B	37	78	23.1	12.5	4	3
A	2	19	48.0	25.6	3	-2	B	4	220	15.1	14.2	3	-2	B	38	275	14.4	13.6	4	4
A	3	155	31.1	28.7	3	-2	B	5	66	19.3	12.0	4	-2	B	39	108	16.0	12.7	3	4
A	4	38	26.5	28.7	4	-2	B	6	212	31.0	14.7	3	-2	B	40	57	4.2	11.5	4	4
A	5	59	61.2	30.5	3	-2	B	7	52	16.0	12.7	3	-2	B	41	52	8.3	13.0	4	4
A	6	31	44.8	27.5	1	-1	B	8	135	18.0	13.2	3	-2	B	42	45	9.9	13.0	4	4
A	7	131	51.0	27.3	3	-1	B	9	39	23.0	19.6	2	-2	B	43	138	11.6	12.7	4	4
A	8	212	31.1	28.7	3	-1	B	10	33	28.9	13.2	3	-2	C	1	42	20.0	16.8	3	-3
A	9	52	32.6	28.8	1	-1	B	11	520	13.5	12.6	3	-2	C	2	156	10.6	14.7	4	-2
A	10	110	35.0	29.6	3	0	B	12	324	16.3	12.7	3	-2	C	3	35	41.6	20.2	1	-2
A	11	168	44.8	27.5	1	0	B	13	23	23.8	12.1	3	-1	C	4	28	13.2	15.8	3	-2
A	12	32	37.3	25.4	3	0	B	14	106	12.2	12.1	1	0	C	5	293	23.0	18.0	1	-2
A	13	44	73.0	29.5	3	0	B	15	106	10.4	12.1	1	0	C	6	305	27.5	18.0	1	-2
A	14	52	32.6	28.8	1	0	B	16	299	35.0	16.8	3	0	C	7	16	22.4	16.8	3	-1
A	15	109	51.0	27.3	3	1	B	17	17	13.1	14.0	4	0	C	8	20	58.8	17.1	2	-1
A	16	27	51.4	31.7	2	1	B	18	17	10.9	24.5	4	0	C	9	43	29.8	18.4	3	0
A	17	232	31.1	28.7	3	1	B	19	33	18.4	12.2	4	0	C	10	364	37.5	18.0	2	0
A	18	197	26.1	30.0	3	1	B	20	110	9.0	12.6	4	0	C	11	69	20.4	18.1	4	0
A	19	26	32.6	28.8	1	1	B	21	132	15.1	14.2	3	0	C	12	42	22.4	16.8	3	0
A	20	75	26.5	28.7	4	1	B	22	324	16.2	15.3	4	0	C	13	108	23.0	18.0	3	1
A	21	31	30.4	25.4	3	2	B	23	74	24.0	14.4	2	0	C	14	47	41.0	19.8	3	1
A	22	232	31.1	28.7	3	2	B	24	43	31.0	14.7	3	0	C	15	47	60.0	17.1	2	2
A	23	197	26.1	30.0	3	2	B	25	65	20.8	14.8	1	0	C	16	55	14.2	15.4	1	2
A	24	26	32.6	28.8	1	2	B	26	42	15.0	13.2	4	1	C	17	27	58.8	17.1	2	2
A	25	75	26.5	28.7	4	2	B	27	165	19.3	12.0	4	1	C	18	162	58.8	17.1	2	2
A	26	206	44.8	27.5	1	3	B	28	168	13.8	13.2	4	1	C	19	394	41.5	17.1	2	2
A	27	108	51.4	31.7	2	3	B	29	258	19.8	14.7	3	1	C	20	55	18.5	15.9	1	2
A	28	232	31.1	28.7	3	3	B	30	244	31.0	14.7	3	1	C	21	63	15.7	14.5	3	3
A	29	118	26.1	30.0	3	3	B	31	110	18.0	11.8	1	2	C	22	63	11.8	14.5	3	3
A	30	103	32.6	28.8	1	3	B	32	121	28.7	18.0	4	2	C	23	106	28.5	18.5	3	3
A	31	34	26.5	28.7	4	3	B	33	73	23.1	12.5	4	2	C	24	17	14.0	20.2	3	4
A	32	44	52.3	27.5	4	4	B	34	44	8.6	11.7	4	3	C	25	364	37.5	18.0	2	4
B	1	19	8.4	12.7	3	-3	B	35	147	24.0	14.4	2	3							
B	2	11	18.2	13.2	2	-3	B	36	38	8.8	11.7	4	3							

The number of flasks, which are to be made is calculated as the number of castings ordered by the customers divided by the number of castings which fit in a single flask. Thus the weight and forming time refer to the whole flask, not to a single casting. Due week is a week which has been agreed with the customer as a term of delivery. A negative number indicates that the remaining castings are already overdue.

There are 3 working shifts for the lines of machine type A and C while there are only 2 working shifts for the lines of machine type B. A common practice in the considered foundry is that only two different castings can be produced during one working shift, so κ_1 and κ_3 are set to 2 and κ_2 is set to 4. The total daily capacity of the furnaces is 21000 kg and a single heat weighs 1400 kg, i.e. at most 15 heats a day are possible. The number of flasks available for all moulding machines during one working shift is limited to 50 big flasks (machine type A), 100 medium (machine type C) and 120 small ones (machine type B).

The goal for optimisation is to create a plan for a week, which consists of 5 working days or for two weeks, consisting of 10 working days.

4 Genetic Algorithm

A weekly plan for moulding machines and a pouring schedule are coded in a single chromosome using integer gene values. First $n*k*(m_1+m_2+..+m_l)$ genes represent the quantity of castings planned for production or equals zero if the production for a particular order during a given shift is not planned. Last $\gamma*k*\max\{m_j\}$ genes represent the number of heats of a particular iron grade. This can be presented as the matrix shown in Figure 1.

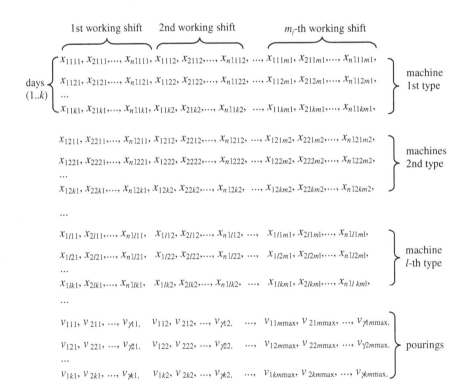

Fig. 1. Moulding plan and pouring schedule coded in a chromosome

4.1 First Variant (GA1)

The proposed chromosome structure is simple and natural. However, there is a lot of zeroes in a chromosome representing a valid plan, regarding the constraint (6). To avoid keeping incorrect individuals in a population, a simple repair algorithm has been introduced. Whenever constraint (6) is violated for one of the machines and working shifts, the smallest lots planned so far are eliminated successively from the plan until the number of different lots which are allowed for production during one working shift is reached. Only the non overdue castings are taken into consideration at the first stage. If the reduction in this stage is not enough the same procedure is applied also for the overdue castings. The scheme of the algorithm can be presented as follows:

```
Step 1. For each day t, working shift z and a machine j:
        K ← SUM(i=1..n_j){x_ijtz}
Step 2. While K>K_j
Step 3. Find a lot, for which the smallest weight of
        castings has been planned:
        x_s ← MIN(i=0..n_j){x_ijtz w_i}
Step 4. Remove the lot x_s from the plan
Step 5. K ← K-1. Go to step 2
```

The above algorithm is used also in the remaining two variants of genetic algorithm, as it improves solutions by 20–30%, on average.

Also a new crossover operator has been introduced. It creates one child from two parents in the following way. A string of genes representing a single shift is chosen randomly in two parents. If the fitness value for the first parent is better than for the second parent, lots from the chosen shift in the first parent are placed in the second parent. If the second parent has better fitness, then the lots from it replace the lots in the first parent. The crossover operator simultaneously alters the pouring schedule for the affected shift.

The irregular mutation in the version proposed by Michalewicz and Janikow [4] has been chosen as a mutation operator with a one modification, which has been applied to it. The probability of increasing a gene value is 0.75 for the overdue castings, while it equals to standard 0.5 for the rest of the castings.

Most of genes in initial population are set to zero and only about 3% of them are set to random values. The genes representing pouring schedule are set randomly in such a way that their sum equals to the limit of the number of heats allowed. Experiments have shown that the solution quality obtained after first 1000 generations had a great impact on the quality of the final solution. That is why the proposed algorithm uses initially 10 populations, starting from 10 different points. The evolution is continued only for the best population after first 500 generations.

The remaining parameters, common for all algorithm variants look as follows:

- population size (fixed): 100 individuals
- number of generations: 10000 for problems with 5 days, 20000 for 10 days problems (+10000 for 10 initial populations)

- selection type: binary tournament with elitism
- crossover type: crossover changing shifts with the probability of 0.8
- mutation type: irregular mutation with the probability of 0.001
- penalty function: sum of squares of constraint violation values multiplied by 10000

4.2 Second Variant (GA2)

In the second variant of genetic algorithm two sets of variables representing moulding plans (x_{ijtz}) and pouring schedule (v_{htz}) are treated as two separate chromosomes.

Irregular mutation operator alters genes only in the first chromosome. For the second chromosome another mutation has been defined. It works as follows. The number of heats of randomly chosen iron grade is decreased by 1 and simultaneously the number of heats of another randomly chosen iron grade within the same working shift is increased by 1. This mutation is used with the probability of 0.05.

All other parameters of the genetic algorithm remain the same as in the first variant.

4.3 Third Variant (GA3)

In the third variant of genetic algorithm the genes in which pouring schedule is coded has been removed from the chromosome structure. Instead of this a second repair algorithm has been used. Its role is to keep moulding plans always acceptable from a pouring schedule point of view. This means there is always enough hot iron for filling all the moulds, which has been prepared. The idea of this algorithm is similar to the first repair algorithm. If the maximum number of heats of a particular iron grade is exceeded than the lot with the minimum weight of castings is removed from the plan. The details of the algorithm are shown below:

```
Step 1. t ← 1
Step 2. For each iron grade h:
        Calculate the summary weight of iron grade h
        (SWₕ) necessary for pouring the moulds prepared
        on day t:
        SWₕ ← SUM(j=1..l,i=1..nⱼ,z=1..mⱼ){(xᵢⱼₜ zwᵢ)(gᵢ=h)}
Step 3. Calculate the number of heats (lw):
        lw ← INT(SUM(h=1..l){SWₕ/W})+1
Step 4. If lw ≤ lwₘₐₓ go to step 9
Step 5. For the day t find a machine j, a shift z and a
        lot i, for which the smallest weight of castings
        has been planned:
        xₛ ← min(j=1..l,i=1..nⱼ,z=1..mⱼ){xᵢⱼₜ zwᵢ}
Step 6. Remove the lot xₛ from the moulding plan
Step 7. Correct lw by the removed lot weight
Step 8. If lw > lwₘₐₓ go to step 4
Step 9. t ← t + 1
Step 10. If t ≤ k go to step 2
```

Experiments have shown that the role of the crossover operator defined earlier is virtually meaningless for the third variant of genetic algorithm. Thus a modified version of it has been introduced for this variant. The lots in randomly chosen shift from the first parent are swapped with the lots in another randomly chosen shift from the second parent. In that way two parents create two children instead of one, as it was in the previous case. However, this crossover plays a role of another mutation operator, rather than crossover itself. It is used with the probability of 0.1.

5   Results and Comparison to Integer Programming

Each variant of genetic algorithm was run for 10 times for fixed1 and *fixed2* test problems, assuming 5 days and 10 days planning horizon. A single run took about 4 minutes for 5-day problems and 8 minutes for 10 days (computer with Pentium 560 processor, 1 GB RAM). The results were then compared with the solutions given by branch-and-bound algorithm, implemented in CPLEX 9.0 mixed integer programming solver. Solving time for CPLEX was limited to the time of 10 runs of a single genetic algorithm. The best results and average results obtained from ten runs of the three genetic algorithm variants and the results generated by CPLEX 9.0 (denoted as bb) are collected in Table 3. The GA1 variant gave only 5 valid solutions for *fixed2* problem with 10 days.

Table 3. Results obtained by the genetic algorithm variants

Problem		GA1	GA2	GA3	bb	GA3-bb
fixed1 with 5 days	best	0.61	1.38	1.92	1.98	3.1%
	avg.	0.21	1.15	1.89		4.6%
fixed1 with 10 days	best	0.52	0.79	1.81	1.89	4.5%
	avg.	0.12	0.50	1.77		6.7%
fixed2 with 5 days	best	-0.35	0.74	1.84	1.96	6.4%
	avg.	-0.48	0.39	1.77		10.0%
fixed2 with 10 days	best	-0.58	-0.15	1.77	1.90	6.9%
	avg.	-1.24	-0.41	1.69		10.7%

The last column in the table shows a relative difference in the objective function value between the third variant of genetic algorithm and the branch-and-bound solution provided by CPLEX 9.0. The best of ten runs solution obtained by the third GA variant is not more than 5% behind the branch-and-bound algorithm for the first problem and less than 7% for the second problem.

Next, the experiments for the problems with the objective function, which consisted only of the first summand in equation (1), i.e. the utilization level of bottleneck aggregates, have been carried out. The same genetic algorithms were tested and only the overdue castings were not treated in a special way by the mutation operator and the repair functions, as it was in the previous case. Table 4 shows the obtained results in the same form as earlier.

Table 4. Results obtained by the genetic algorithm variants for a simplified objective

Problem		GA1	GA2	GA3	bb	GA3-bb
fixed1 with 5 days	Best	1.82	1.89	1.97	1.98	0.8%
	Avg.	1.71	1.76	1.93		2.5%
fixed1 with 10 days	Best	1.04	1.83	1.89	1.90	0.6%
	Avg.	0.72	1.77	1.80		5.1%
fixed2 with 5 days	Best	1.69	1.87	1.96	1.97	0.4%
	Avg.	1.58	1.79	1.92		2.8%
fixed2 with 10 days	Best	1.19	1.85	1.92	1.96	2.3%
	Avg.	0.86	1.65	1.85		5.4%

This time the best of the three genetic algorithm variants remains only less then 1% behind the CPLEX 9.0 algorithm in 3 of 4 test tasks, if the best result from ten runs is taken into consideration. The average solution generated by GA3 is within 3% of the branch-and-bound limit for the 5 days planning problems and within 5.5% for the 10 days instances.

This shows that the combined, but competitive criteria have a significant negative impact on the quality of solutions generated by the genetic algorithms. One of the methods to overcome this problem is to treat all the objective functions independently and use multiobjective evolutionary algorithms. The results of such an approach for a similar production planning problem are described in the recently published author's paper [2].

6 Final Remarks

The results presented in the paper show how much a repair function is important to a genetic algorithm, at least for certain real world problems. Both entering new operators and suiting a chromosome structure to a specific problem can significantly improve the quality of the obtained solutions. Nerveless, it is the introduction of even simple repair functions that lets a genetic algorithm to generate solutions of the high quality.

The third variant of genetic algorithm presented in this paper (GA3) can provide good solutions, which differ by 0.5–7% from the results obtained by the advanced branch-and-bound methods implemented in CPLEX 9.0. However, an integer programming approach cannot always be applied easily, because it requires all the objective functions and constraints to be non-linear. This is not a problem for meta-heuristics such as evolutionary algorithms, for which the models can be written in the natural way.

The optimisation model for operational production planning in a foundry proposed in this paper will be successively complemented with new technological and organizational constraints, which have an impact on the overall production costs. The most interesting seems to be the assessment of the costs of a particular heat sequence, resulting from the costs of changing from one iron grade to another. The possibility of making a given casting from different iron grades (usually higher), if such an operation is acceptable by the customer, will be also introduced into the planning model.

The data for problems *fixed1* and *fixed2* can be downloaded from the author's website at http://www.zarz.agh.edu.pl/jduda/foundry.

Acknowledgment

This study was supported by the State Committee for Scientific Research (KBN) under the Grant No. 0224 H02 2004 27.

References

1. Drexl, A., Kimms, A.: Lot sizing and scheduling – Survey and extensions, European Journal of Operational Research vol. 99, 2 (1997) 221–235
2. Duda, J., Osyczka, A.: Multiple criteria lot-sizing in a foundry using evolutionary algorithms (in:) Coello Coello C.A. et al. (eds.): EMO 2005, Lecture Notes in Computer Science vol. 3410 (2005) 651–663
3. Karimi, B., Fatemi Ghomi, S.M., Wilson, J.M.: The capacitated lot sizing problem: a review of models and algorithms, Omega, vol. 31, 5 (2003) 409–412
4. Michalewicz, Z., Janikow, C.Z.: Genetic algorithms for numerical optimization, Statistics and Computing, vol. 1, 2 (1991) 75–91
5. Pochet, Y.: Mathematical programming models and formulations for deterministic production planning problem (in:) Jünger M., Naddef D. (eds.), Computational Combinatorial Optimization, vol. 2241, Berlin, Springer-Verlag, Berlin (2001)
6. dos Santos-Meza, E., dos Santos, M.O., Arenales, M.N.: A Lot-Sizing Problem in An Automated Foundry. European Journal of Operational Research, vol. 139, 3 (2002) 490–500
7. Voorhis, T.V., Peters, F., Johnson, D.: Developing Software for Generating Pouring Schedules for Steel Foundries. Computers and Industrial Engineering, vol. 39, 3 (2001) 219–234

Estimation of Distribution Algorithms with Mutation

Hisashi Handa

Okayama University, Tsushima-Naka 3-1-1,
Okayama 700-8530, JAPAN
handa@sdc.it.okayama-u.ac.jp
http://www.sdc.it.okayama-u.ac.jp/~handa/index-e.html

Abstract. The Estimation of Distribution Algorithms are a class of evolutionary algorithms which adopt probabilistic models to reproduce the genetic information of the next generation, instead of conventional crossover and mutation operations. In this paper, we propose new EDAs which incorporate mutation operator to conventional EDAs in order to keep the diversities in EDA populations. Empirical experiments carried out this paper confirm us the effectiveness of the proposed methods.

1 Introduction

Recently, Estimation of Distribution Algorithms (EDAs) have been attracted much attention in genetic and evolutionary computation community due to their search abilities [1]. Genetic operators such like crossover and mutation are not adopted in the EDAs. In the EDAs, a new population is generated from the probabilistic model constituted by a database containing the genetic information of the selected individuals in the current generation. Such reproduction procedure by using the probabilistic model allows EDAs to search for optimal solutions effectively. However, it significantly decreases the diversity of the genetic information in the generated population when the population size is not large enough.

In this paper, we discuss on the effectiveness of mutation operation in the case of EDAs. We propose new EDAs which incorporate mutation operator to conventional EDAs in order to keep the diversities in EDA populations. In order to confirm the effectiveness of the proposed approach, Computational simulations on Four-peaks problems, Fc_4 function, and MAXSAT problems are carried out.

Related works are described as follows: The effectiveness of mutation operator in the case of conventional genetic and evolutionary computation has been studied a long time: Ochoa empirically studied a well-known heuristic with respect to mutation: better mutation probability is around $1 / L$ (string length) [2]. The relationship between mutual information and entropy was discussed by Toussaint [3].

In the next section, we will briefly introduce three kinds of the EDAs, which are employed for our experiments. Moreover, we will describe the basic notion of

G.R. Raidl and J. Gottlieb (Eds.): EvoCOP 2005, LNCS 3448, pp. 112–121, 2005.

```
Procedure Estimation of Distribution Algorithm
begin
   initialize $D_0$
   evaluate $D_0$
   until Stopping criteria is hold
      $D_l^{Se} \leftarrow$ Select $N$ individuals from $D_{l-1}$
      $p_l(\mathbf{x}) \leftarrow$ Estimate the probabilistic model from $D_l^{Se}$
      $D_l \leftarrow$ Sampling $M$ individuals from $p_l(\mathbf{x})$
      evaluate $D_l$
   end
end
```

Fig. 1. Pseudo code of Estimation of Distribution Algorithms

Estimation of Distribution Algorithms with mutation, i.e., the proposed method. Then, computational experiments are examined in section 3. Section 4 will conclude this paper.

2   Estimation of Distribution Algorithms

2.1   General Framework of EDAs

The Estimation of Distribution Algorithms are a class of evolutionary algorithms which adopt probabilistic models to reproduce the genetic information of the next generation, instead of conventional crossover and mutation operations. The probabilistic model is represented by conditional probability distributions for each variable (locus). This probabilistic model is estimated from the genetic information of selected individuals in the current generation. Hence, the pseudo-code of EDAs can be written as Fig. 1, where D_l, D_{l-1}^{Se}, and $p_l(\mathbf{x})$ indicate the set of individuals at l^{th} generation, the set of selected individuals at $l-1^{\text{th}}$ generation, and estimated probabilistic model at l^{th} generation, respectively [1]. The representation and estimation methods of the probabilistic model are devised by each algorithm. As described in this figure, the main calculation procedure of the EDAs is that (1) the N selected individuals are selected from the population in the previous generation, (2) then, the probabilistic model is estimated from the genetic information of the selected individuals, (3) a new population whose size id M is sampled by using the estimated probabilistic model, and (4) finally, the new population is evaluated.

In this paper, we discuss the effectiveness of mutation operation in case of UMDA, MIMIC, and EBNA. The difference between these EDAs is the representation and estimation of the probabilistic models. Since our study is relevant to the representation of the probabilistic models, we will briefly describe EDAs with a focus on the representation as follows:

– **UMDA:**   Mühlenbein proposed UMDA (Univariate Marginal Distribution Algorithm) in 1996 [1,7]. As indicated by its name, the variables of the

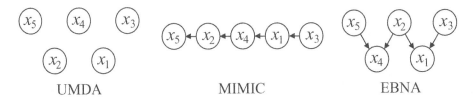

Fig. 2. Probabilistic models for UMDA, MIMIC, and EBNA

probabilistic model in this algorithm is assumed to be independent from other variables. That is, the probability distribution $p_l(\mathbf{x})$ is denoted by a product of univariate marginal distributions, i.e.,

$$p_l(\mathbf{x}) = \prod_{i=1}^{n} p_l(x_i),$$

where $p_l(x_i)$ denotes the univariate marginal distribution $X_i = x_i$ at a variable X_i at generation l.

– **MIMIC:**   De Bonet *et al.* proposed MIMIC [1, 8], a kind of EDAs whose probabilistic model is constructed with bivariate dependency such like COMIT [9]. While the COMIT generates a tree as dependency graph, the probabilistic model of the MIMIC is based upon a permutation π.

$$p_l(\mathbf{x}) = \prod_{j=1}^{n-1} p_l(x_{i_{n-j}} | x_{i_{n-j+1}}) \cdot p_l(x_{i_n}),$$

where the permutation π is represented by $(i_1, i_2, \ldots, i_n)$, and is obtained in each generation. In Fig. 2, the permutation π is set to be $(i_1, i_2, \ldots, i_5) = (5, 2, 4, 1, 3)$ for instance.

– **EBNA:**   Like BOA and LFDA [10, 11], the EBNA (Estimation of Bayesian Networks Algorithms), proposed by Larrañaga *et al.*, adopts Bayesian Network (BN) as the probabilistic model [1, 12]. That is, the probabilistic model used in the EBNA is written as follows: Suppose that S is the network structure of Bayesian Network, then,

$$p_l(\mathbf{x}) = \prod_{i=1}^{n} p(x_i | Pa_i^S),$$

where Pa_i^S denotes a set of parent variables of ith variable. For instance, in Fig. 2, sets of the parent variables Pa_1^S, Pa_3^S of variable x_1 and x_3 are defined as $\{x_2, x_3\}$ and $\oslash$, respectively.

2.2 Estimation of Distribution Algorithms with Mutation

In this paper, we incorporate mutation operators into EDAs. The followings introduce incorporated mutation operators for each of UMDA, MIMIC, and EBNA: In the case of UMDA, we adopt the bitwise mutation operator which is the same as SGA: after each bit is decided in accordance with estimated marginal distribution, the mutation operator independently changes the allele of loci with the mutation probability. That is, the succeeding bit production is not affected by the changes by the mutation operator.

On the other hand, since the probabilistic model of MIMIC is represented by a chain of variables, the changes at a certain locus by mutation operator affects the production of alleles at succeeding loci. The mutation operation for MIMIC is described as follows: After producing at a certain locus x_i, whether mutation operation is carried out is randomly decided with the mutation probability. If the mutation operation is occurred, the allele at the locus x_i is flipped. Now, suppose that above mutation operation is carried out at the last produced locus $x_{i_{j+1}}$ in order π. If the conditional probability $p_l(x_{i_{n-j}}|x_{i_{j+1}})$ for flipped allele is not defined[1] the former value produced at first is used for $x_{i_{j+1}}$.

Finally, mutation operation for EBNA is similar to the one in the case of MIMIC, that is, we should take into consideration for succeeding bit production. Now, we assume that we would like to decide the allele at a certain locus x_i and q variables (loci) in the parent set Pa_i^S are flipped their alleles by past mutation events. The conditional probability $p(x_i|Pa_i^S)$ for flipped alleles is used iff such conditional probability is defined. Otherwise, find a defined conditional probability $p(x_i|Pa_i^S)$ such that the number of flipped alleles is maximum, and use it to produce allele at the current locus.

3 Experiments

3.1 Experimental Settings

This paper examines the effectiveness of mutation operation in the case of EDAs on three kinds of fitness functions, whose explanation is described in the next subsection, Four-peaks function, Fc_4 function, and MAXSAT problems. In this paper, we compare the proposed methods with corresponding conventional methods, that is, UMDA, MIMIC and EBNA. We will represent corresponding proposed method as UMDAwM, MIMICwM, and EBNAwM, respectively. This paper employs $EBNA_{BIC}$ as EBNA. For first two functions, we investigate how many trials these algorithms can achieve to optimal solution effectively. Hence, we adopt two indices to evaluate the effectiveness of algorithms: success ratio (SR) and the number of fitness evaluations until finding optimal solutions (NOE). The SR is defined as the fraction of runs in which find optimal solutions. The NOE in this paper is averaged value over "success" runs. If the SR

---

[1] If the flipped allele is not occurred in selected individuals D_{l-1}^{Se}, we cannot calculate the conditional probability $p_l(x_{i_{n-j}}|x_{i_{j+1}})$ for flipped allele.

Table 1. Genetic parameters for each problem

	Four-Peaks (20 and 40 var.'s)	Four-Peaks (60 and 80 var.'s)	Fc_4	MAXSAT
Mutation Prob.	0.2, 0.1, 0.07, 0.05, 0.02, 0.01	0.02, 0.01, 0.005, 0.002, 0.001	0.2, 0.1, 0.07, 0.05, 0.02, 0.01	0.02, 0.01, 0.005, 0.002, 0.001
No. Indiv.	32, 64, 128, 256, 512, 1024 2048, 4096	1024, 2048 4096, 8192	32, 64, 128, 256, 512, 1024 2048, 4096	256, 512, 1024, 2048, 4096, 8192
No. Fit. Eval.	1,000,000	1,000,000	100,000	200,000

is 0, the NOE is not defined. Hence, lines in graphs in Fig. 3 and Fig. 4 are not plotted for undefined NOE. On the other hand, we examine the solution quality obtained by the proposed methods and conventional methods for MAXSAT problems.

Genetic parameters used in each examination is summarized in Table 1. For each tuple of parameters indicated in the table, trial is examined. The number of trials for each tuple is set to be 30 for Four-peaks and Fc_4 function, and 10 for each problem instance of MAXSAT. We use benchmark problems for MAXSAT which consists of 50 problem instances for each couple of variables and clauses [14][15]. Moreover, for Four-peaks and Fc_4 function, we only plot the best result for the proposed methods over various values of mutation probabilities. Common settings for all problems are as follows: The number of selected individuals N is set to be half of the number of individuals M. We use the truncation selection method, which selects the best N individuals form M individuals, to constitute the selected individuals.

3.2   Test Functions

Four-Peaks Function [8]

$$F_{\text{four-peak}}(T, \mathbf{x}) = \max(\text{head}(x_1, \mathbf{x})) + \max(\text{tail}(1 - x_n, \mathbf{x})) + R(T, \mathbf{x})$$

$$R(T, \mathbf{x}) = \begin{cases} \frac{3}{2}n & \text{if}(\text{head}(x_1, \mathbf{x}) > T))(\text{tail}(1 - x_n, \mathbf{x}) > T) \\ 0 & \text{otherwise}, \end{cases}$$

where $\text{head}(b, \mathbf{x})$ and $\text{tail}(b, \mathbf{x})$ denote the number of contiguous leading bits set to b in $\mathbf{x}$, and the number of contiguous trailing bits set to b in $\mathbf{x}$, respectively. The parameter T is set to be $2/n - 1$ in this paper. There are two optimal solutions: $000\ldots0011\ldots111$ and $111\ldots1100\ldots000$. Furthermore, there are two sub-optimal solutions: $111\ldots1111\ldots111$ and $000\ldots0000\ldots000$ which can be easily achieved to.

Fc_4 [1]

At first, we describe two functions: F^3_{cuban1} and F^5_{cuban1}

$$F^3_{cuban1}(x_1, x_2, x_3) = \begin{cases} 0.595 & \text{for } (x_1, x_2, x_3) = (0,0,0) \\ 0.200 & \text{for } (x_1, x_2, x_3) = (0,0,1) \\ 0.595 & \text{for } (x_1, x_2, x_3) = (0,1,0) \\ 0.100 & \text{for } (x_1, x_2, x_3) = (0,1,1) \\ 1.000 & \text{for } (x_1, x_2, x_3) = (1,0,0) \\ 0.050 & \text{for } (x_1, x_2, x_3) = (1,0,1) \\ 0.090 & \text{for } (x_1, x_2, x_3) = (1,1,0) \\ 0.150 & \text{for } (x_1, x_2, x_3) = (1,1,1) \end{cases}$$

$$F^5_{cuban1}(x_1, x_2, x_3, x_4, x_5) = \begin{cases} 4F^3_{cuban1}(x_1, x_2, x_3) & \text{if } x_2 = x_4 \text{ and } x_3 = x_5 \\ 0 & \text{otherwise.} \end{cases}$$

Then, function Fc_4 is defined as follows:

$$Fc_4(\mathbf{x}) = \sum_{c=1}^{r} F^5_{cuban1}(x_{5c-4}, x_{5c-3}, x_{5c-2}, x_{5c-1}, x_{5c}),$$

where $n = 5r$. This function has only one optimal solution.

MAXSAT

In order to solve the MAXSAT problems, we have to find an assign of values such that the number of satified clauses is maximized. That is, this problem is formulated as the following CNF (Conjunctive Normal Form):

$$\bigwedge_j (\bigvee_{l_i \in cl(j)} l_i),$$

where $cl(j)$ denotes a set of literals which belongs in the jth clauses. Moreover, l_i indicates literals.

3.3   Experimental Results

Fig. 3 depicts the experimental results for the Four-peaks problems. Conventional methods, MIMIC and EBNA, with larger population size could find optimal solutions when dimension = 20. It is difficult to solve for Four-peaks problems by assigning alleles at each locus independently, so that UMDA could not solve the four-peaks problems effectively. The proposed method improve the search ability of the conventional EDAs in the viewpoint of the success ratio. Especially, the proposed method with smaller population size could solve the Four-peaks problems when they are easy problems. Moreover, only MIMICwM could solve for Four-peaks problems with 80 variables.

Next, we carried out experiments on Fc_4 problems as delineated in Fig. 4. The number of fitness evaluations in each run was limited to 100000 so that it

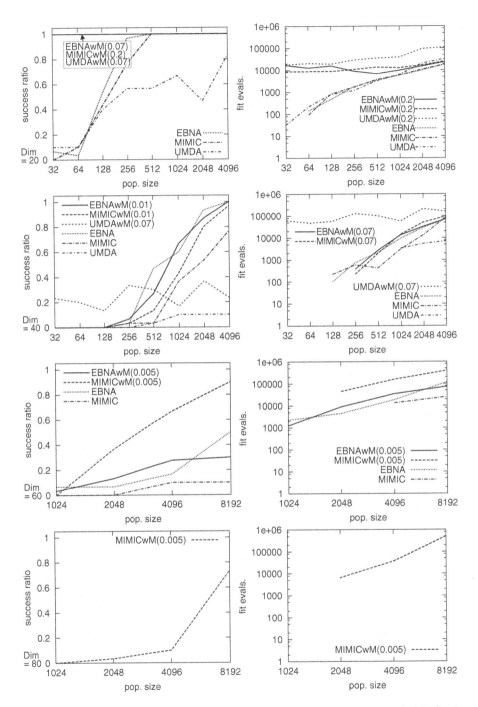

Fig. 3. Experimental results for the Four-peaks problems: Success ratio (LEFT), the number of fitness evaluations until finding optimal solutions (RIGHT); Problem dimension = 20 (UPPER), 40, 60, and 80 (LOWER)

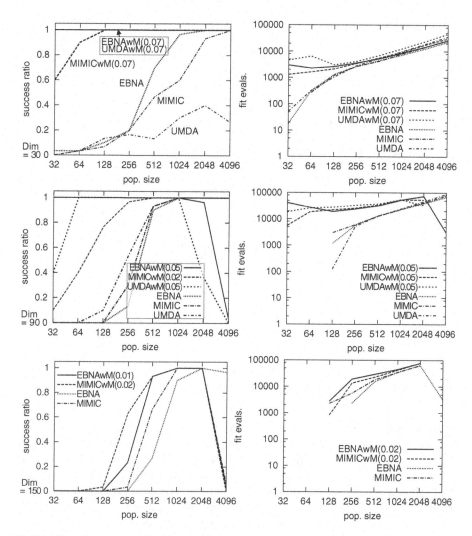

Fig. 4. Experimental results for the Fc_4 function: Success ratio (LEFT), the number of fitness evaluations until finding optimal solutions (RIGHT); Problem dimension = 30 (UPPER), 90 (MIDDLE), and 150 (LOWER)

was impossible for the proposed method whose population size was set to be 4096 to solve the Fc_4 problems with 90 and 150 variables. Except for this, the proposed methods shows better performance in the sense of the success ratio.

Finally, Fig. 5 investigated the quality of acquired solutions on 3-MAXSAT problems with 100 variables. Upper graphs show the results for 500 clauses. On the other hand, lower graphs are the results for 700 clauses. Graphs on the left side and the right side indicates result of MIMIC and EBNA, respectively. For each of the number of population size in all graphs in the figure, 6 lines are plotted: the solid line denotes the conventional method. Other dashed lines

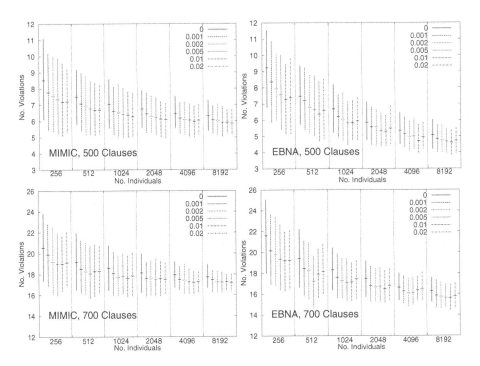

Fig. 5. Experimental results for MAXSAT problems with 100 variables: MIMIC (LEFT) and EBNA (RIGHT); 500 clauses (UPPER) and 700 clauses (LOWER)

represents corresponding mutation probabilities. As mentioned above, there are 50 problem instances for each couple of (variable, clauses). 10 trials are examined for each problem instance. The highest and lowest points indicates the averaged number of unsatisfied clauses for worst and best solutions in 10 trials, respectively. Moreover, the short horizontal lines crossed to corresponding vertical lines means the averaged value over all (500) trials. All solutions used to depict the graphs is acquired when the number of fitness evaluations achieves to 200,000. These graphs reveal that the mutation operator proposed in this paper improves the quality of solutions which are acquired after the convergence.

4   Conclusion

In this paper, we discussed on the effectiveness of mutation in the case of Estimation of Distribution Algorithms from empirical viewpoints. Comparisons on two deceptive functions carried out in section 3 elucidate that (1) the proposed method works well even if the population size M of EDAs is not large enough, and (2) only MIMICwM could solve for the most difficult four-peaks problems applied in this paper. Moreover, the computational results for MAXSAT problems reveal that the mutation operator proposed in this paper improves the quality of solutions after the convergence.

References

1. P. Larrañaga and J. A. Lozano Editors: *Estimation of Distribution Algorithms*, Kluwer Academic Publishers (2002)
2. *Proc. of 2002 Genetic and Evolutionary Computation Conference*, (2003) 495–502
3. Marc Toussaint: The Structure of Evolutionary Exploration: On Crossover, Buildings Blocks, and Estimation-Of-Distribution Algorithms, *Proc. of 2003 Genetic and Evolutionary Computation Conference*, LNCS 2724, **2** (2003) 1444–1455
4. Vose, M.D.: The simple genetic algorithm: foundations and theory. MIT Press (1999)
5. González, C., Lozano J.A., Larrañaga, P.: Mathematical Modeling of Discrete Estimation of Distribution Algorithms. Larrañaga, P. and Lozano, J.A. Eds., *Estimation of Distribution Algorithms*. Kluwer Academic Publishers (2002) 147–163
6. M. Pelikan: Bayesian optimization algorithm: From single level to hierarchy, Ph.D. thesis, University of Illinois at Urbana-Champaign, Urbana, IL. Also IlliGAL Report No. 2002023 (2002)
7. H. Mühlenbein and G. Paaß:From Recombination of genes to the estimation of distributions I. Binary parameters. *Parallel Problem Solving from Nature - PPSN IV* (1996) 178–187
8. J. S. De Bonet *et al.*: MIMIC: Finding optima by estimating probability densities, *Advances in Neural Information Processing Systems* **9** (1996)
9. S. Baluja: Using a priori knowledge to create probabilistic models for optimization *International J. of Approximate Reasoning*, **31(3)** (2002) 193–220
10. M. Pelikan *et al.*: BOA: The Bayesian optimization algorithm, *Proceedings of the Genetic and Evolutionary Computation Conference* **1** (1999) 525–532
11. H. Mühlenbein and T. Mahnig: FDA - a scalable evolutionary algorithms for the optimization of additively decomposed functions, *Evolutionary Computation* **7(4)** (1999) 353–376
12. P. Larrañaga *et al.*: Combinatorial Optimization by Learning and Simulation of Bayesian, *Uncertainty in Artificial Intelligence, Proceedings of the Sixteenth Conference* (2000) 343–352
13. The equation for the response to selection and its use for prediction, *Evolutionary Computation*, **5(3)** (1998) 303–346
14. http://rtm.science.unitn.it/intertools/sat/
15. R. Battiti and M. Protasi: Reactive Search, a history-sensitive heuristic for MAX-SAT, *ACM Journal of Experimental Algorithmics*, **2(2)** (1997)

Property Analysis of Symmetric Travelling Salesman Problem Instances Acquired Through Evolution

Jano I. van Hemert

Centre for Emergent Computing, Napier University, Edinburgh, UK
j.van.hemert@napier.ac.uk

Abstract. We show how an evolutionary algorithm can successfully be used to evolve a set of difficult to solve symmetric travelling salesman problem instances for two variants of the Lin-Kernighan algorithm. Then we analyse the instances in those sets to guide us towards deferring general knowledge about the efficiency of the two variants in relation to structural properties of the symmetric travelling salesman problem.

1 Introduction

The travelling salesman problem (TSP) is well known to be NP-complete. It is mostly studied in the form of an optimisation problem where the goal is to find the shortest Hamiltonian cycle in a given weighted graph [1]. Here we will restrict ourselves to the *symmetric* travelling salesman problem, i.e., distance$(x, y) =$ distance(y, x), with Euclidean distances in a two-dimensional space.

Over time, much study has been devoted to the development of better TSP solvers. Where "better" refers to algorithms being more efficient, more accurate, or both. It seems, while this development was in progress, most of the effort went into the construction of the algorithm, as opposed to studying the properties of travelling salesman problems. The work of [2] forms an important counterexample, as it focuses on determining phase transition properties of, among others, TSP in random graphs, by observing both the graph connectivity and the standard deviation of the cost matrix. Their conjecture, which has become popular, is that all NP-complete problems have at least one order parameter and that hard to solve problem instances are clustered around a critical value of this order parameter.

It remains an open question whether the critical region of order parameters are mainly depending on the properties of the problem, or whether it is linked to the algorithm with which one attempts to solve the problem. However, a substantial number of empirical studies have shown that for many constraint satisfaction and constraint optimisation problems, a general region exists where problems are deemed more difficult to solve for a large selection of algorithms [3, 4, 5].

Often the characterisation of the order parameter includes structural properties [6, 7], which leads to both a more accurate prediction and a better

G.R. Raidl and J. Gottlieb (Eds.): EvoCOP 2005, LNCS 3448, pp. 122–131, 2005.

understanding of where hard to solve problems can be expected. Naturally, this does not exclude that a relationship between an algorithm and certain structural properties can exist. In this study, we shall provide empirical evidence for the existence of such a distinct relationship for two TSP problem solvers, which is of great influence on the efficiency of both algorithms.

In the following section we describe the process of evolving TSP instances. Then, in Section 3 we provide a brief overview of the Lin-Kernighan algorithm, and the variants used in this study. Section 4 contains the empirical investigation on the difficulty and properties of evolved problem instances. Last, in Section 5 we provide conclusions.

2   Evolving TSP Instances

The general approach is similar to that in [8], where an evolutionary algorithm was used to evolve difficult to solve binary constraint satisfaction problem instances for a backtracking algorithm. Here, we use a similar evolutionary algorithm to evolve difficult travelling salesman problem instances for two well known TSP solvers.

A TSP instance is represented by a list of 100 (x, y) coordinates on a 400×400 grid. The list directly forms the chromosome representation with which the evolutionary algorithm works. For each of the 30 initial TSP instances, we create a list of 100 nodes, by uniform randomly selecting (x, y) coordinates on the grid. This forms the first step in the process, depicted in Figure 1. Then the process enters the evolutionary loop: Each TSP instance is awarded a fitness equal to the search effort (defined in Section 3) required by the TSP solver to find a near-optimal shortest tour. Using two-tournament selection, we repeatedly select two parents, which create one offspring using uniform crossover. Every offspring is subjected to mutation, which consists of replacing each one of its nodes with a probability pm, with uniform randomly chosen (x, y) coordinates. This generational model is repeated 29 times, and together with the best individual from the current population (1-elitism), a new population is formed. The loop is repeated for 600 generations.

The mutation rate pm is decreased over the subsequent generations. This process makes it possible to take large steps in the search space at the start, while keeping changes small at the end of the run. The mutation rate is varied using,

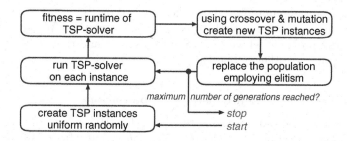

Fig. 1. The process of evolving TSP instances that are difficult to solve

$$pm = pm_{end} + (pm_{start} - pm_{end}) \cdot 2^{\frac{-generation}{bias}},$$

from [9] where the parameters are set as $bias = 2$, $pm_{start} = 1/2$, $pm_{end} = 1/100$, and *generation* is the current generation.

3   Lin-Kernighan

As for other constrained optimisation problems, we distinguish between two types of algorithms, complete algorithms and incomplete algorithms. The first are often based on a form of branch-and-bound, while the latter are equipped with one or several heuristics. In general, as complete algorithms will quickly become useless when the size of the problem is increased, the development of TSP solvers has shifted towards heuristic methods. One of the most renowned heuristic methods is Lin-Kernighan [10]. Developed more than thirty years ago, it is still known for its success in efficiently finding near-optimal results.

The core of Lin-Kernighan, and its descending variants, consists of edge exchanges in a tour. It is precisely this procedure that consumes more than 98% of the algorithm's run-time. Therefore, in order to measure the *search effort* of Lin-Kernighan-based algorithms we count the number of times an edge exchange occurs during a run. Thus, this measure of the time complexity is independent of the hardware, compiler and programming language used. In this study, we use two variants of the Lin-Kernighan algorithm, which are explained next.

3.1   Chained Lin-Kernighan

Chained Lin-Kernighan (CLK) is a variant [11] that aims to introduce more robustness in the resulting tour by chaining multiple runs of the Lin-Kernighan algorithm. Each run starts with a perturbed version of the final tour of the previous run. The length of the chain depends on the number of nodes in the TSP problem.

In [12], a proof is given demonstrating that local optimisation algorithms that are PLS-complete (Polynomial Local Search), can always be forced into performing an exponential number of steps with respect to the input size of the problem. In [13], Lin-Kernighan was first reported to have difficulty on certain problem instances, which had the common property of being clustered. The reported instances consisted of partial graphs and the bad performance was induced because the number of "hops" required to move the salesman between two clusters was set large enough to confuse the algorithm. We are using the symmetric TSP problem, where only full graphs exist and thus, every node can be reached from any other in one "hop".

3.2   Lin-Kernighan with Cluster Compensation

As a reaction on the bad performance reported in [13], a new variant of Lin-Kernighan is proposed in [14], called *Lin-Kernighan with Cluster Compensation* (LK-CC). This variant aims to reduce the computational effort, while maintaining the quality of solutions produced for both clustered and non-clustered instances.

Cluster compensation works by calculating the cluster distance for nodes, which is a quick pre-processing step. The cluster distance between node v and w equals the minimum bottleneck cost of any path between v and w, where the bottleneck cost of a path is defined as the heaviest edge on that path. These values are then used in the guiding utility function of Lin-Kernighan to prune unfruitful regions, i.e., those involved with high bottlenecks, from the search space.

4   Experiments

Each experiment consists of 190 independent runs with the evolutionary algorithm, each time producing the most difficult problem instance at the end of the run. With 29 new instances at each of the 600 generations, this results in running the Lin-Kernighan variant 3 306 000 times for each experiment. The set of problem instances from an experiment is called *Algorithm:Evolved set*, where *Algorithm* is either CLK or LK-CC, depending on which problem solver was used in the experiment.

The total set of problem instances used as the initial populations for the 190 runs is called *Random set*, and it contains $190 \times 30 = 5\,700$ unique problem instances, each of which is generated uniform randomly. This set of initial instances is the same for both Lin-Kernighan variants.

4.1   Increase in Difficulty

In Figure 2, we show the amount of search effort required by Chain Lin-Kernighan to solve the sets of TSP instances corresponding to the different experiments, as

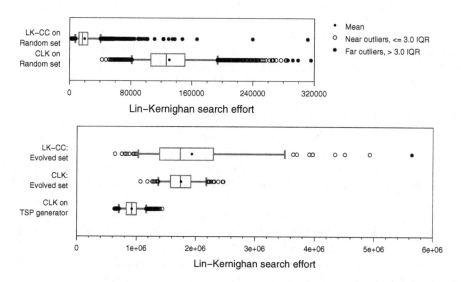

Fig. 2. Box-and-whisker plots of the search effort required by CLK and LK-CC on the Random set (top), and CLK on the TSP generator and on the CLK:Evolved set (bottom) and by LK-CC on the LK-CC:Evolved set (bottom)

well as to the Random set. Also, we compared these results to results reported in [15], where a specific TSP generator was used to create clustered instances and then solved using the Chained Lin-Kernighan variant. This set contains the 50 most difficult to solve instances from those experiments and it is called *TSP generator*.

In Figure 2, we notice that the mean and median difficulty of the instances in the CLK:Evolved set is higher than those created with the TSP generator. Also, as the 5/95 percentile ranges are not overlapping, we have a high confidence of the correctness of the difference in difficulty.

When comparing the difficulty of CLK and LK-CC for both the Random set and the Evolved sets in Figure 2, we find a remarkable difference in the the amount of variation in the results of both algorithms. CLK has much more variation with the Random set than LK-CC. However, for the evolved sets, the opposite is true. We also mention that for the Random set, LK-CC is significantly faster than CLK, while difference in speed for the evolved sets is negligible.

4.2   Discrepancy with the Optimum

We count the number of times the optimum was found by both algorithms for the Random set and for the corresponding Evolved sets. These optima are calculated using Concorde's [16] branch-and-cut approach to create an LP-representation of the TSP instance, which is then solved using Qsopt [17]. We show the average discrepancy between optimal tour length and the length of the tour produced by one of the problem solvers.

For the Random set, CLK has an average discrepancy of 0.004% (stdev: 0.024), and it finds the best tour for 95.8% of the set. For the same set of instances, LK-CC has an average discrepancy of 2.08% (stdev: 1.419), and it finds the best tour for 6.26% of the set.

A similar picture presents itself for the Evolved sets. Here, CLK has an average discrepancy of 0.03% (stdev: 0.098), and find the best tour for 84.7% of the CLK:Evolved set. LK-CC has an average discrepancy of 2.58% (stdev: 1.666), and finds the best tour for 4.74% of the LK-CC:Evolved set.

4.3   Clustering Properties of Problem Sets

To get a quantifiable measure for the amount of clusters in TSP instances we use the clustering algorithm GDBSCAN [18]. This algorithm uses no stochastic process, assumes no shape of clusters, and works without a predefined number of clusters. This makes it an ideal candidate to cluster 2-dimensional spatial data, as the methods suffers the least amount of bias possible. It works by using an arbitrary neighbourhood function, which in this case is the minimum Euclidean distance. It determines clusters based on their density by first seeding clusters and then iteratively collecting objects that adhere to the neighbourhood function. The neighbourhood function here is a spatial index, which results in a run-time complexity of $O(n \log n)$.

Clustering algorithms pose a large responsibility on their users, as every clustering algorithm depends on at least one parameter to help it define what a cluster is. Two common parameters are the number of clusters, e.g., for vari-

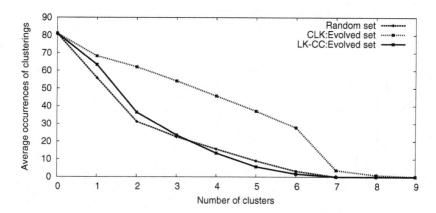

Fig. 3. Average amount of clusters found for problem instances of the Random set and for problem instances evolved against CLK and LK-CC

ants of k-means, and distance measurements to decide when two points are near enough to consider them part of the same cluster. The setting of either of these parameters greatly affects the resulting clustering. To get a more unbiased result on the number of clusters in a set of points we need a more robust method.

To get a more robust result for the number of clusters found in the different sets of TSP instances we repeat the following procedure for each instance in the set. Using the set $\{10, 11, 12, \ldots, 80\}$ of settings for the minimum Euclidean distance parameter for GDBSCAN, we cluster the TSP instance for every parameter setting. We count the number of occurrences of each number of clusters found. Then we average these results over all the TSP instances in the set. The set of minimum Euclidean distance parameters is chosen such that it includes both the peak nd the smallest number of clusters for each problem instance.

We use the above procedure to quantify the clustering of problem instances in the Random set and the two evolved sets. Figure 3 shows that for the Random set and the LK-CC:Evolved set, the average number of clusters found does not differ by much. Instead, the problem instances in the CLK:Evolved set contain consistently more clusters. The largest difference is found for 2–6 clusters.

4.4   Distribution of Segment Lengths

For both problem solvers, we study the difference between the distribution of the segment lengths of resulting tours from both the Random set and the corresponding Evolved set. For both sets, we take the optimal tour of each TSP instance in the set and then, for each tour, observe all the segment lengths. Finally, we count the occurrences of the segment lengths, and average these over the whole set of tours.

Figure 4 shows the average distribution in segment lengths for tours derived from TSP instances from the Random set and for the CLK:Evolved set. We notice that the difference from the Random set to the CLK:Evolved set is the increase of very short segments (0–8) and more longer segments (27–43), and the introduction of long segments (43–58). As the number of segments in a tour is always the

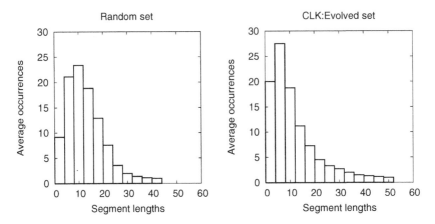

Fig. 4. Average distribution of segment lengths in the resulting Chained Lin-Kernighan tour for TSP instances of the Random set and of the CLK:Evolved set

same, i.e., 100, these increases relate to the decrease of medium length segments (9–26).

Figure 5 shows the average distribution of segment lengths for both problem solvers. For LK-CC, we observe no significant changes in the distribution of segment lengths between the Random set and LK-CC:Evolved set.

When comparing the average distribution of segment lengths of tours in both problem solvers we clearly see a large difference. LK-CC, compared to CLK, uses much longer segments. Those segments most frequently used, in the range of 60–80, never occur at all with CLK. The distribution for LK-CC seems to match a flattened normal distribution, whereas the distribution for CLK is much more skewed, CLK favours the usage of short segments of a length less than 20.

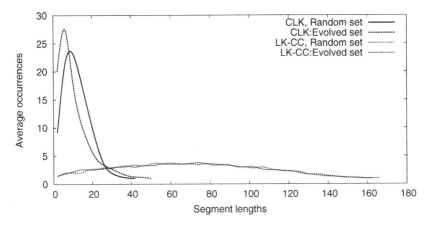

Fig. 5. Average distribution of segment lengths in the resulting tours for both the Random set and Evolved set of problem instances for CLK and LK-CC

4.5   Distribution of Pair-Wise Distances

In Figure 6, we show the average number of occurrences for distances between pairs of nodes. Every TSP instance contains $\binom{100}{2}$ pairs of nodes on the account that it forms a full graph. The distribution of these pair-wise distances mostly resembles a skewed Gaussian distribution. The main exception consists of the cut-off at short segments lengths. These very short distances, smaller than about 4, occur rarely when 100 nodes are distributed over a 400×400 space.

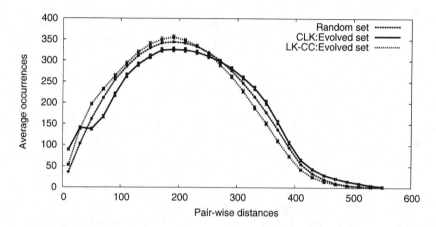

Fig. 6. Distribution of distances over all pairs of nodes in randomly generated and evolved problem instances after 600 generations (CLK and LK-CC), 95% confidence intervals included, most of which are small

For the Chained Lin-Kernighan we notice a change in the distribution similar to that in the previous section. Compared with the Random set, both the number of short segments and the number of long segments increases. Although not to the same extent when observing the distribution of segment lengths. Also, the number of medium length occurrences is less than for the Random set. This forms more evidence for the introduction of clusters in the problem instances. Although this analysis does not provide us with the amount of clusters, it does give us an unbiased view on the existence of clusters, as it is both independent of the TSP algorithms and any clustering algorithm.

Also shown in Figure 6 is the distribution of pair-wise distances for problem instances evolved against the LK-CC algorithm. While we notice an increase in shorter distances, this is matched by an equal decrease in longer distances. Thus, the total layout of the nodes becomes more huddled together.

4.6   Swapping Evolved Sets

We run each variant on the problem instances in the set evolved for the other variant. Table 1 clearly shows that a set evolved for one algorithm is much less difficult for the other algorithm. However, each variant needs significantly

Table 1. Mean and standard deviation, in brackets, of the search effort required by both algorithms on the Random set and both Evolved sets

	CLK	CC-LK
CLK:Evolved set	1 753 790 (251 239)	207 822 (155 533)
CC-LK:Evolved set	268 544 (71 796)	1 934 790 (799 544)
Random set	130 539 (34 452)	19 660 (12 944)

more search effort for the alternative Evolved set than for the Random set. This indicates that some properties of difficulty are shared between the algorithms.

5   Conclusions

We have introduced an evolutionary algorithm for evolving difficult to solve travelling salesman problem instances. The method was used to create a set of problem instances for two well known variants of the Lin-Kernighan heuristic. These sets provided far more difficult problem instances than problem instances generated uniform randomly. Moreover, for the Chained Lin-Kernighan variant, the problem instances are significantly more difficult than those created with a specialised TSP generator. Through analysis of the sets of evolved problem instances we show that these instances adhere to structural properties that directly afflict on the weak points of the corresponding algorithm.

Problem instances difficult for Chained Lin-Kernighan seem to contain clusters. When comparing with the instances of the TSP generator, these contained on average more clusters (10 clusters) then the evolved ones (2–6 clusters). Thus, this leads us to the conjecture that clusters on itself is not sufficient property to induce difficulty for CLK. The position of the clusters and distribution of cities over clusters, as well as the distribution of nodes not belonging to clusters, can be of much influence.

The goal of the author of LK-CC is to provide a TSP solver where its efficiency and effectiveness are not influenced by the structure of the problem [14]. Problem instances evolved in our study, which are difficult to solve for Lin-Kernighan with Cluster Compensation, tend to be condense and contain random layouts. The algorithm suffers from high variation in the amount of search effort required, therefore depending heavily on a lucky setting of the random seed. Furthermore, its effectiveness is much lower than that of CLK, as the length of its tours are on average, further away from the optimum. Thus, it seems that to live up its goals, LK-CC is losing on both performance and robustness.

The methodology described here is of a general nature and can, in theory, be used to automatically identify difficult problem instances, or instances that inhibit other properties. Afterwards, these instances may reveal properties that can lead to general conclusions on when and why algorithms show a particular performance. This kind of knowledge is of importance when one needs to select an algorithm to solve a problem of which such properties can be measured.

Acknowledgements. The author is supported through a TALENT-Stipendium awarded by the Netherlands Organization for Scientific Research (NWO).

References

1. Lawler, E.L., Lenstra, J.K., Rinnooy Kan, A.H.G., Shmoys, D.B.: The Traveling Salesman Problem. John Wiley & Sons, Chichester (1985)
2. Cheeseman, P., Kanefsky, B., Taylor, W.: Where the really hard problems are. In: Proceedings of IJCAI-91. (1991)
3. Beck, J., Prosser, P., Selensky, E.: Vehicle routing and job shop scheduling: What's the difference? In: Proc. of the 13th International Conference on Automated Planning & Scheduling. (2003)
4. Monasson, R., Zecchina, R., Kirkpatrick, S., Selman, B., Troyansky, L.: Determining computational complexity from characteristic phase transitions. Nature **400** (1999) 133–137
5. Hayes, B.: Can't get no satisfaction. American Scientist **85** (1997) 108–112
6. Hogg, T.: Refining the phase transition in combinatorial search. Artificial Intelligence **81** (1996) 127–154
7. Culberson, J., Gent, I.: Well out of reach: why hard problems are hard. Technical report, APES Research Group (1999)
8. van Hemert, J.: Evolving binary constraint satisfaction problem instances that are difficult to solve. In: Proceedings of the IEEE 2003 Congress on Evolutionary Computation, IEEE Press (2003) 1267–1273
9. Kratica, J., Ljubić, I., Tošic, D.: A genetic algorithm for the index selection problem. In Raidl, G., et al., eds.: Applications of Evolutionary Computation. Volume 2611., Springer-Verlag (2003) 281–291
10. Lin, S., Kernighan, B.: An effective heuristic algorithm for the traveling salesman problem. Operations Research **21** (1973) 498–516
11. Applegate, D., Cook, W., Rohe, A.: Chained lin-kernighan for large travelling salesman problems (2000) http://www.citeseer.com/applegate99chained.html.
12. Papadimitriou, C.: The complexity of the Lin-Kernighan heuristic for the traveling salesman problem. SIAM Journal of Computing **21** (1992) 450–465
13. Johnson, D., McGeoch, L.: The traveling salesman problem: a case study. In Aarts, E., Lenstra, J., eds.: Local Search in Combinatorial Optimization. John Wiley & Sons, Inc (1997) 215–310
14. Neto, D.: Efficient Cluster Compensation for Lin-Kernighan Heuristics. PhD thesis, Computer Science, University of Toronto (1999)
15. van Hemert, J., Urquhart, N.: Phase transition properties of clustered travelling salesman problem instances generated with evolutionary computation. In Yao, X., Burke, E., Lozano, J.A., Smith, J., Merelo-Guervós, J.J., Bullinaria, J.A., Rowe, J., Kabán, P.T.A., Schwefel, H.P., eds.: Parallel Problem Solving from Nature (PPSN VIII). Volume 3242 of LNCS., Birmingham, UK, Springer-Verlag (2004) 150–159
16. Applegate, D., Bixby, R., Chvátal, V., Cook, W.: Finding tours in the TSP. Technical Report 99885, Research Institute for Discrete Mathematics, Universität Bonn (1999)
17. Applegate, D., Cook, W., Dash, S., Mevenkamp, M.: Qsopt linear programming solver (2004) http://www.isye.gatech.edu/~wcook/qsopt/.
18. Sander, J., Ester, M., Kriegel, H.P., Xu, X.: Density-based clustering in spatial databases: The algorithm GDBSCAN and its applications. Data Min. Knowl. Discov. **2** (1998) 169–194

Heuristic Colour Assignment Strategies
for Merge Models in Graph Colouring

István Juhos[1], Attila Tóth[2], and Jano I. van Hemert[3]

[1] Dept. of Computer Algorithms and Artificial Intelligence, Univ. of Szeged, Hungary
[2] Department of Computer Science, Univ. of Szeged (JGYTFK), Hungary
[3] Centre for Emergent Computing, Napier University, Edinburgh, UK

Abstract. In this paper, we combine a powerful representation for graph colouring problems with different heuristic strategies for colour assignment. Our novel strategies employ heuristics that exploit information about the partial colouring in an aim to improve performance. An evolutionary algorithm is used to drive the search. We compare the different strategies to each other on several very hard benchmarks and on generated problem instances, and show where the novel strategies improve the efficiency.

1 Introduction

The problem class known as the graph k-colouring problem [1] is defined as follows. Given a graph $G = \langle V, E \rangle$, where $V = \{v_1, \ldots, v_n\}$ is a set of nodes and $E = \{(v_i, v_j) | v_i \in V \wedge v_j \in V \wedge i \neq j\}$ is a set of edges. The objective in the graph k-colouring problem is to colour every node in V with one of k colours such that no two nodes connected with an edge in E have the same colour. Such a colouring is called a *valid colouring*. The smallest number of colours k used to achieve a valid colouring of G is called the *chromatic number* of G, which is denoted by χ.

Most algorithms searching for a solution of a graph k-colouring problem do so by incrementally assigning a colour to a node. Consequently, at every node visited a decision must be made which colour to assign. This choice may prove of vital importance to achieving a valid colouring. Quite a number of strategies for making this decision exist, and each one comes with its own rationale and benefits [2–Chapter 5]. From recent theoretical developments [3] we know that algorithms for finites sets of problems under permutation closure also cannot escape the No Free Lunch Theorem [4], which makes it more important to link properties of problems with algorithms [5].

A great deal of study is devoted to hybrid algorithms, as these have proven to be successful approaches to solving difficult constrained optimisation problems. Popular methodologies are meta-heuristics [6] and, more recently, hyper-heuristics [7] and hybrid meta-heuristics, where the idea is to use combine several heuristics to get a more successful algorithm. This success is measured in both

G.R. Raidl and J. Gottlieb (Eds.): EvoCOP 2005, LNCS 3448, pp. 132–143, 2005.

effectiveness and efficiency, i.e., accuracy in finding solutions and time complexity. Here, we examine the combination of a powerful representation for graph colouring with five different heuristic strategies for colour assignment.

In the next section we explain how solutions of graph colouring problems can be represented using merge models. Then, in Section 3 we explain different heuristic strategies for colour assignment. These strategies are used in an evolutionary algorithm explain in Section 4. The different heuristic strategies are benchmarked in Section 5 together with other algorithms. Finally in Section 6 we draw some conclusions.

2   The Binary Merge Model Representations

Graph colouring algorithms make use of adjacency checking during the colouring process, which has a large influence on the performance. Generally, when assigning a colour to a node, all adjacent nodes must be scanned to check for potential violations. Thus, a number of constraint checks, i.e., checks for equal colours, need to be performed. The exact number of constraint checks performed is bounded by the current number of coloured neighbours and by $|V| - 1$. Using the Binary Merge Model approach, explained next, the number of constraint checks lies between one and the number of colours used up to this point. These bounds arise from the model-induced hyper-graph structure, which guarantees that the algorithms usually performs better .

The *Binary Merge Model* (BMMs) implicitly uses hyper-nodes and hyper-edges (see Figure 1). A hyper-node is a set of nodes that all have the same colour, i.e., they are folded into one node. A hyper-edge connects nodes of which at least one node is a hyper-node. Such a hyper-edge essentially forms a collection of regular edges, i.e., constraints. A hyper-edge only exists if and only if its corresponding nodes are connected by at least two "normal" edges. The BMM concentrates on operations on hyper-nodes and normal nodes, by trying to merge normal nodes with normal nodes and hyper-nodes. Within the context of search effort, which in constraint satisfaction is measured by counting the number of constraint checks, we can save effort if at least one of the nodes is a hyper-node. Then, the number of adjacency checks, i.e, constraint checks, can be reduced as these are performed along hyper-edges instead of normal edges, because one constraint check on a hyper-edge saves at least one, but possible more, constraint checks on the normal edges it incorporates. A more detailed explanation is given in [8], where the model was introduced.

The current colouring of a graph $\langle V, E \rangle$ is stored in a Binary Merge Table (BMT) (for an example, see Figure 2). Every cell (i, j) in the table is binary. The columns refer to the nodes and the rows refer to the colours. A value in cell (i, j) is zero if and only if node $j \in V$ cannot be assigned colour i because of the edges E in the original graph $\langle V, E \rangle$. The initial BMT is the adjacency matrix of the graph, hence a different colour is assigned to each node.

If the graph is not a complete graph, then it might be possible to reduce the number of necessary colours. This corresponds to reducing rows in the BMT. Rows can be reduced by repeatedly using a Binary Merge Operation, which attempt to merge two rows. If a merge is possible, i.e., no violations are introduced, the number of colours is decreased by one. Otherwise, the number of colours remains the same. A merge is successful only when two nodes are not connected by a normal edge or a hyper-edge. An example of both a successful and unsuccessful merge is shown in Figures 1 and 2.

Definition 1. *The Binary Merge Operations $\cup$ merges an initial row r_j into an arbitrary (initial or merged) row r_i if and only if $(j,i) = 0$ (i.e. the hyper-node x_j is not connected to the node x_i) in the BMT. If rows r_i and r_j can be merged then the result is the union of them.*

Formally, let I be the set of initial rows of the BMT and R be the set of all possible $|V|$ size rows, i.e. binary vectors. Then an merge operation is defined as

$$\cup : R \times I \to R$$
$$r'_j := r_j \cup r_i, \quad r'_j, r_j \in R, \quad r_i \in I, \text{ or by components}$$
$$r'_j(l) := r_j(l) \vee r_i(l), \quad l = 1, 2, \ldots, |V|$$

With regard to the time complexity of the binary operation, it is proportional to a binary OR operation on a register of l bits. If l is the number of bits in one such operation and, under assumption that the time complexity of that operation is one, the merge of two rows of length n by l length parts takes $\lceil n/l \rceil$ to complete. If k is the number of rows left in the BMT, then the number of merge operations is $|V| - k$, where $k \in \{\chi, \ldots, |V|\}$.

2.1   Permutation Merge Model

Finding a minimal colouring for a graph k-colouring problem using the binary merge table representation requires finding the sequence of merge operations that leads to that colouring. This can be represented as a sequence of candidate reduction steps using the greedy approach described above. The permutations of this representation form the *Permutation Merge Model* [8].

The result of colouring a graph after two or more merge operations depends on the order in which these operations were performed. Consider the hexagon in Figure 1(a) and its corresponding BMT in Figure 2. Now let the sequence $P_1 = 1, 4, 2, 5, 3, 6$ be the order in which the rows are considered for the merge operations and consider the following merging procedure. Take the first two rows in the sequence, then attempt to merge row 4 with row 1. As these can be merged the result is $1 \cup 4$ (see Figure 1(b)). Now take row 2 and try to merge this with the first row, i.e. $(1 \cup 4)$. This is unsuccessful, so row 2 remains unaltered. The merge operations continue with the next rows 5 and 3, and finally, with 6. The allowed merges are $1 \cup 4$ and $2 \cup 5$. This sequence of merge operations results in the 4-colouring of the graph depicted in Figure 1(c). However, if we use the sequence $P_2 = 1, 4, 2, 6, 3, 5$ then the result will be only a 3-colouring, as shown in Figure 1(e) with the merges $1 \cup 4$, $2 \cup 6$ and $3 \cup 5$. The defined merge is

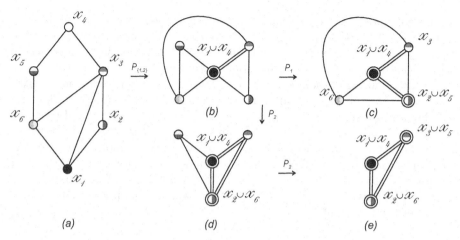

Fig. 1. Examples of the result of two different merge orders $P_1 = 1, 4, 2, 5, 3, 6$ and $P_2 = 1, 4, 2, 6, 3, 5$. The double-lined edges are hyper-edges and double-lined nodes are hyper-nodes. The P_1 order yields a 4-colouring (c), but with the P_2 order we get a 3-colouring (e).

greedy, i.e. it takes a row and tries to find the first row from the top of the table that it can merge. The row remains unaltered if there is no suitable row. After performing the sequence P of merge operations, we call the resulting BMT the *merged* BMT.

(a)	x_1	x_2	x_3	x_4	x_5	x_6
r_1	0	1	1	0	0	1
r_2	1	0	1	0	0	0
r_3	1	1	0	1	0	1
r_4	0	0	1	0	1	0
r_5	0	0	0	1	0	1
r_6	1	0	1	0	1	0

(b)	x_1	x_2	x_3	x_4	x_5	x_6
$r_1 \cup r_4$	0	1	1	0	1	1
r_2	1	0	1	0	0	0
r_3	1	1	0	1	0	1
r_5	0	0	0	1	0	1
r_6	1	0	1	0	1	0

(c)	x_1	x_2	x_3	x_4	x_5	x_6
$r_1 \cup r_4$	0	1	1	0	1	1
$r_2 \cup r_5$	1	0	1	1	0	1
r_3	1	1	0	1	0	1
r_6	1	0	1	0	1	0

(d)	x_1	x_2	x_3	x_4	x_5	x_6
$r_1 \cup r_4$	0	1	1	0	1	1
$r_2 \cup r_6$	1	0	1	0	1	0
r_3	1	1	0	1	0	1
r_5	0	0	0	1	0	1

(e)	x_1	x_2	x_3	x_4	x_5	x_6
$r_1 \cup r_4$	0	1	1	0	1	1
$r_2 \cup r_6$	1	0	1	0	1	0
$r_3 \cup r_5$	1	1	0	1	0	1

Fig. 2. Binary Merge Tables corresponding to the graphs in Figure 1.

In practice the graphs start out uncoloured, the colouring is then constructed by colouring the nodes in steps. We deal with the sub-graphs of the original graph defined by the colouring steps. The related binary merge tables contain partial information about the original one. Let the original graph with its initial BMT be defined by Figure 3(a) on which the colouring will be performed. Taking the $x_1, x_4, x_2, x_6, x_3, x_5$ order of the nodes into account for colouring G, then using the ordering $P_1 = 1, 4, 2, 6, 3, 5$, an attempt will be made to merge rows. After the greedy colouring of the nodes x_1, x_4, x_2 there is a related partial or sub-BMT along with the (sub-)hyper-graph. These are depicted in Figure 3(b). The 1st and the 4th row are merged together, but the 2nd cannot be merged with the $1 \cup 4$ merged row, thus the 2nd row remains unaltered in the related sub-BMT.

From here on, we concentrate on how to extract valuable information from the sub-structures to get an efficient colour assignment strategy for the the nodes, which takes into account the current state of the colouring. This as opposed to the usually greedy manner, which is blind in this sense, i.e., it does not consider the current environment.

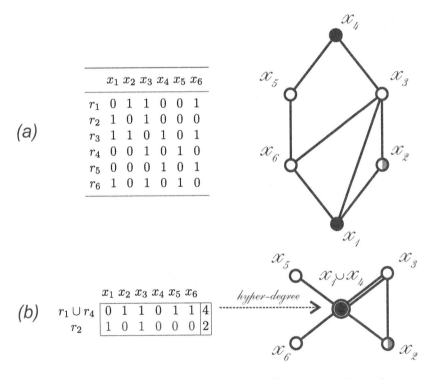

Fig. 3. The left side shows the partial colouring of the G graph according to the x_1, x_4, x_2 greedy order and the adjacency matrix of the graph. The right one shows the partial or sub-BMT related to this colouring with its co-structures and sub-BMT induced hyper-graph.

3   Heuristic Strategies for Colour Assignment

Finding the appropriate node order is important, which will be left to the evolutionary algorithm described in Section 4. However, the choice of which colour to assign also has much influence on the success of the final algorithm. The greedy strategy finds the first available colour for assigning it to the node currently being coloured. It does so by trying to merge the corresponding $r_{P(i)}$ row of the BMT to the previously merged rows following the natural order $1, 2, \ldots, i - 1$.

Instead of using a simple greedy procedure, we will employ a number of more sophisticated ways for assigning colours. The way in which the merge model reduces the number of colours leaves open the way in which the colours are assigned to various groups of nodes. We take advantage of this by employing different heuristic strategies for the assigning colours to nodes. Formally, let $x_{P(i)}$ be the next node to be coloured, where i is the index of the node in the permutation P. Next, we provide several strategies for assigning a colour to $x_{P(i)}$ using the information provided by the current merge model. The first two strategies use only information about the already coloured structure, i.e., about the sub-BMT not dealing with the current node. The remaining, novel strategies, use information about the current node and its context so far, and then try to exploit this information by using it to avoid getting stuck later on.

3.1   Hyper-node Cardinality

This strategy attempts to merge row i with its preceding rows by favouring hyper-nodes with high cardinality. A hyper-node's cardinality is defined by the number of normal nodes it encompasses. The strategy consists of first colouring the hyper-node with the highest cardinality. In other words, choosing the colour that colours the most nodes and gives valid colouring. Although this strategy has a greedy component to it, its ability to use knowledge on-line, i.e., while searching for a solution, may give it an edge over the simple greedy method.

3.2   Hyper-node Constrainedness

While the previous heuristic supposes that the hyper-node cardinality determines its constrainedness, this one expresses it in a direct way examining the context of the considered hyper-node. With this heuristic, we favour the most constrained hyper-nodes. The intuition is to avoid the possibility that the least constrained nodes pick up too many irrelevant nodes. This method also works for the BMT, where we count all the connecting hyper-edges, so calculating the hyper-degree (see Figure 3). This information can easily be obtained by summarising the rows, i.e., by making the first order norm $||r_{P(i)}||_1$ of them in the sub-BMT.

3.3   Suitable Matches

Up until now, we did not consider the characteristics of the structure of $x_{P(i)}$. The previous strategy used only the constrainedness of hyper-nodes, and here we shall use the $x_{P(i)}$ constraints as well to find a suitable match for merging rows. We say that the m-th vector of the sub-BMT $r'(m)$ is the most suitable candidate for merging with $r_{P(i)}$ if they share the most constraints. The dot product of two vectors provides the number of shared constraints. Thus, by reverse sorting all the sub-BMT vectors on their dot product with $r_{P(i)}$, we can reduce the number of colours by merging $r_{P(i)}$ with the most suitable match.

These approach include implicitly the least constraining value heuristic [2–Chapter 5], but provide additional one. Try to find that hyper-nodes (group of the nodes) which has the most number of common neighbors. Thus, reducing implicitly the structure of the graph in a way which is explicit described in [9].

3.4   Topological Similarity

The dot product, as described above, provides a measure for the similarity of the vectors. If we normalise these vectors by their length, the result is a measure for similarity in a topological sense. As the normalised dot products gives the cosines of the angles of the vectors, higher cosines corresponds to vectors located nearer. This strategy exploits this idea by collecting vectors that are spatially near to each other. By performing merge operations on these collections we get convex combinations of vectors. Thus, the result of a merge remains in the span of the merged vectors. The idea behind this is to carefully combine similar groups of colours, thereby building up a solid colouring of the graph that leaves enough room for further merging of groups, i.e., rows in the BMT.

4   Evolutionary Algorithm to Guide the Models

We have two goals. The first is to find a successful order of the nodes and the second is to find a successful order for assigning colour. While the order of the node can be represented by a fix length permutation, the order for colour assignment needs a variable length representation. We turn to the heuristics described above to guide the colour assignment dynamically. For the first goal, we must search the permutation search space of the model described in Section 2.1, which is of size $n!$. Here, we use an evolutionary algorithm to search through the space of permutations. The genotype consists of the permutations of the nodes, i.e., rows of the BMT. The phenotype is a valid colouring of the graph after using a colour assignment strategy on the permutation to select the order of the binary merge operations.

An intuitive way of measuring the quality of an individual p in the population is by counting the number of rows remaining in the final BMT. This equals to the number of colours $k(p)$ used in the colouring of the graph, which needs to be minimised. When we know that the optimal colouring is χ then we may normalise this fitness function to $g(p) = k(p) - \chi$. This function gives a rather low diversity of fitnesses of the individuals in a population because it cannot distinguish between two individuals that use an equal number of colours. This problem is called the fitness granularity problem. We address it by introducing a new fitness, which relies on the heuristic that one generally wants to avoid highly constraint nodes and rows in order to have a higher chance of successful merges at a later stage. It works as follows. After the final merge the resulting BMT defines the colour groups. There are $k(p) - \chi$ over-coloured nodes, i.e., merged rows. Generally, we use the indices of the over-coloured nodes to calculate the number of nodes that need to be minimised. But these nodes are not necessarily responsible for the over-coloured graph. Therefore, we choose to count the nodes that are in the smallest group of nodes with the same colour. In the context of the merge model, this corresponds to hyper-nodes with the smallest cardinality. To cope better with the fitness granularity problem we should also deal with the constraints causing high constrainedness. The final fitness function is then defined as follows. Let $\zeta(p)$ denote the number of constraints, i.e., ones, in the rows of the final BMT that belong to the $k(p) - \chi$ hyper-nodes having the smallest cardinality. The fitness function becomes $f(p) = (k(p) - \chi)\zeta(p)$.

Note that here the cardinality of the problem is known, and used as a stopping criterium $(f(p) = 0)$ to determine the efficiency of the algorithm. For the case where we do not know the cardinality of the problem, this approach can be used by leaving out the normalisation step.

5   Empirical Comparison

We use a generational model with 2-tournament selection and replacement, where we employe elitism of size one. The stop condition is that either an individual p exists with $f(p) = 0$ or that the maximum number of generations of 6 000 generations is reached (twice as many as in a previous study due to the harder problems here). The latter means that the run is unsuccessful, i.e., the optimal colouring is not found. This setting is used in all experiments. The initial population is created with 100 random individual. Two variation operators are used to provide offsprings. First, the 2-point order-based crossover (OX2) [10–in Section C3.3.3.1] is applied. Second, the other variation operator is a simple swap mutation operator, which selects at random two different items in the permutation and then swaps. We use simple operators to make sure that any success gained, stems from the heuristic strategies for colour assignment. The probability of using OX2 is set to 0.4 and the probability for using the simple swap mutation is set to 0.6. These values are take from a previous study [8].

5.1   Means of Comparisons

The performance of an algorithm is expressed in its effectiveness and its efficiency in solving a problem instance. The first is measured using the success ratio, which is the amount of runs where an algorithm has found the optimum divided by the total number of runs. The second is measured by keeping track of how many constraint checks are being performed on average, for a successful run. This measure is independent of hardware and programming language as it counts the number of times an algorithm requests information about the problem instance, e.g., it checks if an edge exists between two nodes in the graph. This check, or rather the number of times it is performed, forms the largest amount of time spend by any constraint solver. A *constraint check* is defined for each algorithm as checking whether the colouring of two nodes is allowed (satisfied) or not allowed (violated). An evolutionary algorithm is of stochastic nature. Therefore, we always perform ten independent runs with different random seeds for each problem instance. Results are averaged over these runs and, where appropriate, over multiple instances with equal characteristics.

5.2 Benchmarks

We compare the five different strategies on a number of benchmark problems from the "The Second DIMACS Challenge" [11], which is a standard competition repository. For demonstration purposes we choose the extremely difficult Leighton graphs [12]. In a previous study [13], these graphs took 10–20 hours to be solved by specialised evolutionary solvers, due to the structure induced during their creation. Also, the failure of the well-known heuristic DSATUR of Brélaz [14] confirms the difficulty of these problem. They are random graphs on $n = 450$ nodes with an edge density of 0.25. A graph is constructed by first generating cliques of varying sizes in such a way that the pre-specified value of χ is not violated. Here we use $\chi = 15$. They are identified by `le450-15x`, where x is one of a, b, c, or d, to differentiate the individual instances.

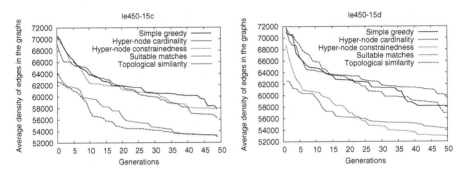

Fig. 4. Convergence graphs of the fitness for all five strategies on two hard problems

In Figure 4, the fitness of the best individual in the population is presented for each generation. These results are averaged over ten runs. This provides insight into the convergence when employing the different heuristic strategies. Not much difference exists between the simple greedy strategy and both the hyper-node cardinality and hyper-node constrainedness. However, we notice a significant faster convergence for both the suitable matches and the topological similarity. For the graph `le450-15c`, the convergence of the topological similarity is slightly faster than that of the suitable matches. Observing the starting phase for the `le450-15d`, the topological similarity shows large improvements, however after the good starting the suitable matches catches it up.

5.3   In the Phase Transition

Using the well known graph k-colouring generator of Culberson [15], we generate a test suite of 3-colourable graphs with 200 nodes. The edge density of the graphs is varied in a region called the phase transition. This is where hard to solve problem instances are generally found, which is shown using the typical easy-hard-easy pattern. The graphs are all equipartite, which means that in a solution each colour is used approximately as much as any other. The suite consists of nine groups where each group has five instances, one each instance we perform ten runs and calculate averages over these 50 runs. The connectivity is changed from 0.020 to 0.060 by steps of 0.005 over the groups.

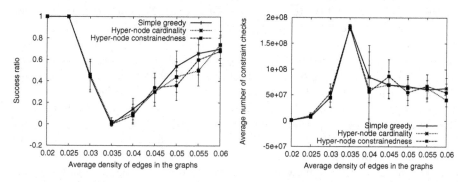

Fig. 5. Success ratio and average constraint checks to a solution for the simple greedy strategy, the hyper-node cardinality strategy, and the hyper-node constrainedness strategy (with 95% confidence intervals; for 0.035, the success ratio is too low thus we include all runs)

Figure 5 shows the performance measured by success ratio and by average constraint checks performed for the simple greedy strategy and the two strategies that restrict their on-line heuristics to the current node under consideration for colouring. No significant improvement is made over the simple greedy method.

The two novel strategies that employ knowledge over the colouring of the graph made so far are shown in Figure 6 together with the simple greedy strategy. Here we clearly notice an improvement in both efficiency and effectiveness over the simple greedy strategy. Especially, the search effort needed for denser graphs is much lower. Furthermore, the confidence intervals for this range are small and non-overlapping. These two approaches give a much robuster algorithm for solving graph k-colouring.

6   Conclusions

We have combined a powerful representation for the graph k-colouring problem, called the permutation merge model, with several heuristic strategies for colour

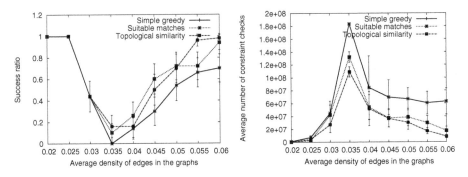

Fig. 6. Success ratio and average constraint checks for the simple greedy strategy, the suitable matches, and the topological similarity (with 95% confidence intervals for 0.035, the success ratio is too low thus we include all runs)

assignment. Four novel strategies use information about the current state of the colouring of the graph to infer where problems can be expected in a future stage of the colouring process. The aim of exploiting this knowledge is to improve performance by increasing the efficiency and effectiveness of the evolutionary algorithm that uses the permutations merge model.

By comparing the different strategies on several hard to solve problems, we showed how employing on-line heuristics improves the convergence speed of the evolutionary algorithm. Furthermore, the two novel strategies, by exploiting the suitability of matches and the topological similarity, showed more potential then the two strategies that restrict to using knowledge about the current node only.

In order to get a strong comparison, we compared all the strategies on a suite of generated problem instances that encompasses the phase transition. This way we ensure a comparison on very hard to solve problems. This confirmed the results on the benchmarks, as the two novel strategies are more effective, i.e., had a higher success ratio, on the right side of the phase transition. Also, they were far more efficient, and more consistent in their efficiency.

Acknowledgements

This work was supported by the Hungarian National Information Infrastructure Development Program through High Performance Supercomputing as the project CSPAI/1066/2003-2004. The third author is supported by a TALENT-Stipendium awarded by the Netherlands Organization for Scientific Research (NWO).

References

1. Jensen, T., Toft, B.: Graph Coloring Problems. Wiley-Interscience Series in Discrete Mathematics and Optimization. John Wiley & Sons, Inc (1995)

2. Russell, S., Norvig, P.: Artificial Intelligence: A Modern Approach. second edn. Prentice Hall Series in Artificial Intelligence. Englewood Cliffs, New Jersey (1995)
3. Schumacher, C., Vose, M., Whitley, L.: The no free lunch and problem description length. In Spector, L., Goodman, E.D., Wu, A., Langdon, W.B., Voigt, H.M., Gen, M., Sen, S., Dorigo, M., Pezeshk, S., Garzon, M.H., Burke, E., eds.: Proceedings of the Genetic and Evolutionary Computation Conference (GECCO-2001), Morgan Kaufmann (2001) 565–570
4. Wolpert, D., Macready, W.: No free lunch theorems for optimization. IEEE Transactions on Evolutionary Computation 1 (1997) 67–82
5. Culberson, J.: On the futility of blind search: An algorithmic view of "No Free Lunch". Evolutionary Computation 69 (1998) 109–128
6. Glover, F., Kochenberger, W., Gary, A.: Handbook of Metaheuristics. Volume 57 of International Series in Operations Research and Management Science. Kluwer (2003)
7. Burke, E., Kendall, G., Soubeiga, E.: A tabu-search hyper-heuristic for timetabling and rostering. Journal of Heuristics 9 (2003) 451–470
8. Juhos, I., Tóth, A., van Hemert, J.: Binary merge model representation of the graph colouring problem. In Gottlieb, J., Raidl, G., eds.: Evolutionary Computation in Combinatorial Optimization. (2004) 124–134
9. Cheeseman, P., Kanefsky, B., Taylor, W.M.: Where the Really Hard Problems Are. In: Proceedings of the Twelfth International Joint Conference on Artificial Intelligence, IJCAI-91, Sidney, Australia. (1991) 331–337
10. Bäck, T., Fogel, D., Michalewicz, Z., eds.: Handbook of Evolutionary Computation. Institute of Physics Publishing Ltd, Bristol and Oxford University Press (1997)
11. Johnson, D., Trick, M.: Cliques, Coloring, and Satisfiability. American Mathematical Society, DIMACS (1996)
12. Leighton, F.T.: A graph colouring algorithm for large scheduling problems. J. Res. National Bureau Standards 84 (1979) 489–503
13. Dimitris Fotakis, Spyros Likothanassis, S.S.: An evolutionary annealing approach to graph coloring. In: Proceedings of Applications of Evolutionary Computing, EvoWorkshops 2001. (2001) 120–129
14. Brélaz, D.: New methods to color the vertices of a graph. Communications of the ACM 22 (1979) 251–256
15. Culberson, J.: Iterated greedy graph coloring and the difficulty landscape. Technical Report TR 92-07, University of Alberta, Dept. of Computing Science (1992)

Application of the Grouping Genetic Algorithm to University Course Timetabling

Rhydian Lewis and Ben Paechter

Centre for Emergent Computing, Napier University, Edinburgh EH10 5DT, UK
{r.lewis, b.paechter}@napier.ac.uk

Abstract. University Course Timetabling-Problems (UCTPs) involve the allocation of resources (such as rooms and timeslots) to all the events of a university, satisfying a set of hard-constraints and, as much as possible, some soft constraints. Here we work with a well-known version of the problem where there seems a strong case for considering these two goals as separate sub-problems. In particular we note that the satisfaction of hard constraints fits the standard definition of a grouping problem. As a result, a grouping genetic algorithm for finding feasible timetables for "hard" problem instances has been developed, with promising results.

1 Introduction

The university course-timetabling problem (UCTP)[1] is the task of assigning the events of a university (lectures, tutorials, etc) to rooms and timeslots in such a way as to minimise violations of a predefined set of constraints. This version of the problem is already well known and in the last few years has become somewhat of a benchmark in a problem area that is notorious for having a multitude of different definitions. Specifically, given a set of events E, the task is to assign every event a room from a set R and timeslot from a set T (where $|T|=45$, comprising 5 days of nine timeslots). The problem is made taxing by the fact that various pairs of events *clash* - i.e. they can't be scheduled in the same timeslots because one or more student may be required to attend them both, making it analogous to the well known NP-hard graph colouring problem. There are also other complications - not all rooms are suitable for each event (it may be too small to accommodate the students or might not have the facilities the event requires), and cases of *double booking* (where a particular room is given more that one event in a timeslot) are strictly disallowed. A violation of any of these three so-called hard-constraints makes a timetable *infeasible*. The total number of possible assignments (timetables) is therefore $|E|^{|R|.|T|}$ and it can be easily appreciated that in anything but trivial cases, the vast majority of these contain some level of infeasibility.

---

[1] Defined by B. Paechter for the International Timetabling Competition 2002. More details, and example problem instances are at [19].

G.R. Raidl and J. Gottlieb (Eds.): EvoCOP 2005, LNCS 3448, pp. 144–153, 2005.

In addition to finding feasibility, it is usual in timetabling problems to define a number of soft constraints. These are rules that, although not imperative in their satisfaction, should be avoided, if possible, in order to show some consideration to the people who will have to base their working lives around it. In this particular UCTP these are (1) no student should be scheduled to sit more than three events in consecutive timeslots on the same day, (2) students should not be scheduled just one event in a day and (3) events should not be scheduled in the last timeslot of a day.

This UCTP has been studied by Rossi-Doria *et al.* [14] as a means for comparing different metaheuristics. A conclusion of this substantial work is that the performance of any one metaheuristic with respect to satisfying hard constraints and soft constraints might be different; i.e. what may be a good approach for finding feasibility may not necessarily be good for optimising soft constraints. The authors go on to suggest that hybrid algorithms comprising two stages, the first to find feasibility, the second to optimise soft constraints whilst staying in feasible regions of the search space might be the more promising approach. This hypothesis was reinforced when the International Timetabling Competition [19] was organised in 2002 and people were invited to design algorithms for this problem - as it turned out, the best algorithms presented used this two-stage approach [1, 4, 13], using various constructive heuristics to first find feasibility, followed by assorted local improvement algorithms to deal with the soft constraints.

It seems then that we have a case for this two-stage approach, but although there is substantial work pertaining to the optimisation of soft constraints whilst preserving feasibility [1, 4, 11, 13], there is still a major issue of concern: How we can ensure that we have a good chance of finding feasibility in the first place? Indeed, this (sub)problem is still NP-hard [10] and should not be treated lightly. Therefore in "harder" instances, where methods such as those in [1], [4], [11] and [13] might start to fail, some sort of stronger search algorithm is required.

1.1   Grouping Genetic Algorithms and Their Applicability for the UCTP

Grouping genetic algorithms (GGAs) may be thought of as a special type of genetic algorithm specialised for *grouping problems*. Falkenauer [9] defines a grouping problem as one where the task is to partition a set of objects U into a collection of mutually disjoint subsets u_i of U, such that $\cup u_i = U$ and $u_i \cap u_j = \varnothing$, $i \neq j$, and according to a set of problem-specific constraints that define valid and legal groupings. The NP-hard *bin packing* problem is a well used example - given a finite set of "items" of various "sizes", the task is to partition all of the items into various "bins" (groups) such that (1) the total size of all the items in any one bin does not exceed the bin's maximum capacity and (2) the number of bins used is minimised (a *legal* and *optimal* grouping respectively).

It was bin packing that was first used in conjunction with a GGA by Falkenauer [7]. Here, the author argues that when considering problems of this ilk, the use of classical genetic operators in conjunction with typical representation schemes[2] (as

---

[2] Such as the standard-encoding representations where the value of the ith gene represents the group that object i is in, and the indirect order-based representations that use a decoder to build solutions from permutations of the objects.

used for example with timetabling in [3] and [14]) are highly redundant due to the fact that the operators are *object-oriented* rather than *group-oriented*, resulting in a tendency for them to recklessly break up building blocks that we might otherwise want promoted. Falkenauer concludes that when considering grouping problems, the representations and resulting genetic operators need to be defined such that they allow the *groupings* of objects to be propagated, as it is these that are the innate building blocks of the problem, and not the particular positions of any one object on its own.

With this in mind, a standard GGA methodology is proposed in [9]. There has since been applications of these ideas to a number of grouping problems, with varying degrees of success. Examples include the equal piles problem [8], graph colouring [5, 6], edge colouring [12] and the exam-timetabling problem [6]. To our knowledge, there is yet to have been an application of a GGA towards a UCTP[3], although it is fairly clear that, at least for finding feasibility, it *is* a grouping problem - in this case, the set of events represents the set of objects to partition and the groups are defined by the timeslots. A feasible solution is therefore one where all of the $|E|$ events have been partitioned into $|T|$ feasible timeslots $t_1,\dots,t_{|T|}$, where a feasible timeslot t_i $1 \leq i \leq |T|$ is one where none of the events in t_i conflict, and where all the events in t_i can be placed in their own suitable room.

Note that soft constraints are not considered in this definition. There are two reasons for this. Firstly, in this UCTP violations of soft constraints (1) and (2) arise as a result of factors such as timeslot ordering and the occurrence of sequences of events with common students across adjacent timeslots. Thus they are in disagreement with the more general definition of a grouping problem [9]. Secondly, if we *were* to take soft constraints into account at this stage they would need to be incorporated into the fitness function (which we'll define in section 2.1). But taking soft-constraints into account, while at the same time searching for feasibility (as used in [3] and [14] for example), might actually have the adverse effect of leading the search away from attractive (and 100% feasible) areas of the search space, therefore compromising the main objective of the algorithm.

2   The Algorithm

Similarly to the work presented in [4], [11], [15] and [17], in our approach each timetable is represented by a two dimensional matrix M where rows represent rooms and columns represent timeslots; thus if $M(a,b)=c$, then event c is to occur in room a and timeslot b. If, on the other hand, $M(a,b)$ is blank, then no event is to be scheduled in room a during timeslot b. In our method, the timeslots are always kept feasible and the number of timeslots in each timetable is allowed to vary. We therefore open new timeslots when events cannot be feasibly placed in any existing one, and the aim of the algorithm is to reduce this number down to the required amount $|T|$ (remembering that in this case $|T|=45$).

---

[3] We note that UCTPs are generally considered to be different problems to exam timetabling problems due to various differences that we will not go into here, but are detailed in [16] for example.

Fig. 1 shows how we perform recombination to construct the first offspring timetable using parents P_1 and P_2, and randomly selected crossover points $x_{1,...,4}$. To form the second offspring the roles of the parents and the crossover points is reversed. The mutation operator we use follows a typical GGA mutation scheme - remove a small number (specified by the mutation rate mr) of randomly selected timeslots from a timetable and reinsert the events contained within them using the rebuild scheme (see below). We also use an inversion operator that randomly selects two columns in the timetable and swaps all the columns between them.

An initial population of timetables is constructed using the recursive rebuild scheme defined below. The same scheme is also used to reconstruct partial timetables that occur during recombination and mutation (with subtle differences regarding the breaking of ties - see table 1. Note too that event selection for mutation-rebuild is also different). The heuristics we use with this scheme are variations on those used in [1] and [11] and have already shown themselves to be powerful with this sort of problem.

Specifically, the rebuild scheme takes an empty or partial timetable tt and a set U of unplaced events. It then assigns all $u \in U$ a room and timeslot to produce a complete timetable, opening new timeslots where necessary. Note that for the initial population generator $U=E$. Let S represent the set of timeslots in tt and let P represent the set of *places* in tt - i.e. $P=R \times S$.

Rebuild (tt, U)

1. If $U=\varnothing$ end, else if $(|S|<|T|)$ open $(|T|-|S|)$ new slots, else open $\left\lceil \frac{|U|}{|R|} \right\rceil$ new slots.
2. PlaceEvents (tt, U).
3. Rebuild (tt, U).

PlaceEvents (tt, U)

1. Pick $u \in U$ with the smallest number of possible places to which it can be feasibly assigned in tt. Break ties with H_1 (see table 1). For mutation, just choose any u randomly.
2. Pick the feasible place for u in tt that the least number of other events in U want. Break ties with H_2 (see table 1).
3. Remove u from U and insert it into tt at the chosen place.
4. If there are still events in U with feasible places, go back to step 1.

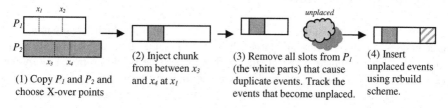

Fig. 1. How recombination is performed in this algorithm

2.1   Judging Criteria and the Fitness Function

When looking at the final output of the algorithm, it seems reasonable to assume that what we are ultimately interested in is the timetable's *distance to feasibility*, if indeed feasibility has not been found. In timetabling this can be measured in various ways such as the level of student inconvenience, the number of broken constraints, the number of extra timeslots used, etc. Of course, what *is* chosen depends first and foremost on user preference. In our approach we choose to use the number of unplaced events. This is calculated by doing the following: Recall that $|T|$ represents the target number of timeslots that we wish to use (i.e. 45), and $|S|$ represents the current number of timeslots. Additionally, let F_i indicate the number of events placed in timeslot i, where $1 \leq i \leq |S|$, and let $|S|'$ represent the number of *extra* timeslots being used i.e. $|S|'=(|S|-|T|)$. The distance to feasibility is the total number of events in the $|S|'$ timeslots to which F_i is minimal – i.e. the $|S|'$ timeslots with the least events in them.

During the algorithm's run however, the distance to feasibility need not be the only measure that we use to determine fitness. Indeed, if other information is present then it makes sense to use it if it is thought that it can help guide the search towards more promising areas of the search space. Consequently, we use a fitness function somewhat akin to the one proposed for graph colouring by Eiben *et al.* [5]: we calculate the number of extra timeslots $|S|'$ being used and the distance to feasibility. The fitness function is the total of these two values.

Table 1. Showing the various ways that ties are broken in the rebuild scheme for the three genetic operators (described in section 2). Note that H_1 for mutation is not applicable. As explained in the text, for this operator the order that unplaced events are inserted back into the timetable is entirely random

	Heuristic - H_1	*Heuristic - H_2*
Recombination	Choose the event that conflicts with the most others	Choose the place that defines the emptiest slot
Initial population	Choose randomly	Same as above
Mutation	N/A	Choose randomly

3   Experimental Analysis

We created sixty test instances using an instance generator, which we separate into three classes: small, medium and large[4]. It is known that all have at least one feasible timetable. These instances were created with no reference to the proposed GGA but are deliberately intended to be troublesome for finding feasibility. This was mainly achieved by simple experiments whereby instances were created and run on two

---

[4] For the small instances $|E|$ = 200 to 225 and $|R|$ = 5 or 6. For the medium instances $|E|$ = 390 to 425 and $|R|$ = 10 or 11. For the large instances $|E|$ = 1000 to 1075 and $|R|$ = 25 to 28. Other parameters for the instances can be found online at the URL at the end of section 4.

existing constructive algorithms [1, 11] that attempt to use stochastic heuristics to place all events feasibly. With many instances, these algorithms could only place about 80% of events (and sometimes even less) before running out of ideas and getting totally stuck. We therefore tended to take these as the instances to use in our experiments. Indeed, between them these algorithms could not find feasibility in 52 of the 60 instances.

For all the experiments we used a PC Pentium 4 2.40GHz with 512MB RAM. For the evolution scheme, a steady-state population of size ps was used: at each step, two parents are selected using binary tournament selection with parameter ts. Two offspring are then created with recombination rate rr. These are then mutated and inserted back into the population, in turn, over the least fit individual. If there is more than one least-fit individual we choose between these randomly. Also at each step, ir individuals are also chosen randomly and inversion is applied. During the run we keep track of the fittest solution found so far according to the fitness function defined in section 2.1. This is the algorithm's final output. In all experiments we set ps=50, ts=0.9, mr=3, and ir=4.

For our first set of experiments[5] we tested the algorithm on all sixty problem instances. We introduced time limits of 30, 200 and 800 seconds for the instance sets small, medium and large respectively, and set the recombination rate to 0.5. In this case, even with the strict time limits and the fact that we performed minimal parameter tuning, the algorithm found feasibility in 23 of the 60 instances – fifteen more than the algorithms presented in [1] and [11]. Additionally, there is the obvious advantage that this new algorithm is able to produce a number of different solutions in the same run.

In the experiments we also noticed that, in some cases (10 in small, 3 in medium and 1 in large), solutions were actually found in the initial populations! Although this might lead the prudent reader to suspect that the test instances are suspiciously easy, we argue that if anything this just goes to highlight the strength of our rebuild heuristics. Indeed this could be somewhat expected - similar heuristics have already shown themselves able to find feasibility in one go with other well-known instances [1, 11] and so there is no reason why they shouldn't occasionally do the same here.

Our next experiments attempted to address an issue that we consider to be of particular interest with this algorithm - the consequences that the recombination operator has on the number of new individuals that can be produced within the time limit, and the effects that this has on movement in the search space. It is well acknowledged that the general goal of recombination is to aid the search by allowing useful building blocks from multiple parents to be combined into new, different and hopefully fitter offspring. However, as can be imagined in this algorithm, recombination is more expensive than mutation, so if it is not doing its intended job at an acceptable level then its presence is questionable. If, on the other hand, recombination *is* aiding the search, there will still be a trade-off between the amount of recombination that is used, and the resultant number of new individuals that can be produced within the time limit.

---

[5] Full results can be found online at the URL defined at the end of section 4.

To investigate this, we conducted experiments using different recombination rates with all other parameters, including the time limits, remaining as previously defined. For all problem instances we ran five separate trials using recombination rates 0.0, 0.25, 0.5, 0.75 and 1.0.

Table 2 shows clearly that as we increase the recombination rate, the mean number of new individuals produced within the time limit falls. The effect that this characteristic has on the resultant movements in the search space within the time limit is illustrated in fig 2. This graph shows the mean fitness of populations for each of the twenty medium instances, over time. The effect of using a high level of recombination is shown well here - a rate of 1.0 for instance, at least for the first half of the run, gives the slowest progression through the search space per second, whilst a rate of 0.0 offers the most. Clearly, if the time limits for these instances were shorter, then using no recombination would seem the more sensible choice. What is also noticeable however is that the higher rates of 0.75 and 1.0 still seem to be making positive movements in the search space towards the end of the runs, whilst the other three rates (which use less recombination) seem to be levelling off. Indeed, the lines for both 0.75 and 1.0 both cross the other three towards the end of the run and, if the time limit were increased, look like they could go on to make further positive movements.

Table 2. Showing the mean number of new individuals the algorithm is able to produce within the specified time limits for the different recombination rates

	Rr=0.0	Rr=0.25	Rr=0.5	Rr=0.75	Rr=1.0
Small	24382	19750	14542	11600	7662
Medium	62752	45658	32262	26160	10454
Large	67594	38424	18148	4072	1052

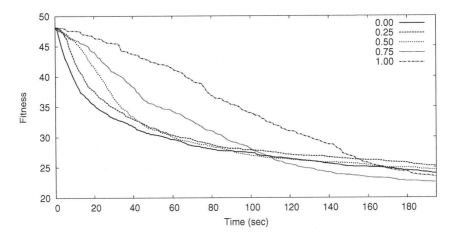

Fig. 2. Mean fitness of populations for the twenty medium instances, per second, for the different recombination rates

This is exactly what is shown in fig 3. Here, rather than concern ourselves with CPU time, we look at the way the fitness changes according to the number of timetable evaluations performed. This measure is frequently considered instead of CPU time when looking at the performance of an evolutionary algorithm [3] as evaluation can often be the most expensive operation. Indeed, timetabling problems in particular are prone to this, as often there might be an abundance of different real-world constraints that need to be checked, resulting in very complex and expensive evaluation functions. Although this is not the case for the relatively simple UCTP used here, this still seems a reasonable criterion to study, as it might be the case that as we add extra realism to the problem, the fitness function might become so intricate that the relative expense of the recombination operator becomes negligible. In this graph then, it can be seen quite clearly that as we increase the recombination rate, both the amount of positive movement through the search space per evaluation *and* the quality of the final solution increases (reflected in the steepness of the curve and the lower levelling-off point respectively). It can also be seen that as we decrease the recombination rate, these characteristics lessen in a uniform fashion. Thus it would seem that recombination is indeed doing its intended job.

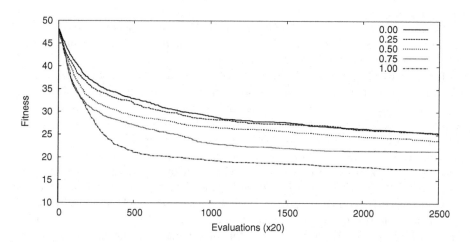

Fig. 3. Mean fitness of populations for the twenty medium instances, per evaluation, for the different recombination rates

4   Conclusions and Further Work

We have presented an algorithm for university-course timetabling that combines powerful constructive heuristics with GGA methodology. To our knowledge this is the first such algorithm aimed at this problem domain. Our initial experiments with sixty new "hard" problem instances have shown that results are promising with regards to the number of cases where we have found feasibility, although we do not yet claim these results to be state of the art. Further experiments in this paper have shown that the use of recombination does seem to aid the search towards better solutions, but if the time limit imposed is highly restricted it should probably be used in smaller amounts. Finally we round off this paper with some remarks about possible future work and various other issues that might be of interest.

A place where some improvements might be found is the fitness function. Currently the search landscape defined by our function can sometimes seem a little stepped in nature - especially when we are close to finding feasibility. In these cases the jump from near-feasibility to full-feasibility might well be a lucky one. However, other *smoother* fitness functions that concentrate more on favouring solutions with good combinations of events in timeslots might show to improve the search. We have recently been conducting experiments using the fitness function

$$f = \frac{\sum_{i=1}^{|S|} (F_i / |R|)^2}{|S|} \tag{1}$$

where F_i is as defined in section 2.1. Although this is more archetypal of this type of algorithm [6, 7, 8] it actually seems to give significantly worse results than our current function under the same test conditions. Further work might reveal other promising avenues.

Other improvements may also be seen through the introduction of some sort of smart-mutation [3] and/or various local search operators [14]. Whether or not these will improve the algorithm's results is also pending further work.

It is also worth noting that not all cases of UCTPs will have these same classical grouping characteristics as this one. In some problem definitions, all timeslots might not be the same because certain resources might be unavailable in certain predefined timeslots. Secondly, some cases may incorporate hard-constraints that pan across the timeslots such as the specification that one event must take place before another etc. in which case the *ordering* of the timeslots might also become important.

Finally, and perhaps most noticeably, we have not addressed for the time being the important task of optimising the soft constraints. The two-stage approach that we support here dictates that only once feasibility is found should soft constraints be considered. However, whether effective searches can still be made in the more restricted, feasible-only search space for these "hard" instances is yet to be investigated. In the meantime, the sixty problem instances and full tables of results are available at www.emergentcomputing.org/timetabling/harderinstances.htm.

References

1. H. Arntzen and A. Løkketangen. A Tabu Search Heuristic for a University Timetabling Problem. (2003), Proceedings of the Fifth Metaheuristics International Conference MIC 2003, Kyoto, Japan. An older version of the paper is also available at (accessed Dec 2004) http://www.idsia.ch/Files/ttcomp2002/arntzen.pdf
2. E. Burke, D. Elliman and R. Weare, Specialised Recombinative Operators for Timetabling Problems. (1995) In Proceedings of the AISB (AI and Simulated Behaviour) Workshop on Evolutionary Computing, Springer LNCS 993, pp. 75-85.

3. D. Corne, P. Ross, H-L Fang, The Practical Handbook of Genetic Algorithms, Applications, Volume 1. (1995). Edited by Lance Chambers, CRC Press, pp 219-276.
4. M. Chiarandini, K. Socha, M. Birattari, and O. Rossi-Doria. An effective hybrid approach for the university course timetabling problem.(2003) Technical Report AIDA-2003-05, FG Intellektik, FB Informatik, TU Darmstadt, Germany.
5. A.E. Eiben, J.K. van der Hauw, and J.I. van Hemert. Graph Coloring with Adaptive Evolutionary Algorithms (1998). Journal of Heuristics, 4(1):25-46.
6. W. Erben, A grouping Genetic Algorithm for Graph Colouring and Exam Timetabling (2000). Proceedings of the Practice and Theory of Automated Timetabling III, Springer LNCS 2079, pp132-156.
7. E. Falkenauer. A New Representation and Operators for GAs Applied to Grouping Problems (1994). Evolutionary Computation, Vol. 2 Issue 2, Summer 1994 pp123-144.
8. E. Falkenauer. Solving equal piles with the grouping genetic algorithm (1995) Proceedings of the 6th Int. Conf. on Genetic Algorithms. Morgan Kaufmann, pp 492-497.
9. E. Falkenauer. Genetic Algorithms and Grouping Problems (1999). John Wiley and Sons Ltd.
10. M. R. Garey and D. S. Johnson. Computers and Intractability – A guide to NP-completeness. (1979). W. H. Freeman and Company, San Francisco.
11. R. Lewis and B. Paechter. New Crossover Operators for Timetabling with Evolutionary Algorithms (2004). In proceedings of the 5th International Conference on Recent Advances in Soft Computing RASC2004. ISBN 1-84233-110-8, pp189-194. A copy is also available at http://www.soc.napier.ac.uk/publication/op/getpublication/publicationid/7207469
12. S. Khuri, T. Walters and Y. Sugono. A Grouping Genetic Algorithm for Coloring the Edges of Graphs (2000), Proceedings of the 2000 ACM/SIGAPP Symposium on Applied Computing, ACM Press, pp 422-427.
13. P. Kostuch. The University Course Timetabing Problem with a 3-stage approach (2004). In E. Burke and M. Trick (eds.) Proceedings of the 5th International Conference on the Practice and Theory of Automated Timetabling, pp 251-266.
14. O. Rossi-Doria, M. Samples, M. Birattari, M. Chiarandini, J. Knowles, M. Manfrin, M. Mastrolilli, L. Paquete, B. Paechter, T. Stützle. (2002). A comparison of the performance of different metaheuristics on the timetabling problem. In E. Burke and W. Erben (eds.) Proceedings of the Practice and Theory of Automated Timetabling III, Springer LNCS 2740, pp329-351.
15. B. Paechter, H. Luchian, A. Cumming and M. Petriuc. Two Solutions to the General Timetable Problem Using Evolutionary Algorithms (1994). In Proceedings of the IEEE World Congress in Computational Intelligence, pp 300 305.
16. A Schaerf. A Survey of Automated Timetabling. (1995) Centrum voor Wiskunde en Informatica (CWI) report CS-R9567, Amsterdam, The Netherlands. A revised version appeared in Artificial Intelligence Review 13(2), 87-127.
17. K. Socha, J. Knowels, M. Sampels. A MAX-MIN Ant System for the University Course Timetabling Problem (2002). In Dorigo, M., Di Caro, G., Sampels , M. (eds.), Proceedings of the 3rd International Workshop on Ant Algorithms (ANTS'2002), Springer LNCS 2463, pp 1-13.
18. J. M. Thompson and K. Dowsland. A Robust Simulated Annealing Based Examination Timetabling System (1998). Computers and Operations Research 25 pp 637- 648 ISSN 0305-0548.
19. International Timetabling Competition - http://www.idsia.ch/Files/ttcomp2002, accessed Dec 2003.

Self-Adapting Evolutionary Parameters: Encoding Aspects for Combinatorial Optimization Problems

Marcos H. Maruo[1], Heitor S. Lopes[1], and Myriam R. Delgado[1]

Centro Federal de Educação Tecnológica do Paraná -CEFET/PR
Av. Sete de Setembro, 3165 – 80230-901 Curitiba, Brazil
{ maruo, hslopes, myriam}@cpgei.cefetpr.br

Abstract. Evolutionary algorithms are powerful tools in search and optimization tasks with several applications in complex engineering problems. However, setting all associated parameters is not an easy task and the adaptation seems to be an interesting alternative. This paper aims to analyze the effect of self-adaptation of some evolutionary parameters of genetic algorithms (GAs). Here we intend to propose a flexible GA-based algorithm where only few parameters have to be defined by the user. Benchmark problems of combinatorial optimization were used to test the performance of the proposed approach.

1 Introduction

The two major definitions in applying any heuristic search algorithm to a particular problem are the representation and the evaluation (fitness) function. When using an evolutionary algorithm (EA) it is also needed to specify how candidate solutions will be changed to generate new solutions. This encompasses the definition of genetic operators (mainly mutation and crossover) suited to the encoding, and a selection method to enforce the survival-of-the-fittest evolutionary rule. Each of these components may have parameters, for instance: the probability of mutation, the tournament size of selection, or the population size. Values of these parameters greatly determine the quality of the solution found and the efficiency of the search [1]. Frequently, choosing suitable parameter values, is problem-dependent and requires previous experience of the user. Since this can be a time-consuming task, considerable effort has been applied to develop good heuristics for it, so as to avoid trial-and-error [2]. Despite its crucial importance, there is no consistent methodology for the determination of the running parameters of an EA, which are, most time, arbitrarily set within predefined ranges.

Globally, we distinguish two major forms of setting parameter values: parameter *tuning* and parameter *control*. The first means the commonly practised approach that tries to find good values for the parameters *before* running the algorithm, and then tuning the algorithm using these values, which remain fixed during the run. A general drawback of the parameter tuning approach, regardless of how the parameters are tuned, is based on the observation that a run of

G.R. Raidl and J. Gottlieb (Eds.): EvoCOP 2005, LNCS 3448, pp. 154–165, 2005.

an EA is an intrinsically dynamic, adaptive process. The use of rigid parameters that do not change their values is thus in contrast with this spirit [1]. Therefore, parameter control is an alternative. In this approach, a run is started with initial values for the parameters, and then they are dynamically changed *during* the run.

This paper aims to analyze the parameter control in contrast with the parameter tuning technique while solving combinatorial optimization problems, taking into account three important issues:

- Does real-valued encoding take any benefit for a discrete problem using parameter control in a GA?
- How the performance of the GA is affected by the adaptation in their parameters?
- Is there any advantage for the user using parameter control in a GA?

In this work the parameter control technique is based on self-adaptation of several parameters associated with the evolutionary process. The main goal here is to produce a flexible GA, in which only few running parameters need to be defined by the user.

2   Parameter Control

In classifying parameter control techniques of an evolutionary algorithm, many aspects can be taken into account: *What* is changed? *How* the change is made? The *scope/level* of change and the *evidence* upon which the change is carried out.

According to Eiben et al. [1], the change can be categorized into three classes:

- *Deterministic* parameter control: this take place when the value of a parameter is altered by some deterministic rule.
- *Adaptive* parameter control: this take place when there is some form of feedback from the search that is used to determine the direction and/or the magnitude of the change to the parameter.
- *Self-adaptive* parameter control: the idea of "evolution of the evolution" can be used to implement the self-adaptation of parameters. Here the parameters to be adapted are encoded into the chromosome and undergo the action of genetic operators. The better values of these encoded parameters lead to better individuals which, in turn, are more likely to survive and produce offspring and hence propagate these better parameter values.

Some authors have introduced a different terminology based on the level of change [3, 4] and in how the change is made [5]. A more detailed discussion of parameter control can be found in [6].

The straightforward way to control parameters in a deterministic way is by using parameters that may change over time, that is, by replacing a parameter p_{stat} by a function $p_{dyn}(t)$, where t is the generation counter. However, this process presents some disadvantages: the difficulty in designing an optimal function

$p_{dyn}(t)$, and the fact that this function does not take into account any clue of the actual progress in solving the problem. Hence, it is thus seemingly natural to use an evolutionary algorithm not only for finding solutions to a problem, but also for tuning the (same) algorithm to the particular problem. Technically speaking, it is tried to modify the values of parameters during the run of the algorithm by taking the actual search progress into account. As discussed in [1], there are two ways to do this. The first way is to use some heuristic rule which takes feedback from the current state of the search and modifies the parameter values accordingly (adaptive parameter control), such as the credit assignment process presented by [7]. A second way is to incorporate parameters into the chromosomes, thereby making them subject to evolution (self-adaptive parameter control), like the approach presented in [8].

2.1   Mutation Parameters Control

De Jong recommended $pm = 0.001$ [9], the meta-level GA used by Grefenstette [10] indicated $p_m = 0.01$, while Schaffer et al. [11] came up with $p_m \in [0.005, 0.001]$. Folowing the earlier work of Bremmermann [12], Mühlenbein derived a formula for p_m which depends on the length of the bitstring (L), namely $p_m = \frac{1}{L}$ should be a generally "optimal" static value for p_m. This rate was compared with several fixed rates by Smith and Fogarty [13] who found that $p_m = \frac{1}{L}$ outperformed other values for p_m in their comparison. The same was found by Bäck [14] using gray coding. However, as pointed by [15, 13], there is an increasing body of evidence that the optimal rate of mutation is not only different for every problem encoding but will vary with evolutionary time according to the state of the search and the nature of the fitness landscape being explored.

These ideas have been applied to a generational GA by adding a further 20 bits to the problem genotype, which were used to encode the mutation rate [16]. The results showed that in generational setting the mechanism proved competitive with a genetic algorithm using a fixed (optimal) mutation rate, provided that a high selection pressure was maintained (this is referred to as "extinctive" selection). In [17] they proposed two simple adaptive mutation rate control schemes and show their feasibility in comparison with several other fixed and adaptive schemes applied to combinatorial optimization problems.

2.2   Crossover Parameters Control

Effectiveness of crossover has been frequently discussed in the literature, and some interesting results were reported by De Jong [9]. More recently, Schaffer and Eshelman [18] empirically compared mutation and crossover and concluded that the latter is capable of exploring epistatic problems more efficiently, in contrast with the mutation alone.

There are several types of crossover, but GAs use more frequently only one- or two-point crossovers. However, there are some situations when using a multi-point crossover can be beneficial [19, 20]. Then, an interesting option is the uniform crossover, that produces, in average, $\frac{L}{2}$ combinations in L-long strings [21, 20]. Besides the empirical research, many efforts have been directed to the theoretical comparisons between different crossover types [22, 23]. However, conclusions are not general enough to foresee which crossover is the best for a given problem. For instance, such theoretical approaches do not consider population size, although this parameter can affect directly the utility of crossover [24]. Furthermore, there are evidences that the utility of mutation operators can also be affected by the population size: it seems to be more useful than crossover when the population is small, and crossover seems to be more effective when the population is large [25].

Spears [8] proposed a self-adapting mechanism that chooses between two crossover types: two-point and uniform crossovers. An extra bit is added to the chromosome indicating which type of crossover it will be used for this particular individual. Descendants will inherit the crossover type from parents. Some experiments indicate that the GA using adaptive crossover had performance equal or better than a classic GA for a set of test problems [11].

When we use multi-parent operators [26], a new parameter is included: the number of parents used in the crossover. Eiben [27] presents an adjustment mechanism for the recombination arity based on competing sub-populations. In particular, the population is divided into disjoint sub-populations, and each one uses a crossover with different arity. These sub-populations evolve independently during a time-window and then they interchange information by allowing migrations between them. Migration favors those sub-populations that evolved better within the time-window, allowing them to be increased, accepting migrants. Conversely, sub-populations that evolved worse, loose individuals and decrease in size. This method achieved similar performance than the conventional GA using a 6-parents crossover. However, this algorithm does not succeed to identify clearly the best operators, regarding the population sizes, thus agreeing with Spears' experiments [8].

3 Proposed Approach: Self-Adapting Parameters

As discussed before, parameter control in evolutionary algorithm is a poorly structured, ill-defined, complex problem [1], and then, self-adapting appears as an interesting alternative. Most works in recent literature discuss the adaptation of just one evolutionary parameter at a time (e.g., probability of crossover and mutation). In this work it is aimed to self-adapt concomitantly several parameters associated with the evolutionary process. It is also evaluated their influence on the performance for different encodings. The main objective is to develop a flexible GA with few user-defined parameters.

In self-adapting GA parameters, regardless of the adopted encoding, the chromosome must be modified to accommodate genes encoding parameters' values that will be fine-tuned during the evolutionary process. Considering the parameters that will be changed, each individual is codified into a chromosome with $n+p$ genes, where n and p are, respectively, the number of genes that encode the

	Solution	Pr_mut	Pr_cross	cross_type	mut_nonunif	tourn_size
1	Solution	Pr_mut	Pr_cross	cross_type	mut_nonunif	tourn_size
2	Solution	Pr_mut	Pr_cross	cross_type	mut_nonunif	tourn_size
	.	.				.
	.					
N	Solution	Pr_mut	Pr_cross	cross_type	mut_nonunif	tourn_size

Fig. 1. Self-adapting: encoding aspects

problem solution and those that encode parameter values. In the population N individuals, they are encoded with a single chromosome, as shown in Figure 1.

In figure 1 the strategic parameters that will be adjusted by means of evolution are: mutation rate (Pr_mut), crossover rates (Pr_cross), crossover type ($cross\_type$), mutation step (for real encoding) ($mut\_nonunif$), and the tournament size ($tour\_size$), resulting in $p = 5$ genes. Mutation rate, mutation step and crossover rate are applied at individual level, whereas the tournament size and crossover type are determined by the mean among all individuals and are applied at population level. To compare the behavior of parameters adaptation in continuous and discrete search spaces, two encoding schemes will be analyzed: binary and real encoded chromosomes.

During the evolutionary process, genetic operators treat indistinctly all genes of the chromosome, despite what it encodes (part of the solution or any strategic parameter). For crossover, a N-length random vector is generated, where N is the population size and each element of the vector is within $[0, 1]$. Each value of this vector is compared with the crossover rate encoded in an individual. If it is lower, the individual is selected for crossover. For mutation, a $N \times (n + p)$ matrix is generated with random values. For each individual there is a corresponding line of $n + p$ random numbers in the matrix, which are compared with the mutation rate encoded in the individual. If it is lower, the associated gene of the individual will undergo mutation. Both crossover type and tournament size are chosen based on the frequency of the corresponding value in the population. The most frequent values are accepted as parameter for the next generation. It is important to note that with this encoding, whenever an individual has the genes that encode the strategic parameters modified by genetic operators (but not the solution genes), the corresponding fitness does not need to be re-calculated.

4   Experiments, Results and Discussion

This paper aims to analyze the performance of the self-adapting approach when applied to a benchmark of a well-known combinatorial optimization problem: the multiple knapsack problem (MKP), that is a generalization of the simple 0/1 knapsack problem [28]. The 0/1 KP involves selecting from among various items those that will be most profitable, given the knapsack has limited capacity. The 0/1 MKP involves m knapsacks of capacities $c_1, c_2, \cdots, c_m$. Every object selected must be placed in all m knapsacks, although neither the weight of an object o_j nor its profit is fixed, and, probably, they will have different values in each knapsack (for additional details see [29]).

In our implementation, each gene that encodes the problem solution should indicate the presence or absence of an item in all knapsacks. For binary encoding, each gene corresponds to a single bit, and for real encoding, each gene corresponds to a number in the range [0..1] that is rounded to the closest integer, as shown in figure 2.

a)   b)

Fig. 2. Solution encoding using: a) binary encoding, b) real encoding

When using binary encoding, the search space (ss) is defined by the number of items i as $ss = 2^i$. Although the number of knapsacks does not influence the search space, a large number of knapsacks implies in more complexity in computing the fitness function.

For analyzing the behavior of the algorithm in different difficulty levels, simulations were done for nine MKP problems, detailed in Table 1.

Table 1. Characteristics of MKP problems used in this work

Problem	Items	Knapsacks	Optimal solution
Weing7	105	2	1095445
Pb6	40	30	776
Pet6	39	5	10618
Weish18	70	5	9580
Weish22	80	5	8947
Weish26	90	5	9584
Flei	20	10	2139
Hp2	35	4	3186
Sent01	60	30	7772

In our experiments, the value of GA parameters were based on those recommended by De Jong [9], and were set as follows:

- Number of generations: 2000;
- Population size (N): 100 individuals;

– Crossover rate: $Pr\_cross \in [0.4, 1]$ for adaptive approach and $Pr\_cross = 0.6$ for fixed approach;
– Mutation rate: $Pr\_mut \in [0.0001, 0.3]$ for adaptive approach and $Pr\_mut = 0.001$ for fixed approach;
– Tournament size: $s \in \{2, \ldots, 5\}$ for adaptive approach and $s = 5$ for fixed approach;
– Crossover type: one-point and uniform for adaptive approach and one-point for fixed approach;
– Stop criterion: Maximum number of generations.

Table 2 compares the performance of our self-adapting parameters algorithm (self-adapt) and fixed parameters algorithm (fixed) for all problems described in Table 1. As shown in Figure 2, for real encoding, rounding to the closest integer is used to transform a real-valued chromosome to a binary string.

Self-adaptation is based in the expectation that the best strategic parameters will be able to produce more adapted individuals. Table 2 shows the results of self-adaptation in both cases, real and binary encoding. The problems were evaluated considering 500 runs [1] and the problem files were taken from [31]. For each problem instance, we evaluated the distribution tendency [2]. Besides, we also presented the best individual found in all runs (Best) and when this individual has been found (Gen_Best).

Table 2 clearly shows that the self-adapt approach out-performed the fixed approach, independently of the encoding scheme. Therefore, we can answer the first question pointed in the Introduction: "Does real-valued encoding take any benefit for a discrete problem using parameter control in a GA?". For the discrete problems dealt in this work real encoding does seems to be less appropriate than binary encoding. The hypothesis that making transitions between values of genes smoother before rounding (using real values instead of discrete ones) can facilitate GA to find better solutions was not confirmed.

To better investigate the effects of self-adaptation in the system performance, Table 3 shows normalized results [3] considering binary encoding and different levels of system autonomy. Results from [32] that self-adapts mutation rates are also presented.

It is important to point out that, due to the penalties applied to the fitness function, all the solutions obtained from approaches with adaptation were feasible. It is a strong restriction that could be relaxed in the future to improve the system performance. In [32], for instance, they reported the elimination of all unfeasible solutions. As can be noted, the version with the highest level of autonomy

---

[1] Different random seeds were generated using L'Ecuyer with Bays-Durham shuffling [30]
[2] Measures of central tendency are measures of the location of the middle or the center of a distribution. For symmetric distributions, mean and median are the same. In general, the mean will be higher than the median for positively skewed distributions and less than the median for negatively skewed distributions.
[3] The optimal known value (profit=1095445) for the Weing7 problem is the reference value 1.00

Table 2. Results for multi knapsack 0/1 optimization problems

Problem	Encoding	Adaptation	Mean	Median	SD	Best	Gen(Best)
Weing7	Binary	Self-Adapt	1095264	1092466	48892.98	1095445	231
		Fixed	1075147	1071868	48891.31	1091327	1721
	Real	Self-Adapt	1079962	1083937	49065.43	1079880	63
		Fixed	1034557	1038552	50443,06	1094957	9
Pb6	Binary	Self-Adapt	714.0699	729	62.27475	776	24
		Fixed	407.0459	405	104.9299	657	57
	Real	Self-Adapt	681.1497	704	79.87887	776	17
		Fixed	470.5289	468	91.4316	745	3751
Pet6	Binary	Self-Adapt	10468.01	10504	474.841	10618	34
		Fixed	8749.058	8271	992.6105	10396	1328
	Real	Self-Adapt	10460.21	10496	475.6101	10618	19
		Fixed	10152.5	10201	497.6526	10584	5526
Weish18	Binary	Self-Adapt	9510.048	9548	433.288	9580	37
		Fixed	8102.982	8157	534.998	9109	1915
	Real	Self-Adapt	9218.266	9282	463.818	9580	61
		Fixed	7535.108	7556	641.971	8938	46
Weish22	Binary	Self-Adapt	8834.535	8857	400.092	8947	924
		Fixed	7435.112	7458	514.995	8589	1407
	Real	Self-Adapt	8239.800	8334	540.689	8929	45
		Fixed	5735.299	5764	810.009	7629	37
Weish26	Binary	Self-Adapt	9493.311	9539	428.442	9584	97
		Fixed	8115.074	8130	515.070	9117	1865
	Real	Self-Adapt	8716.764	8840	657.194	9533	55
		Fixed	6080.405	6120	867.144	8429	62
Flei	Binary	Self-Adapt	2067.920	2068	94.8643	2139	8
		Fixed	1944.948	1956	102.612	2059	60
	Real	Self-Adapt	2052.653	2059	98.2860	2139	1
		Fixed	1918.399	1922	117.593	2139	1
Hp2	Binary	Self-Adapt	3080.966	3089	149.404	3186	18
		Fixed	2855.920	2868	158.267	3119	13
	Real	Self-Adapt	3038.818	3048	148.173	3169	19
		Fixed	2894.940	2910	161.950	3157	29
Sent01	Binary	Self-Adapt	7622.100	7698	383.915	7772	44
		Fixed	4559.451	4573	816.850	6698	1939
	Real	Self-Adapt	7111.149	7295	650.846	7772	36
		Fixed	3843.950	3758	973.846	6505	42

(self-adapt approach) out-performed the other approaches with a high degree of confidence (see t-Student test column [4]), confirming that the self-adaptation is a good alternative to release user from arbitrarily defining evolutionary parameters. By answering the second question "How the performance of the GA is

---

[4] t-Student tests were performed comparing all the approaches with the fixed one.

Table 3. Normalized results for Weing7 problem

Adaptation	Mean	Median	SD	Best	Gen(Best)	t-Student
Self-Adapt	0.99983	0.99728	0.044633	1.000000	231	1.0e-190
Kimbrough [32]	0.99934	-	0.000364	-	-	-
Mutation	0.99906	0.99638	0.044590	1.000000	375	2.8e-183
Crossover/crossover-type	0.98897	0.98581	0.044439	0.99866	1591	1.64e-40
Tournament size	0.98235	0.97883	0.044704	0.99719	945	0.27
Fixed	0.98147	0.97848	0.044631	0.99624	1721	-

affected by the adaptation in their parameters?", results show that almost all forms of adaptation resulted in a GA with better performance when compared with the fixed one. Also, the self-adaptation of mutation plays a main role when compared with the remaining parameters.

Regarding the third question: "Is there any advantage for the user using parameter control in a GA?", the better results found with self-adapting parameters point out a clear advantage. Nevertheless, it could be argued that the final results using self-adaptation of (only) mutation are not striking better than using self-adaptation of all parameters together. In fact this is not really the point because it should be taken into account the fact that having less parameters to adjust, user can focus in understanding/tuning the remaining ones.

To analyze the behavior of evolutionary operators during the process, Figure 3 depicts the evolution of crossover and mutation rates of the self-adaptive method. It is also shown the evolution of fitness for both methods [5].

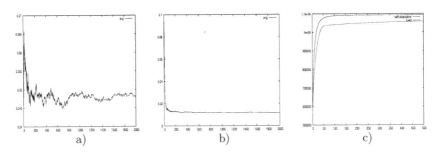

a)   b)   c)

Fig. 3. Evolution of: a) Crossover Rate, b) Mutation Rate, c) Fitness

It can be noted in Figure 3 that the oscillation of crossover rate is higher during the early stages of the evolutionary process, and observe the convergence

---

[5] Since the fitness evolution has not shown significant improvement after 500 generations, the following generations were omitted from the graphics

of mutation rates to small values at the end of evolution. This fact confirms the hypothesis that using fixed parameters during the evolutionary process is not adequate for all the problems solved by GAs.

5   Conclusions

This paper presented a method for self-adapting several parameters associated with the evolutionary process of a GA. After answering the questions asked in the Introduction, we conclude that the proposed approach has shown a good performance with a high degree of autonomy, where only few evolutionary parameters have to be defined by the user. Also, we verified that, for the MKP instances tested, the search in a continuous space resulted in a worse performance when compared with the search in a binary space. So, the possible benefits associated with the fact of producing more robust algorithms (which can perform independently in continuous or discrete spaces) are reduced in performance.

Finally, simulation results confirmed that self-adaptation is an interesting alternative in the search for parameterless heuristic optimization systems, since it is able to generate models that explore the space of parameters looking for the best ones at the same time as well as it searches for the problem's solution. Results obtained encourage further research towards a fully self-adaptive GA, relieving user from the burden of adjusting parameters for each problem.

Acknowledgements. The authors would like to thank the anonymous referees for their helpful suggestions. Thanks to Fundação Araucária/Funcefet-PR 018/2004 for the financial support, for CNPq for a grant to M.H.Maruo, and CNPq for a research grant to H.S.Lopes, process 350053/03-0.

References

1. Eiben, A.E., Hinterding, R., Michalewicz, Z.: Parameter control in evolutionary algorithms. IEEE Transactions on Evolutionary Computation **3** (1999)
2. Rojas, I., Gonzalez, J., Pomares, H., Merelo, J., Castillo, P., Romero, G.: Statistical analysis of the main parameters involved in the design of a genetic algorithm. IEEE Transactions on Systems Man and Cybernetics **32** (2002) 31–37
3. Angeline, P.J.: Adaptive and self-adaptive evolutionary computation. In M., P., Attikiouzel, Y., Marks, R., D., F., Fukuda, T., eds.: Computational Intelligence, A Dynamic System Perspective, IEEE Press (1995) 152–161
4. Hinterding, R., Michalewicz, Z., Eiben, A.E.: Adaptation in evolutionary computation: A survey. In: Proceedings of the 4th IEEE International Conference on Evolutionary Computation. (1997) 65–69
5. Smith, J., Fogarty, T.C.: Operator and parameter adaptation in genetic algorithms. Soft Computing **1** (1997) 81–87
6. Eiben, A.E., Smith, J.E.: Introduction to Evolutionary Computing. Springer (2003)
7. Davis, L., ed.: Handbook of Genetic Algorithms. Van Nostrand Reinhold, New York, New York (1991)

8. Spears, W.M.: Adapting crossover in evolutionary algorithms. In: Proceedings of the Fourth Annual Conference on Evolutionary Programming, San Diego, CA (1995)

9. De Jong, K.: The Analysis of the behaviour of a Class of Genetic Adaptive systems. PhD thesis, Department of Computer Science, University of Michigan (1975)

10. Grefenstette, J.J.: Optimization of control parameters for genetic algorithms. IEEE Transactions on Systems, Man, and Cybernetics 16 (1986) 122–128

11. Schaffer, J.D., Morishima, A.: An adaptive crossover distribution mechanism for genetic algorithms. In: Proceedings of the Second International Conference on Genetic Algorithms on Genetic algorithms and their application, Lawrence Erlbaum Associates, Inc. (1987) 36–40

12. Bremermann, H.J., Rogson, M., Salaff, S.: Global properties of evolution processes. In Pattee, H.H., Edlsack, E.A., Fein, L., Callahan, A.B., eds.: Natural Automata and Useful Simulations. Spartan Books, Washington D.C. (1966) 3–41

13. Smith, J., Fogarty, T.C.: Self-adaptation of mutation rates in a steady-state genetic algorithm. Proceedings of IEEE International Conference on Evolutionary Computation (1996) 318–323

14. Bäck, T.: Evolutionary algorithms in theory and practice: evolution strategies, evolutionary programming, genetic algorithms. Oxford University Press (1996)

15. Fogarty, T.C.: Varying the probability of mutation in the genetic algorithm. In: Proceedings of the 3rd International Conference on Genetic Algorithms, Morgan Kaufmann Publishers Inc. (1989) 104–109

16. Bäck, T.: Self-adaptation in genetic algorithms. In Varela, F.J., Bourgine, P., eds.: Proc. of the 1st European Conf. on Artificial Life, Cambridge, MA, MIT Press (1992) 227–235

17. Thierens, D., Goldberg, D.E.: Mixing in genetic algorithms. In: Proceedings of the 5th International Conference on Genetic Algorithms, Morgan Kaufmann Publishers Inc. (1993) 38–47

18. Schaffer, J.D., Eshelman, L.J.: On crossover as an evolutionary viable strategy. In Belew, R., Booker, L., eds.: Proceedings of the Fourth International Conference on Genetic Algorithms. (1991) 61–68

19. Eshelman, L.J., Caruana, R., Schaffer, J.D.: Biases in the crossover landscape. In: Proceedings of the 3rd International Conference on Genetic Algorithms, Morgan Kaufmann Publishers Inc. (1989) 10–19

20. Sywerda, G.: Uniform crossover in genetic algorithms. In: Proceedings of the third international conference on Genetic algorithms, Morgan Kaufmann Publishers Inc. (1989) 2–9

21. Spears, W.M., De Jong, K.: On the virtues of parameterized uniform crossover. In Belew, R.K., Booker, L.B., eds.: Proc. of the Fourth Int. Conf. on Genetic Algorithms, San Mateo, CA, Morgan Kaufmann (1991) 230–236

22. De Jong, K., Spears, W.M.: A formal analysis of the role of multi-point crossover in genetic algorithms. Annals of Mathematics and Artificial Intelligence 5 (1992) 1–26

23. Spears, W.M.: Crossover or mutation? In Whitley, L.D., ed.: Foundations of Genetic Algorithms 2. Morgan Kaufmann, San Mateo, CA (1993) 221–237

24. De Jong, K., Spears, W.M.: An analysis of the interacting roles of population size and crossover in genetic algorithms. In Schwefel, H.P., Männer, R., eds.: Parallel Problem Solving from Nature - Proceedings of 1st Workshop, PPSN 1. Volume 496., Dortmund, Germany, Springer-Verlag, Berlin, Germany (1991) 38–47

25. Spears, W.M., Anand, V.: A study of crossover operators in genetic programming. In Ras, Z.W., Zemankova, M., eds.: Proceedings of the Sixth International Symposium on Methodologies for Intelligent Systems ISMIS 91, Springer-Verlag (1991) 409–418

26. Eiben, A.E.: 3.7. In: Multi-parent Recombination. IOP Publishing Ltd and Oxford University Press (1995)

27. Eiben, A.E.: Multiparent recombination in evolutionary computing. Advances in evolutionary computing: theory and applications (2003) 175–192

28. Mumford, C.L.: Comparing representations and recombination operators for the multi-objective 0/1 knapsack problem. In: CEC2003: Proceedings of The IEEE Conference on Evolutionary Computation, IEEE. (2003) 854–861

29. Thierens, D.: Adaptive mutation rate control schemes in genetic algorithms. In: CEC2002: Proceedings of The IEEE Conference on Evolutionary Computation, IEEE. (2002) 980–985

30. Press, W.H., Teukolsky, S.A., Vetterling, W.T., Flannery, B.P.: Numerical Recipes in C: The Art of Scientific Computing. Cambridge University Press (1992)

31. Khuri, S., Bäck, T., Heitkötter, J.: SAC 94: suite of 0/1 multiple knapsack problem instances, http://elib.zib.de/pub/packages/mp-testdata/ip/sac94-suite (1994)

32. Kimbrough, S.O., Lu, M., Wood, D.H., Wu, D.J.: Exploring a two-market genetic algorithm. In: Proceedings of the Genetic and Evolutionary Computation Conference, Morgan Kaufmann Publishers Inc. (2002) 415–422

Population Training Heuristics

Alexandre C.M. Oliveira[1] and Luiz A.N. Lorena[2]

[1] Universidade Federal do Maranhão - UFMA, Depto. de Informática,
S. Luís MA, Brasil
`acmo@deinf.ufma.br`
[2] Instituto Nacional de Pesquisas Espaciais - INPE, Lab. Associado de Computação e
Matemática Aplicada, S. José dos Campos SP, Brasil
`lorena@lac.inpe.br`

Abstract. This work describes a new way of employing problem-specific
heuristics to improve evolutionary algorithms: the Population Training
Heuristic (PTH). The PTH employs heuristics in fitness definition, guiding the population to settle down in search areas where the individuals
can not be improved by such heuristics. Some new theoretical improvements not present in early algorithms are now introduced. An application
for pattern sequencing problems is examined with new improved computational results. The method is also compared against other approaches,
using benchmark instances taken from the literature.

Keywords: Hybrid evolutionary algorithms; population training; MOSP;
GMLP.

1   Introduction

Evolutionary algorithms are efficient to explore a wide search space, converging
quickly to local minima. However, their lack of exploiting local information is
a well-known drawback to reach global minima. Evolutionary operators to incorporate knowledge about problem particularities have encapsulated heuristics
and local search procedures. Such procedures basically consist in searching for
better solutions in the set of candidate solutions (*neighborhood*) that can be
obtained from a given solution by heuristic moves. An individual improved by
heuristic, in general, is replaced as soon as a better individual is obtained. The
more individuals are heuristically improved, the more the heuristic leads the
population to incorporate the desired features.

Due to the computational cost of some heuristic procedures, a challenge in
such hybrid methods is to define efficient strategies to cover all search space,
applying local search only in actually promising neighborhoods. Elitism plays
an important role towards achieving this goal, once the best individuals can represent such promising neighborhoods. But the elite can be sharing the same few
neighborhoods and then the heuristic moves does not improve the population.

The Population Training Heuristic (PTH) proposes a way of leading the population to acquire desired characteristics. All individuals are evaluated by two

G.R. Raidl and J. Gottlieb (Eds.): EvoCOP 2005, LNCS 3448, pp. 166–176, 2005.

functions: a function computing *what the individual is* and another estimating *what it should be*. The later does not take account of a presumed potential of the individual, but its deficiency by not being what it should be.

The evolutionary process is driven by a problem-specific heuristic (called training heuristic), employed in the fitness function formulation. The *well-adapted* individuals are those who can not be improved by the training heuristic. They are *what they should be*, i.e., the best inside the neighborhood generated by the training heuristic and tend to stay in the population longer times.

The Constructive Genetic Algorithm (CGA) was proposed in [1] to Location Problems, and applied to other problems as Timetabling [2]. The CGA presents a number of new features compared to a traditional genetic algorithm, such as a dynamic-sized population composed of *schemata* (incomplete solutions) and *structures* (complete solutions). Each individual (structure or schema) has a fitness evaluation based on two functions, f and g (*fg-fitness*), that are built considering specific aspects of the problem at hand in a way that an individual with $|f - g| \approx 0$ corresponds to an optimal solution. A further optimization objective is introduced to guide the search to find structures: to maximize g. Thus, no matter the nature of the problem (minimization or maximization), the original problem is transformed in a bi-objective problem (BOP): $f - g$ minimization (optimal phase) and g maximization (constructive phase) [1].

The CGA was inspiration to PTH, especially by the double evaluation of individuals. In PTH, the *fg-fitness* leads population to subsearch spaces where no improvement can be reached by applying the training heuristic, probably optimal solutions depending on how much less *approximative* is the heuristic. To avoid costly fitness evaluation, light heuristics are used. Local search mutations are included in the evolutionary process to make a fine tuning of the well-adapted individuals.

Some early CGA applications, since they employ training heuristics, can be considered as based on PTH fundamentals [3], [4]. Such applications rank the population by a *constructive ranking* that considers simultaneously a constructive and optimal phases, as in the CGA original form. The constructive approach of PTH is still called Constructive Genetic Algorithm, but denoted by CGA^H, to avoid misunderstanding with the original CGA, which no heuristic training had been employed yet [1], [2].

On the other hand, the *non-constructive ranking*, proposed in this work, only focuses the optimal phase, concerning to the adaptation of individuals to training heuristic. The main reason for this alternative form of ranking is to aggregate more flexibility to the approach, not necessarily coding individuals as incomplete solutions. The non-constructive approach is called Population Training Algorithm (PTA) and it was firstly applied to numerical optimization [5].

This work introduces theoretical improvements for PTH and also presents new results for pattern sequencing problems not found in early works [3], [4], [5]. The remainder of this paper is organized as follows. Section 2 presents the general guidelines of PTH, consolidating the formulation needed to future applications. In section 3, an application in pattern sequencing problems is modeled. The conclusions are summarized in the section 4.

2   General Guidelines of **PTH**

The PTH can be defined by the tuple $\{P, \Theta, f, H, \wp, \delta\}$, where P is a population sampled from the coded search space S, hence individuals $s_k \in P \subset S$. Individuals are generated by a set of evolutionary specific operators Θ and evaluated by an objective function f that maps S in $\Re$. The training heuristic H is defined by the pair $\{\varphi^H, g\}$, where g heuristically evaluates each solution generated by the neighborhood relationship φ^H. The neighborhood relationship φ^H can be understood as set of solutions which can be obtained from an original one, s_k, by heuristic moves (H moves):

$$\varphi^H(s_k) = \{s_k, s_{v1}, s_{v2}, \ldots, s_{vl}\} \tag{1}$$

where $l + 1$ is the length of the neighborhood of s_k, including itself.

The heuristic knowledge about a problem is then used to define g. A typical g, adopted in this work, is the objective function f calculated over $\varphi^H(s_k)$. Thus, for minimization problems:

$$g(s_k) = f(s_{vb}) = \min\{f(s_k), f(s_{v1}), f(s_{v2}), \ldots, f(s_{vl})\} \tag{2}$$

The best neighbor of s_k is denoted by s_{vb}. The concept of proximity, $\wp$, is concerned with the effort necessary to reach s_{vb} from s_k by H moves. More proximity means more adaptation of s_k to the heuristic that generated s_{vb}. Depending on the coding being employed in the application, some distance metrics may be used, such as hamming for binary-coded, euclidean for real-coded and heuristic distance [6]. In this application, the fitness distance between s_k and s_{vb} is adopted:

$$\wp(s_k, s_{vb}) = |f(s_k) - f(s_{vb})| \tag{3}$$

Independently of the nature of the problem (minimization or maximization), the $\wp(s_k, s_{vb})$, adopted in this work, always reflects how much s_{vb} is better than s_k. However, if another distance metric was used, $\wp(s_k, s_{vb})$ would mean just the adaptation to the training heuristic. Finally, the entire population is ranked by a function δ that considers the overall individual adaptation. The constructive and non-constructive rankings are, respectively:

$$\delta_{cons}(s_k) = \frac{d \cdot G_{max} - |f(s_k) - g(s_k)|}{d \cdot [G_{max} - g(s_k)]} \tag{4}$$

$$\delta_{ncons}(s_k) = d \cdot [G_{max} - g(s_k)] - |f(s_k) - g(s_k)| \tag{5}$$

Considering G_{max} an estimate of the upper bound for all possible values of the function g (even function f), the interval $G_{max} - g(s_k)$, gives the fitness distance between individual s_k and the upper bound. This distance is used in two distinct ways. In the constructive ranking, to estimate the completeness of the individuals, penalizing the schemata. In the non-constructive ranking, such interval just considers the objective function evaluation, once in minimization problems, the greater is $G_{max} - g(s_k)$, the better is the individual.

The constant d is used to equilibrate both ranking equations (generally, about $1/G_{max}$). In the beginning of the evolution, the upper bound G_{max} can be analytically calculated, considering the problem instance, or estimated by sampling. For maximization problems, a lower bound G_{min} is introduced in the constructive ranking:

$$\delta_{ncons}^{max}(s_k) = d \cdot [g(s_k) - G_{min}] - |f(s_k) - g(s_k)| \qquad (6)$$

The constructive ranking considers simultaneously a constructive and optimal phases, as in the CGA original form. The non-constructive ranking, on the other hand, only focuses the optimal phase. The main reason for these two forms of ranking is to aggregate more flexibility to the approach, as the possibility of employing distinct heuristics, evolutionary operators and, especially, other solution codings: incomplete solutions (schemata) not always may be naturally incorporated by the evolutionary process. A good example of such applications is the numerical optimization coded in real numbers [5].

3   Applications in Pattern Sequencing Problems

The problems treated in this section can be classified as pattern sequencing problems. Pattern sequencing problems may be stated by a matrix with integer elements where the objective is to find a permutation of rows or patterns (client orders, or gates in a VLSI circuit, or cutting patterns) minimizing some objective function [7]. Objective functions considered here differ from traveling salesman like problems because the evaluation of a permutation can not be computed by using values that only depend on adjacent patterns.

The PTA is modeled for two similar pattern sequencing problems found in the literature: Minimization of Open Stacks Problem (MOSP) and Gate Matrix Layout Problem (GMLP). Theoretical aspects are basically the same for both problems. The difference between them resides only in their enunciation. A more detailed description of the MOSP is emphasized in next sections. The particularities of the GMLP are occasionally presented, when needed.

3.1   Theoretical Issues of the MOSP

The MOSP appears in a variety of industrial sequencing settings, where distinct patterns need to be cut and each one may contain a combination of piece types. For example, consider an industry of woodcut where pieces of different sizes are cut of big foils. Pieces of equal sizes are heaped in a single stack that stays *opened* until the last piece of the same size is cut. A stack is considered *opened* while there exist pieces of the same size to be cut. A MOSP consists of determining a sequence of cut patterns that minimizes the maximum of open stacks (MOS) during the cutting process. Typically, this problem is due to limitations of physical space, so that the accumulation of stacks can cause the temporary need of removal of one or other stack, delaying the whole process.

The data for a MOSP are given by an $I \times J$ binary matrix P, representing patterns (rows) and pieces (columns), where $P_{ij} = 1$, if pattern i contains piece j, and $P_{ij} = 0$ otherwise. Patterns are processed sequentially, opening stacks whenever new piece types are processed and closing stacks of pieces that do not have any items else to be cut. The sequence of patterns being processed determines the number of stacks that stays open at same time. Another binary matrix, here called the open stack matrix Q, can be used to calculate the MOS for a certain pattern permutation. It is derived from the input matrix P, by the following rules:

- $Q_{ij} = 1$ if there exists x and $y | \pi(x) \leq i \leq \pi(y)$ and $P_{xj} = P_{yj} = 1$;
- $Q_{ij} = 0$, otherwise;

where $\pi(b)$ is the position of pattern b in the permutation.

The Q shows the consecutive-ones property [8] applied to P: in each column, "0" 's between "1" 's are replaced by "1" 's. The sum of "1" 's, by row, computes the number of open stacks when each pattern is processed. Figure 1 shows an example of matrix P, its corresponding matrix Q, and the number of open stacks to each pattern processed. When pattern 1 is cut, 3 stacks are opened. No stack else is opened for patterns 2 and 3, but pattern 4 requires 5 open stacks. At most, 7 stacks ($MOS = \max\{3, 3, 3, 5, 6, 7, 7, 5, 3\} = 7$) are needed to process the permutation $\pi_0 = \{1, 2, 3, 4, 5, 6, 7, 8, 9\}$.

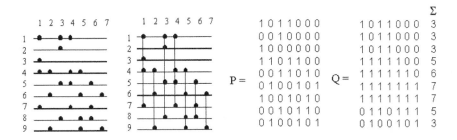

Fig. 1. MOSP (or GLMP) instance: original and corresponding matrix

Recently, several aspects of the MOSP and other related problems, as the GMLP, have been presented, including the NP-hardness of them [9], [10], [11]. The GMLP goal is to arrange a set of gates (horizontal wires), which are interconnected by nets (vertical wires), such that the number of tracks is minimized. This can be achieved by placing non-overlapping nets in the same track. The same example of Figure 1 can be seen as a GMLP instance. The number of open stack is equivalent to the number of overlapping nets.

3.2   PTA Modeling

A very simple representation is implemented for the MOSP and GMLP: a direct alphabet of symbols (natural numbers) represents the pattern (or gate) permutation. Each label is associated to a row of binary numbers, representing the piece type presence in each pattern. A permutation of rows is called structure and consists of a candidate solution for an instance.

A second objective for the MOSP have been used by early works: to close the stacks as soon as possible, allowing that the customer's requests be available with minimum delay [11], [3]. The second objective is to minimize the time that the stacks stay open (TOS) and it can be calculated by the sum of all "1" 's in Q. The TOS is particularly useful for increasing the fitness distinction among individuals. The function f reflects the total cost of a given permutation and considers the primary (MOS) and secondary (TOS) objectives:

$$f(s_k) = I \cdot J \cdot \max_{i \in I} \left\{ \sum_{j \in J} Q_{ij} \right\} + \sum_{i \in I} \sum_{j \in J} Q_{ij} \tag{7}$$

where the product $I \cdot J$ is a weight to reinforce MOS part of the cost.

A dynamic-sized population was implemented and controlled by an adaptive rejection threshold that eliminates the ill-adapted individuals, i.e., structures such that $\alpha \geq \delta(s_k)$. The adaptive rejection threshold, α, is initialized at the beginning of the process with the rank of the worst individual in population. During the evolutionary process, α is updated with adaptive increments, considering the current range of the rank values in the population, the population size, and the remaining number of generations. The adaptive increment of α is:

$$\alpha = \alpha + Step \cdot |P| \cdot \frac{(\delta_{bst} - \delta_{wst})}{RG} \tag{8}$$

where δ_{bst} and δ_{wst} are, respectively, the best and the worst rank of structures in current population, $|P|$ is the current population size, RG is the remaining number of generations, and $Step$ is an adjustment parameter, used to give more or less speed to the evolutionary process.

At the beginning, the population tends to grow up, generally, accepting all new individuals. After some generations, α determines the adaptation values that can be kept in population and the ill-adapted individuals are eliminated. Whenever no improvement is obtained, the population eventually can collapse, becoming empty. Therefore, the correct adjustment of $Step$ (generally, a value about 0.001 is used) is needed to avoid premature emptying of the population.

3.3   The Training Heuristics

The 2-Opt is a well-known improvement heuristic based on k-changes over a complete initial solution. Typically, a 2-change of a permutation consists of deleting 2 edges and replacing them by 2 other edges to form a new permutation. It can be obtained by breaking the permutation in 2 reference points and inverting the order in the middle subpermutation. For example, $\{1-2-3-4-5-6-7-8\}$, breaking in 3 and 6 becomes $\{1-2-3-6-5-4-7-8\}$.

The 2-Opt-like heuristic is employed in function g for training the population. Each one of the 2-changes generates a neighbor structure that is evaluated, looking for the best objective function value. At the end, up to $0.5 \cdot (I^2 - I)$ neighbor structures are evaluated .

Another heuristic used in this work, called the Faggioli-Bentivoglio's heuristic, is based on the constructive heuristic described in [12]. The basic idea of this heuristic is to build a complete solution, minimizing the differences among the patterns. An initial group of patterns (in this work, the first $N/2$ patterns), in a given structure, is accepted as start patterns. The neighborhood is defined as all structures that begin with the start group of patterns and minimize the difference to the subsequent patterns, according to a three stage criterion.

At the first stage, the patterns that open as few new stacks as possible are chosen. A stack is opened when the new sequenced pattern contains a piece type that is not yet stacked, i.e., the i_{th} item presents a $0 - 1$ transition, from previous pattern to next. At the second stage, the pattern that removes the greatest number of stack is chosen among the patterns previously selected. A stack is removed when the new sequenced pattern ends a piece type that is being stacked, i.e., the i_{th} item presents a $1 - 0$ transition, from previous pattern to next. At the last stage, the pattern that continues the production of the greatest number of stacked pieces is chosen among the patterns previously selected. The production continues when the new sequenced pattern contains a piece type that is already stacked, i.e., the i_{th} item presents a $1 - 1$ transition, from previous pattern to next. If these three rules lead to more than one pattern to be inserted in sequence, one of them is selected at random.

3.4 Evolutionary Operators

The structures in the population are kept in descending order, according to the ranking in equation 5. Thus, well-adapted individuals appear in first places on the population, being privileged for evolutionary operations.

Two structures are selected for recombination. The first is called the base (s_{base}) and it is randomly selected out from the first positions in the population (generally, about 20% from the population). The second structure is called the guide (s_{guide}) and is randomly selected out of the entire population. They are recombined by a variant of the Order Crossover (OX) called Block Order Crossover (BOX) [13]. The parent(A) and parent(B) are mixed into only one offspring, by copying blocks of both parents, at random. Pieces copied from a parent are not copied from other, keeping the offspring feasible (Figure 2a).

A local search mutation is applied to each new structure generated with a certain probability (generally, about 20%). This procedure is very important to get intensification moves around a solution. The local search mutation explores a search tree, considering several 2-Opt neighborhoods, not only one as fitness function g (Figure 2b).

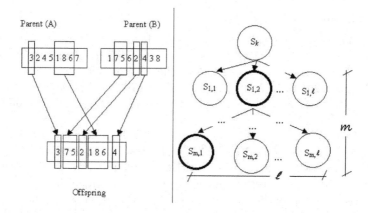

Fig. 2. (a) Block order crossover and (b) local search mutation tree

Some neighbors (generally, $l = 20$ neighbors) are evaluated in each tree level and the best one is held on to be used as starting point to next tree level. Successive neighborhoods are generated until a pre-defined maximum number of neighborhoods (generally, $m = 20$ neighborhoods).

The PTA pseudo-code, shown as follows, is based on traditional genetic algorithms. The stop criteria is either when the best-known solution is found, or after a certain number of objective function calls, to be set depending on the number of patterns (length) of the instance at hand. Once the stop criteria is reached the searching process has failed to find the best-known (optimal) solution.

```
{PTA pseudo-code}
RandomlyInitialize (P_α);
for all s_k ∈ P_α do
    Compute f(s_k), g(s_k), δ(s_k); {equations 7, 2, 5}
end for
α := δ_wst;
while not stopCriteria do
    while numberOfCrossovers do
        SelectionBaseGuide (s_base, s_guide);
        CrossoverBOX (s_base, s_guide) giving s_new;
        if mutationCondition then
            LocalSearchMutation (s_new);
        end if
        Compute f(s_new), g(s_new), δ(s_new);
        Update (s_new) in P_α;
    end while
    α := AdaptiveIncrement (α); {equation 8}
    for all s_k ∈ P_α e δ(s_k) < α  do
        Delete (s_k) from P_α;
    end for
end while
```

3.5   Computational Tests

A pool of 300 MOSP instances and one GMLP instance were chosen for tests, taken from [11], [12]. The MOSP instances have different number of patterns ($I \in \{10, 15, 20, 25, 30, 40\}$), each one of them with different number of piece types ($J \in \{10, 20, 30, 40, 50\}$). The GMLP instance, called $w4$, has 141 gates (patterns) and 202 nets (piece types) and it is the largest instance of pattern sequencing found in literature [11].

The PTA population size was 50 for the MOSP instances and 100 for GMLP instance. Two versions of PTA (PTA^{2opt} and PTA^{fag}), respectively using the 2-Opt-like and Faggioli-Bentivoglio's heuristics for training, were built to evaluate how different heuristics interfere in the algorithm performance.

The best two approaches presented in [12]: a) a tabu search method (TS) based on an optimized move selection process; and b) a generalized local search method (GLS) that employs the Faggioli-Bentivoglio's procedure in a simplified tabu search that only accepts improving moves [12]. Besides, another two methods are included in this comparison: c) the Constructive Genetic Algorithm (CGA^{2opt}); and d) the Collective method (COL). The CGA^{2opt} employs 2-Opt-like heuristic as training heuristic [3], [4]. The COL method explores distance measures among permutations and employs 2-exchange local search to drive the search of a simulated annealing like algorithm [11].

Table 1 shows the best MOS averages obtained by the methods in each instance group. Each pair (I, J) is an instance group with ten instances and different solutions. A comparison is made putting together the results shown in [11], [12] and the new results found by the PTH approaches in this work.

Both versions PTA^{fag} and PTA^{2opt} found the same results and were referred as PTA. The TOS is not considered in the other works and was excluded from the comparison. For this test, CGA^{2opt} and PTA were run 10 times. The PTA have found the best overall average of solutions for the instance groups. The CGA^{2opt} appears with the second best performance, failing in achieving the best average in 2 instance groups (20×40 and 25×40). The hardest instances to find the best-known solution were $p2040n6$, $p2540n3$.

Other performance aspects are focused in Table 2: average of the MOS found (AS), the number of times that the best solution was found (NS), the average of

Table 1. Performance comparison with another approaches per instance groups

I	J	COL	TS	GLS	CGA	PTA	I	J	COL	TS	GLS	CGA	PTA	I	J	COL	TS	GLS	CGA	PTA
10	10	5.5	5.5	5.5	5.5	5.5	20	10	7.5	7.7	7.5	7.5	7.5	30	10	7.8	7.8	7.8	7.8	7.8
	20	6.2	6.2	6.2	6.2	6.2		20	8.5	8.7	8.6	8.5	8.5		20	11.2	11.2	11.2	11.1	11.1
	30	6.1	6.1	6.2	6.1	6.1		30	9.0	9.2	8.9	8.8	8.8		30	12.2	12.6	12.2	12.2	12.2
	40	7.7	7.7	7.7	7.7	7.7		40	8.6	8.6	8.7	8.6	**8.5**		40	12.1	12.6	12.4	12.1	12.1
	50	8.2	8.2	8.2	8.2	8.2		50	7.9	8.0	8.2	7.9	7.9		50	11.2	12.0	11.8	11.2	11.2
15	10	6.6	6.6	6.6	6.6	6.6	25	10	8.0	8.0	8.0	8.0	8.0	40	10	8.4	8.4	8.4	8.4	8.4
	20	7.2	7.2	7.5	7.2	7.2		20	9.8	9.8	9.9	9.8	9.8		20	13.0	13.1	13.1	13.0	13.0
	30	7.3	7.4	7.6	7.3	7.3		30	10.6	10.7	10.6	10.5	10.5		30	14.5	14.7	14.6	14.5	14.5
	40	7.2	7.3	7.4	7.2	7.2		40	10.4	10.7	10.6	10.4	**10.3**		40	15.0	15.3	15.3	14.9	14.9
	50	7.4	7.6	7.6	7.4	7.4		50	10.0	10.1	10.2	10.0	10.0		50	14.6	15.3	14.9	14.6	14.6

objective function calls (FC). A parallel memetic algorithm (PMA) taken from
[14] was included in this comparison. The PMA presents a new 2-exchange local
search with a reduction scheme, which discards useless swaps, avoiding unnec-
essary objective function calls. Table 2 shows the comparison among PTA^{fag},
PTA^{2opt}, CGA^{2opt} and PMA in 10 trials for GMLP instance $w4$.

Table 2. Comparison between PTA and CGA^{2opt} and PMA for instance $w4$

	AS	NS	FC		AS	NS	FC
PTA^{2opt}	28.6	2	8,488,438	CGA^{2opt}	28.0	3	6,537,706
PTA^{fag}	28.3	2	9,330,802	PMA	29.4	2	9,428,591

Observing Table 2, CGA^{2opt} has obtained the best AS and NS for $w4$.
PTA^{fag} and PTA^{2opt} are slightly similar in AS, but the later seems to perform
less function calls. All approaches based on population training were better than
PMA. Despite the Faggioli-Bentivoglio's procedure seemingly should perform
less function calls than 2-Opt-like heuristic, this can not be observed in FC.
Indeed, it was expected a superior FC for versions employing 2-Opt. This fact
can be explained perhaps by the mutation procedure: the mutation would dominate the number of function calls and the training heuristic was not relevant for
FC. Another possibility is that 2-Opt-like training heuristic would improve the
algorithm performance so that it could compensate its computational cost.

The comparison among the methods here presented are based only in the average performance because the other works found in literature does not mention
nothing about the variability of their models. Table 3 shows the average and
standard desviation in 20 trials for the hard MOSP instances $p2040n6$, $p2540n3$.

Table 3. Average and standard desviation for MOSP instances $p2040n6$ and $p2540n3$

Instances (solution)		PTA^{2opt}	PTA^{fag}	CGA^{2opt}	CGA^{fag}
p2040n6	(8, 0)	$8,7 \pm 0,5$	$8,7 \pm 0,5$	$8,9 \pm 0,3$	$8,9 \pm 0,3$
p2540n3	(10, 0)	$10,7 \pm 0,5$	$10,8 \pm 0,4$	$10,7 \pm 0,5$	$10,9 \pm 0,3$

Statistical tests showed that the differences in averages are not statisticaly
significant for MOSP instances $p2040n6$ and $p2540n3$. For GMLP instance $w4$,
averages obtained by CGA^{2opt} was significantly better than those obtained by
PTA^{fag} and PTA^{2opt}.

4   Conclusion

In the Population Training Heuristic (PTH), proposed in this paper, the evolutionary process is driven by a training heuristic, employed in the fitness definition. The population is led to settle down in search areas where the individuals
can not be improved by such heuristic. In this work, the general guidelines for

PTH are introduced and new versions employing a non-constructive ranking are presented.

The algorithms based on PTH showed the best performance when compared against other approaches found in literature. Both constructive and non-constructive approaches were able to reach the known optimal solutions. The 2-Opt-like training heuristic has presented better results concerning the computational cost. For further work, it is intended to implement a multi-heuristic version with subpopulations trained by different heuristics, evolving in parallel, for multi-objective problems.

References

1. Lorena, L.A.N., Furtado, J.C.: Constructive genetic algorithm for clustering problems. *Evolutionary Computation.* (2001) 9(3): 309-327.
2. Ribeiro Filho, G., Lorena, L.A.N.: A constructive evolutionary approach to school timetabling, In: Applications of Evolutionary Computing, Boers, E.J.W., Gottlieb, J., Lanzi, P.L., Smith, R.E., Cagnoni, S., Hart, E., Raidl, G.R., Tijink, H., (Eds.) - *Springer LNCS 2037*(2001). 130-139.
3. Oliveira, A.C.M., Lorena, L.A.N.: A constructive genetic algorithm for gate matrix layout problems. *IEEE Trans. on Computer-Aided Designed of Integrated Circuits and Systems*(2002) 21(8): 969-974.
4. Oliveira, A.C.M., Lorena, L.A.N.: 2-Opt population training for minimization of open stack problem. Advances in Artificial Intelligence - XVI Brazilian Symposium on Artificial Intelligence. Guilherme Bittencourt e Geber L. Ramalho (Eds). Springer LNAI 2507. (2002) 313-323.
5. Oliveira, A.C.M., Lorena, L.A.N.: Real-coded evolutionary approaches to unconstrained numerical optimization. Advances in Logic, Artificial Intelligence and Robotics. Jair Minoro Abe and João I. da Silva Filho (Eds). (2002) 10-15.
6. Reeves, C.R. Landscapes, operators and heuristic search. Annals of Operations Research, v. 86, p. 473490, 1999.
7. Fink, A., Voss, S.: Applications of modern heuristic search methods to pattern sequencing problems, Computers and Operations Research, (1999) 26(1): 17-34.
8. Golumbic, M.: Algorithmic graph theory and perfect graphs. *Academic Press*, New York (1980).
9. Möhring, R.: Graph problems related to gate matrix layout and PLA folding, *Computing* (1990) 7: 17-51.
10. Kashiwabara, T., Fujisawa, T.: NP-Completeness of the problem of finding a minimum clique number interval graph containing a given graph as a subgraph, *In Proc. Symposium of Circuits and Systems*(1979).
11. Linhares, A.: Industrial pattern sequencing problems: some complexity results and new local search models. Doctoral Thesis, INPE, S. José dos Campos, Brazil (2002).
12. Faggioli, E., Bentivoglio, C.A.: Heuristic and exact methods for the cutting sequencing problem, *European Journal of Operational Research*(1998) 110: 564-575.
13. Syswerda, G.: Schedule optimization using genetic algorithms. *Handbook of Genetic Algorithms*,Van Nostrand Reinhold, New York (1991) 332-349.
14. Mendes, A., Linhares, A.: A multiple population evolutionary approach to gate matrix layout, Int. Journal of Systems Science, Taylor & Francis Eds, (2004), 35(1): 13-23.

Scatter Search Particle Filter to Solve the Dynamic Travelling Salesman Problem

Juan José Pantrigo, Abraham Duarte, Ángel Sánchez, and Raúl Cabido

Universidad Rey Juan Carlos, c/ Tulipán s/n
28933 Móstoles, Spain
{j.j.pantrigo, a.duarte, an.sanchez}@escet.urjc.es

Abstract. This paper presents the Scatter Search Particle Filter (SSPF) algorithm and its application to the Dynamic Travelling Salesman Problem (DTSP). SSPF combines sequential estimation and combinatorial optimization methods to improve the execution time in dynamic optimization problems. It allows obtaining new high quality solutions in subsequent iterations using solutions found in previous time steps. The hybrid SSPF approach increases the performance of general Scatter Search (SS) metaheuristic in dynamic optimization problems. We have applied the SSPF algorithm to different DTSP instances. Experimental results have shown that SSPF performance is significantly better than classical DTSP approaches, where new solutions of derived problems are obtained without taking advantage of previous solutions corresponding to similar problems. Our proposal reduces execution time appreciably without affecting the quality of the estimated solution.

1 Introduction

Dynamic optimization problems are characterized by an initial problem definition and a collection of "events" over the time. An event defines some changes on the data of the problem [1]. Therefore the optimization method needs from adaptive strategies for these changing conditions. In these problems, a key question is how to use information found in previous time steps to obtain high quality solutions in subsequent ones, without restarting the computation from scratch.

Dynamic optimization problems play an important role in industrial applications, such as transportation, telecommunications and manufacturing [1]. Surprisingly, compared to the amount of research undertaken on static optimization problems, relatively little work has been devoted to dynamic problems [2][3].

Unlike static problems, dynamic ones often lack well defined optimization functions, standard benchmarks or criteria for comparing solutions [1][4][2][3]. Nowadays, main used strategies have been specific heuristics [1] and manual procedures [5][2]. Traditionally, metaheuristics using constructive methods such as Ant Systems [6][7] and population-based metaheuristics such as evolutionary algorithms [8] have been applied to dynamic problems.

G.R. Raidl and J. Gottlieb (Eds.): EvoCOP 2005, LNCS 3448, pp. 177–189, 2005.
© Springer-Verlag Berlin Heidelberg 2005

The filtering problem deals with updating the present state of knowledge and predicting with drawing inferences about the future state of the system [9]. Sequential Monte Carlo algorithms (also called particle filters) are a special class of filters in which theoretical distributions on the state space are approximated by simulated random measures (also called particles) [9].

We propose a new approach, called Scatter Search Particle Filter (SSPF) to solve the Dynamic Travelling Salesman Problem. SSPF combines sequential estimation (Particle Filter) [9][10] and metaheuristic methods (Scatter Search) [11] in two different stages. In the Particle Filter (PF) stage, a particle set is propagated and updated to obtain a new particle set. In the Scatter Search (SS) stage, some solutions from the particle set are selected and combined to obtain new optimized solutions.

This paper considers a dynamic variant of the Travelling Salesman Problem [12] in which "distances" among cities vary over the time. In dynamic problems, meta-heuristics based approaches tend to restart the search method after events or to use the best solutions found in previous time steps. In the first approach, the derived problem is processed as unrelated with respect to its origin problem. It implies that useful information can be rejected, increasing the computation time. As a consequence, a reasonable trade-off between the analysis of the prior problem solution and actual problem computational effort must be found. Experimental results have shown that the proposed algorithm has successfully been applied to different DTSP instances.

2   The Dynamic Travelling Salesman Problem

The Travelling Salesman Problem (TSP) consists of finding the shortest tour connecting a fixed number of locations (cities), visiting each city exactly once [12]. This problem can be represented by a graph $G = \{V, E, W\}$, where V is a set of vertex representing the cities, E is a set of edges which model the paths connecting cities and W is a symmetric matrix of weights. We suppose that there is an edge jointing every pairs of cities. Weighs $w_{ij} \in W$, attached to edges $(i,j) \in E$ represent the distance between cities $i, j \in V$. The TSP can be described as the problem of finding a Hamiltonian circuit with minimum length in the graph G [8]. The TSP belongs to the NP-hard class [13]. It is one of the most considered problems in Combinatorial Optimization and several approaches have been used to solve it. In the public library TSPLIB [14] sample instances for the static TSP can be found.

The Dynamic Travelling Salesman Problem (DTSP) is a generalization on TSP where G is time-dependent. This problem has got several practical applications such as traffic jams [12] or fluctuating set of active machines [15]. Two different varieties of Dynamic Travelling Salesman Problem exist in the literature. The first one consists of inserting or deleting cities into a given problem instance [16][17]. A different approach [12] consists of keeping constant the number of cities, allowing distance changing among them. It can be applied to describe traffic jams and motorways. In this context, good solutions may not be optimal after changes and the salesman needs to be re-routed. In this paper, we focus on the second approach.

Ant Systems (AS) and Evolutionary Computation (EC) have been the most common applied metaheuristics to solve the DTSP. These techniques should be able to adjust their solutions under changing environments. In [12][15][17] different AS implementations applied to DTSP can be found. The main reason for using AS to dynamic problems is based on the pheromone concept. Pheromone can be exploited as reinforcement, and therefore, as a way to transferring knowledge. When a change is detected a partial decomposition-reconstruction procedure is performed over old solutions [1]. This procedure determines which elements of ant's solutions must be discarded to satisfy the feasibility of the new conditions.

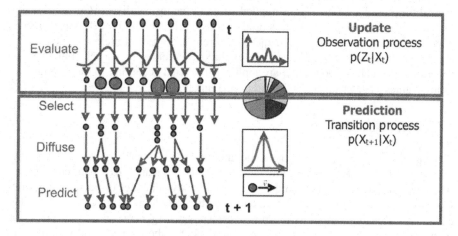

Fig. 1. Particle Filter scheme

3 Particle Filter Framework

Sequential Monte Carlo algorithms (also called Particle Filters) are filters in which theoretical distributions on the state space are approximated by simulated random measures (called particles) [9]. The state-space model consists of two processes: (i) an observation process $p(Z_t|X_t)$, where X denotes the system state vector and Z is the observation vector, and (ii) a transition process $p(X_t|X_{t-1})$. Assuming that observations $\{Z_0, Z_1,..., Z_t\}$ are known, the goal is to recursively estimate the posterior pdf $p(X_t|Z_t)$ and the new system state $\{\chi_0, \chi_1, ... , \chi_t\}$ at each time step. In Sequential Bayesian Modelling framework, posterior pdf is estimated in two stages:

(i) Evaluation: posterior pdf $p(X_t|Z_t)$ is computed at each time step by applying Bayes theorem, using the observation vector Z_t:

$$p(X_t \mid Z_t) = \frac{p(Z_t \mid X_t)p(X_t \mid Z_{t-1})}{p(Z_t)} \qquad (1)$$

(ii) Prediction: the posterior pdf $p(X_t|Z_{t-1})$ is propagated at time step t using the Chapman-Kolmogorov equation:

$$p(X_t \mid Z_{t-1}) = \int p(X_t \mid X_{t-1})p(X_{t-1} \mid Z_{t-1})dX_{t-1} \qquad (2)$$

A predefined system model is used to obtain an updated particle set.

In Figure 1 an outline of the Particle Filter scheme is shown. The aim of the PF algorithm is the recursive estimation of the posterior pdf $p(X_t|Z_t)$, that constitutes the complete solution to the sequential estimation problem. This pdf is represented by a set of weighted particles $\{(x_t^0, \pi_t^0) \dots (x_t^N, \pi_t^N)\}$.

PF starts by setting up an initial population X_0 of N particles using a known pdf. The measurement vector Z_t at time step t, is obtained from the system. Particle weights Π_t are computed using a weighting function. Weights are normalized and a new particle set X_t^* is selected. As particles with larger weight values can be chosen several times, a diffusion stage is applied to avoid the loss of diversity in X_t^*. Finally, particle set at time step $t+1$, X_{t+1}, is predicted using the motion model. A pseudocode of a general PF is detailed in [10][18]. Therefore, Particle Filters can be seen as algorithms handling the particles time evolution. Particles in Particle Filters move according to the state model and are multiplied or died according to their weights or fitness values as determined by the likelihood function [9].

4 Scatter Search

Scatter Search (SS) [19][20] is a population-based metaheuristic that provides unifying principles for recombining solutions based on generalized path construction in Euclidean spaces. In other words, SS systematically generates disperse set of points (solutions) from a chosen set of reference points throughout weighted combinations. This concept is introduced as the main mechanism to generate new trial points on lines jointing reference points. SS metaheuristic has been successfully applied to several hard combinatorial problems. A recent method review can be found in [20].

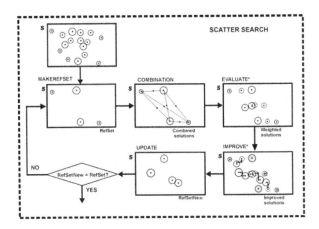

Fig. 2. Scatter Search scheme

In Figure 2 an outline of the SS is shown. SS procedure starts by choosing a subset of solutions (called *RefSet*) from a set S of initial feasible ones. The solutions in *RefSet* are the h best solutions and the r most diverse ones of S. Then, new solutions are generated by making combinations of subsets (pairs typically) from *RefSet*. The resulting solutions, called trial solutions, can be infeasible. In that case, repairing methods are used to transform these solutions into feasible ones. In order to improve the solution fitness, a local search from trial solutions is performed. SS ends when the new generated solutions do not improve the quality of the *RefSet*.

5 Scatter Search Particle Filter

In our opinion, dynamic optimization problems deal with optimization techniques, but also with prediction tasks. This is due to the fact that the optimization method for changing conditions needs from adaptive strategies. Therefore, one key aspect is how to efficiently use important information found in previous time steps in order to find high quality solutions for new derived problems instances.

Usually in metaheuristics, two approaches can be used depending on the problem change rate. If it is high, each problem is tackled as a different one, so the computation is restarted from scratch. If change rate is low, the last solution or a set of the best solutions found are used as starting point in the new search. For instance, Genetic Algorithms use the previous population as initial set in the next time step. On the other hand, Ant Colony Optimization uses the previous pheromone deposition in each node as initial pheromone distribution of subsequent steps. The same idea can be extended to other metaheuristics. In Scatter Search, the *RefSet* obtained in the previous time step can be used as a new *RefSet* for the next one or *RefSet* could be improved with diverse solutions. Nevertheless, it is very important to make a decision of which information is propagated to the next time step. This is because it is possible that the search algorithm get stuck near local optimum. As a consequence, a reasonable trade-off between both restart from scratch and restart from previous optimum must be found. Therefore, it could not be appropriate to use optimization procedures in the prediction stage.

Analogously, sequential estimation algorithms like particle filters are well-suited in prediction stages, but they are not good enough for solving dynamic optimization problems. Optimization strategies performed with this kind of algorithms are usually very computationally inefficient.

Then, dynamic optimization problems need from both optimization and prediction tasks. The key question is how to hybridize these two kinds of algorithms to obtain a new one which combines both techniques. In this way, a novel hybrid algorithm called Scatter Search Particle Filter (SSPF) is presented to solve the Dynamic TSP.

5.1 Scatter Search and Particle Filter Hybridization

The Scatter Search Particle Filter (SSPF) algorithm is introduced in this paper to be applied to dynamic optimization problems. SSPF integrates both Scatter Search (SS) and Particle Filter (PF) frameworks in two different stages:

- In the Particle Filter stage, a particle set is propagated and updated to obtain a new one. This stage is focused on the evolution in time of the best solutions found in previous time steps. The aim for using PF is to avoid the loss of needed diversity in the solution set.
- In the Scatter Search stage, a fixed number of solutions from the particle set are selected and combined to obtain better ones. This stage is devoted to improve the quality of a reference subset of good solutions in such a way that the final solution is also improved.

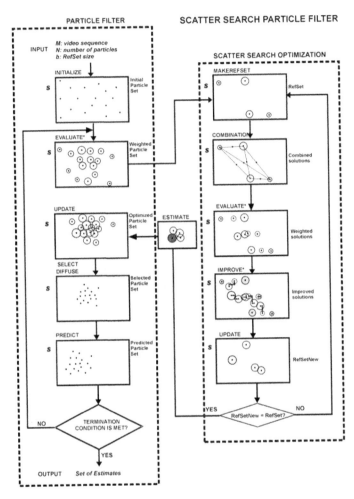

Fig. 3. Scatter Search Particle Filter scheme. Weight computation is required during EVALUATE and IMPROVE stages (*)

Figure 3 shows a graphical template of the SSPF algorithm. Dashed lines separate the two main components in the SSPF scheme: PF and SS optimization, respectively. SPF starts with an initial population of N particles drawn from a known pdf (Figure 3: INITIALIZE). Each particle represents a possible solution of the problem. Particle weights are computed using a weighting function (Figure 3: EVALUATE). SS stage is later applied to improve the best obtained solutions of the particle filter stage. A Reference Set (*RefSet*) is created selecting a subset of b ($b<<N$) particles from the particle set (Figure 3: MAKEREFSET). This subset is composed by the $b/2$ best solutions and the $b/2$ most diverse ones of the particle set. New solutions are generated and evaluated, by combining all possible pairs of particles in the *RefSet* (Figure 3: COMBINE and EVALUATE). In order to improve the solution fitness, a local search from each new solution is performed (Figure 3: IMPROVE). Worst

solutions in the *RefSet* are replaced when there are better ones (Figure 3: UPDATEREFSET). SS stage ends when new generated solutions *RefSetNew* do not improve the quality of the *RefSet*. Once the SS stage is finished, the "worst" particles in the particle set are replaced with the *RefSetNew* solutions (Figure 3: INCLUDE). Then, a new population of particles is created by selecting the individuals from particle set with probabilities according to their weights (Figure 3: SELECT and DIFFUSE). Finally, particles are projected into the next time step by following the update rule (Figure 3: PREDICT).

5.2 Scatter Search Particle Filter Main Features

The SSPF leads the search process to a region of the search space in which it is highly probable to find new better solutions than the initial computed ones. PF increases the performance of general SS in dynamic optimization problems by improving the quality of the diverse initial solution set S. In order to obtain the solution set $S(t+1)$ PF performs two tasks over the set $S(t)$: (i) selecting the best solutions and (ii) predicting new solutions from the best ones. Firstly, the selection procedure selects particles with larger weight values more likely than those with lower weights. Secondly, PF performs a prediction procedure over these best solutions to obtain the set $S(t+1)$. In this way, PF tackles with problem changes in time by predicting the best solution time evolution. As results, solutions in $S(t+1)$ will be closer to global optimum than another ones obtained randomly. On the other hand, a diffusion procedure is applied to the selected solutions to include diversity in the set $S(t+1)$.

Therefore, SSPF adapts computational load to problem constraints, by reducing the number of required evaluations of the particle weighting function. In this way, solutions in *RefSet* will be selected from a better solutions set. This is the main reason why SSPF reduces the required number of evaluations for the fitness function, and hence the computational load.

SS and PF are related in such a way that when the SS improves its performance, the PF performance also improves and vice versa. PF allows parameter tuning in order to adjust the quality and the diversity of the set S, used by SS. On the other hand SS improves the quality of the particle set allowing the better estimation of the pdf, by including *RefSet* solutions in the set S. This fact yields to a highly configurable algorithm. The main considered SSPF algorithm parameters are:

- The *size of the particle set N* is the number of particles in the particle set. These should be enough particles to support a set of diverse solutions, avoiding the loss of diversity in the particle set. Therefore, N influences on the performance of the SS stage. The value of N depends on the problem instance complexity.
- The *size of the reference set b* is the number of solutions in the *RefSet*. A typical b used in the literature is $b = 10$ [20][22].
- The diffusion stage is applied to avoid the loss of diversity in S. It is performed by applying a random displacement with *maximum amplitude A*. This amplitude is a measure of the diversity produced in the new particle set. Therefore, A influences the performance of the SS by tuning the diversity of the initial solution set, and hence, the diversity of the *RefSet*.

The SSPF is presented to be applied to hard dynamic optimization problems. In this kind of problems, it is usual to perform some preliminary experimentation in order to achieve the appropriate parameter values.

6   SSPF Implementation to DTSP Solving

In the Scatter Search Particle Filter (SSPF) implementation, solutions (particles) are represented as paths over cities. The number of particles N in the particle set S is chosen according to the problem size. Concretely, N varies from 100 in the 25-cities problem to 1000 in the 100-cities problem. The *RefSet* is created by selecting the 5 best solutions and the 5 most diverse ones in S.

DTSP is considered as an R-permutation problem [20], because relative positioning of the elements is more important than absolute positioning. Therefore, to find the most diverse solutions, the distance metric for *R-permutation problems* was used [20]. The distance between two solutions p and q for R-permutation problems is defined as:

$$d(p,q) = \text{number of times } p_{i+1} \text{ doesn't immediately follow } p_i \text{ in } q \text{ for } i=1,\dots,n\text{-}1 \quad (3)$$

Voting method [20] has been used as combination procedure over all pairs of solutions in the *RefSet*. In this procedure, each reference solution votes for its first sector not included in the combined solution. The voting determines the element to be assigned to the next free position in the combined solution.

The *2-opt method* [21] was employed as improvement stage in the SS scheme (Figure 3). Given a solution, consider all pairs of edges connecting four different cities are considered. Removing two edges from the solution tour, there is a unique way of reconnecting the two remaining paths such that a new tour is obtained. If the new tour is shorter, then it replaces the old tour and the procedure is repeated until no such improvement is produced.

7   Experimental Results

To analyze the performance of the proposed algorithm, implementations of three evolutionary methods have been developed: Scatter Search, Evolutionary Algorithm and Scatter Search Particle Filter. The experiments were evaluated in an Intel Pentium 4 at 1.7 GHz and 256 MB RAM. All algorithms were coded in MATLAB 6.1, without optimization and by the same programmer. Different methods were applied to several instances of DTSP and results were compared. The following sections are devoted to describe used data, algorithms and obtained results.

7.1   Problem Instances

Unfortunately, as far as the authors know, there are no benchmarks for the DTSP. Thus, we generated synthetic and standard-based data. Synthetic data are composed by four different graph sequences, using 25, 50, 75 and 100 cities. Each sequence is composed by 10 different graphs. To know the value of the optimum, cities are

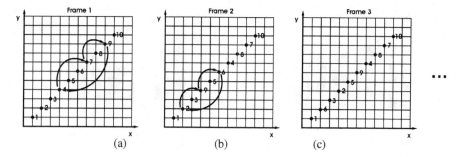

Fig. 4. Graph sequence generation process

located in the Euclidean plane along the diagonal as shown in Figure 4. In the first frame, cities are located in lexicographic order. Subsequent frames are generated by performing exchanges of cities in groups of three (see Figure 4). The average probability of node exchange in the considered graph sequences was $p_{change} = 0.15$.

Standard-based data are built as dynamic version of benchmarks from the public-domain library TSPLIB [14]. In particular, they are dynamic modifications of BAYG29, BERLIN51 and ST70 instances. We built each sequence starting from the original graph. Subsequent 4 graphs are obtained from the previous one introducing a perturbation in the actual location of each city according to a Gaussian distribution.

7.2 Algorithms Description

Different versions of Scatter Search and Evolutionary Algorithms were implemented. All of them use as stopping criteria "10^6 evaluations of fitness function or none improvement in the population". Solutions are coded as tours connecting all cities.

The two Scatter Search implementations were called SS1 and SS2. SS1 considers that each graph in the sequence is totally decoupled from the previous ones. Therefore, computation is restarted from scratch after graph changes. Basically, SS1 is based on Campos implementation described in [20][22].

In the second implementation (SS2) the graphs are supposed quite related. Thus, the *RefSet* obtained in the actual time step is used as a new *RefSet* for the next one. SS parameters *PopSize* and *b* for both implementations SS1 and SS2 were set to 100 and 10 respectively, as recommended in [22]. To obtain comparable results, we use the same *RefSet* composition, combination and improvement methods as in SSPF implementation (5-best and 5-diverse solutions in *RefSet*, voting method and 2-opt).

The implementation of the Evolutionary Algorithm (EA) performs the main stages of a standard genetic algorithm, including an improvement stage. The algorithm use voting method and 2-opt. EA parameters were set to *PopSize* = 100, crossover probability $p_c = 0.25$ and mutation probability $p_m = 0.01$ as recommended in [23]. Finally, improvement probability was set to $p_c = 0.25$.

7.3 Computational Testing

Experimental results are organized in three sections. In the first one the approach goodness in synthetic data is justified. Next section is devoted to the comparison between the SSPF and the two SS and the EA implementations.

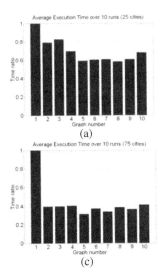

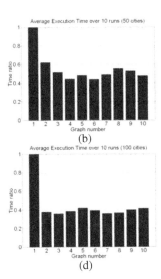

Fig. 5. Fitness function evaluations per graph in (a) 25, (b) 50, (c) 75 and (d) 100-cities problems

7.3.1 SSPF in Synthetic Data

Experimental results obtained by applying SSPF to synthetic data are presented in this section. In table 1, mean value of the execution time for the first graph is compared to the mean value of the execution time for the rest of the graph sequence.

The proposed strategy, based on Particle Filter and Scatter Search hybridization seems to be more advantageous than the classical SS one, in which an execution from scratch is performed. In this table, the column *Ratio* represents the average time SSPF improvement with respect to the corresponding time of the SS1 solution. As it can be seen, ratio between execution times is always in favor of SSPF algorithm.

Figure 5 shows the average fitness function evaluation per frame, over 10 runs of the same graph sequence. Each one is composed by 10 similar graphs. In this figure, relative execution time is represented for each frame. Elapsed time for the 2nd to 10th graphs (SSPF improvement) significantly reduces elapsed time for the first graph (SS approach) in all instances. Results show that SSPF achieves the best solution in all instances. Moreover, it is faster than SS1 implementation without loss of quality.

Table 1. Average fitness function evaluations values over 10 runs for each graph sequence

Nº of cities	Size of RefSet	Average Time SS	Average Time SSPF	Ratio
25	100	0.3×10^6	0.2×10^6	0.69
50	100	2.1×10^6	1.1×10^6	0.55
75	500	6.8×10^6	3.0×10^6	0.44
100	1000	14.7×10^6	6.6×10^6	0.44

7.3.2 SSPF Vs SS and EA in Standard-Based Data

This section presents a comparison between SSPF and two different implementations of SS. Results obtained by these algorithms (SS1, SS2, EA and SSPF) over all standard-based data (BAYG29, BERLIN52 and ST70) are presented in figure 6. Because of initial conditions and initial procedures performed are the same in SS1, SS2 and SSPF, solutions found in the first graph is exactly the same one. As the EA approach is different to the other ones, the solution and the time required to found this solution in the first graph are also unlike.

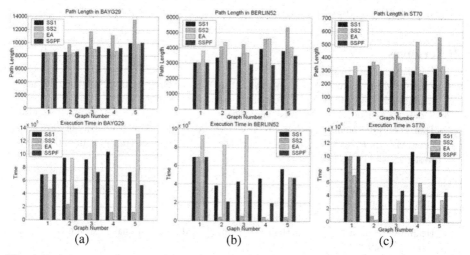

Fig. 6. Path length (upper row) and fitness function evaluations (lower row) using SS1, SS2, EA1 and SSPF in (a) BAYG29, (b) BERLIN52 and (c) ST70

Quality of the estimation performed by SS1 and SSPF are similar in subsequent graphs. However, execution time is significantly lower in SSPF approach, as explained in previous sections. In the SS2 implementation, the search procedure is trapped in a local optimum (maybe in the neighbourhood of the previous optimum). This yields SS2 achieves the lowest execution time, but with very poor quality. Finally, EA finds good quality solutions, but the time required to obtain them is larger than using SSPF. Table 2 resumes the main results obtained using different approaches. Execution time and path length demonstrate the performance of SSPF.

Table 2. Average execution time and path lengths of SS1, SS2, EA and SSPF over all instances

	SS1		SS2		EA		SSPF	
Cities	Length	Time	Length	Time	Length	Time	Length	Time
BAYG29	0.86×10^6	0.91×10^4	0.25×10^6	1.09×10^4	1.03×10^6	0.89×10^4	0.58×10^6	1.09×10^4
BERLIN52	5.07×10^6	3.51×10^3	1.75×10^6	$4..27 \times 10^3$	6.37×10^6	4.12×10^3	3.79×10^6	3.11×10^6
ST70	9.65×10^6	302.97	2.84×10^6	427.58	3.97×10^6	331.15	5.72×10^6	272.15

8   Conclusions

The main contribution of this work is the development of the Scatter Search Particle Filter (SSPF) algorithm. SSPF hybridizes the Scatter Search metaheuristic and the Particle Filter framework to solve dynamic problems. We have successfully applied the proposed SSPF algorithm to the Dynamic Travelling Salesman Problem (DTSP). Experimental results have shown that SSPF appreciably increases the performance of derived Scatter Search and Evolutionary Algorithm methods in a challenging dynamic optimization problem (DTSP), without losing quality in the estimation procedure. This improvement is more significant as the size of the problem increases.

References

1. Randall, M.: Constructive Meta-heuristics for Dynamic Optimization Problems. Technical Report. School of Information Technology. Bond University (2002)
2. Sadeh, N., Kott, A.: Models and Techniques for Dynamic Demand-Responsive Transportation Planning. Tech. Rept. TR-96-09, Carnegie Mellon University (1996)
3. Dror, M., and Powell, W.: Stochastic and Dynamic Models in Transportation. Operations Research, 41 (1993) 11-14
4. Beasley, J., Krishnamoorthy, M., Sharaiha, Y. and Abramson, D.: The displacement Problem and Dynamically Scheduling Aircraft Landings. Working paper, Available online at http://graph.ms.ic.ac.uk/jeb/displace.pdf (2002)
5. Beasley, J., Sonander, J. and Havelock, P.: Scheduling Aircraft Landings at London Heathrow using a Population Heuristic, Journal of the Operational Research Society, 52 (2001) 483-493
6. Glover, F., Kochenberger, G. A.: Handbook of metaheuristics. Kluwer (2002)
7. Dorigo, M., Gambardella, L.M.: Ant colony system: A cooperative learning approach to the traveling salesman problem. IEEE Trans. on Evolutionary Computation, 1(1) (1997) 53–66
8. Zhang-Can H.; Xiao-Lin H.; Si-Duo C.: Dynamic traveling salesman problem based on evolutionary computation. Proc. of Evolutionary Computation Conf, 2 (2001) 1283 - 1288
9. Carpenter, J., Clifford, P., Fearnhead, P.: Building robust simulation based filters for evolving data sets. Tech. Rep., Dept. Statist., Univ. Oxford, Oxford, U.K. (1999)
10. Arulampalam, M., et al.: A Tutorial on Particle Filter for Online Nonlinear/Non-Gaussian Bayesian Tracking. IEEE Trans. On Signal Processing, 50 (2) (2002) 174–188
11. Blum, C., Roli, A.: Metaheuristics in Combinatorial Optimization: Overview and Conceptual Comparison. ACM Computing Surveys, 35 (3) (2003) 268 - 308
12. Eyckelhof, C.J., Snoek, M.: Ant Systems for A Dynamic DSP: Ants Caught in a Traffic Jam. Proc. of ANTS02 Conference (2002)
13. Karp, R.M.: Reducibility among Combinatorial Problems. R. Miller and J. Thatcher (eds.): Complexity of Computer Computations. Plenum Press (1972) 85-103
14. Reinelt, G.: TSPLIB. University of Heidelberg. Available online at http://www.iwr.uni-heidelberg.de/groups/comopt/software/TSPLIB95/ (1996)
15. Guntsh, M., Middendorf, M.: Applying Population based ACO to Dynamic Optimization Problems. In Ant Algorithms, Proceedings of Third International Workshop ANTS 2002, LNCS 2463 (2002) 111-122

16. Guntsh, M., Middendorf, M., Schmeck, H.: An Ant Colony Optimization Approach to Dynamic TSP. In Proc. GECCO-2001 Conference, San Francisco, CA: Morgan Kaufmann Publishers (2000) 860-867

17. Guntsh, M., Middendorf, M.: Pheromone Modification Strategies for Ant Algorithms applied to Dynamic TSP. Lecture Notes in Computer Science, 2037 (2001) 213-222

18. Pantrigo, J.J., Sánchez, A., Gianikellis, K., Duarte, A.: Path Relinking Particle Filter for Human Body Pose Estimation. Lecture Notes in Computer Science 3138 (2004) 653-661

19. Glover, F.: A Template for Scatter Search and Path Relinking. LNCS 1363 (1997) 1-53

20. Laguna, M., Marti, R.: Scatter Search methodology and implementations in C. Kluwer Academic Publisher (2003)

21. Vizeacoumar, F.: TSP Implementation. Project report Combinatorial Optimization CMPUT – 670

22. Campos, V., Laguna, M. , Martí, R.: Scatter Search for the Linear Ordering Problem. *New Ideas in Optimization*. McGraw-Hill (1999)

23. Michalewitz, Z.: Genetic Algorithms + Data Structures = Evolution Programs Springer-Verlag, 1996.

The Use of Meta-heuristics to Solve Economic Lot Scheduling Problem

Syed Asif Raza and Ali Akgunduz

Concordia University, 1455 de Maisonneuve Blvd. W.
Montreal, Quebec H3G 1M8 Canada
{sraza, akgunduz}@me.concordia.ca

Abstract. Economic lot scheduling problem has been an important topic in production planning and scheduling research for more than four decades. The problem is known to be NP-hard due to it's combinatorial nature. In this paper, two meta-heuristics algorithms - Tabu Search and Simulated Annealing - are proposed. To investigate the effect of control parameters to the performance of tabu search and simulated annealing algorithms, a general factorial design of experiment study is used. Two Neighborhood Search heuristics that differ in rounding off scheme of the production frequencies are also tested. Experimental study shows that both tabu search and simulated annealing algorithms outperform two best known solution methods - Dobson's Heuristic and Hybrid Genetic Algorithm.

1 Introduction

Economic Lot Size Problem (ELSP) deals with the production assignment of several products sharing a common production facility in order to minimize the total cost. It is a constraint optimization problem where production scheduling is done in such a way that all products are manufactured and their demands are satisfied during the planning period. There have been many articles published in last forty years covering a wide range of possible solutions to ELSP. The problem has many applications in production planning and scheduling (see Moon et al. [1]). The earliest contribution to ELSP is due to Elion [2], Rogers [3], Maxwell [4], Hanssmann [5] and Bomberger [6]. A Lower Bound (LB) on cost was developed by Bomberger [6]. The LB is tight because it incorporates machine sharing constraint in calculation, whereas the previous lower bound was an Independent Solution (IS) obtained by ignoring all constraints. Elmaghraby [7] presents a comprehensive review of the research up to late 1970s. In previous studies, three different approaches are used to solve ELSP : *common cycle*, *basic period* and *time varying lot size*. In all approaches it is mostly desired to generate cyclic schedules and Zero Switch Rule (ZSR) is also considered in most studies. ZSR enforces the condition that a product will only be produced if it's inventory reaches to zero. ZSR does not guarantee optimality as it is shown in some cases (see Maxwell [4] and Delporte and Thomas [8]). A common period

G.R. Raidl and J. Gottlieb (Eds.): EvoCOP 2005, LNCS 3448, pp. 190–201, 2005.

approach is the simplest to implement in which all products are manufactured for the same period. The basic period approach allows different cycle times for different products; however the cycle time of each product has to be an integer multiple of the basic period. Research shows that both approaches, common cycle and basic period, have limitations. The common cycle approach may produce solutions that are far from LB (see Moon et al. [1]). The basic period approach tends to find better solutions than the common period approach; however finding a feasible solution is NP-hard (Hsu [9]). On the other hand, the time varying lot size approach is more flexible in solving ELSP than the other two approaches as it allows different cycle times for products. Maxwell [4] and Delporte and Thomas [8] started using time varying lot size approach to solve ELSP, then Dobson [10] showed that under this approach any production sequence can be converted into a feasible sequence and, thus proposed a heuristic solution, known as Dobson's Heuristic (DH). In general, the ELSP is known as an NP-hard problem (see Hsu [9], Gallego and Shaw [11]). Although it is typically assumed that the demand rate, production rate, set-up time and set-up cost are deterministic, known and product dependent they nevertheless independent of production sequence. Dobson [12] in extension to his work, considered the sequence dependent set-up times. ELSP model is extended by considering some given parameters as decision variables and relaxing some of the assumptions. Allen [13] considered the production rate as a decision variable in ELSP model. Several authors, Silver [14], Moon et al. [15], Gallego [16], Khouja [17] and Moon and Christy [18] concluded that slowing down production rates is more profitable in an under-utilized production facility. Silver [19], Viswanathan and Goyal [20] studied the ELSP under shelf life constraint. Under this constraint inventory is subjected to expiration while in stock. A feasible production schedule is obtained by using either production rate or cycle time as decision variable. Issues related to set-up time reduction were consider by Gallego and Moon [21], Hwang et al. [22] and Moon [23]. The concept of stabilization is another extension to ELSP. During stabilization period the production rate gradually increases and it is suggested by Moon et al. [24]. Silver [25] presented a comprehensive survey of heuristic solution methods. Khouja et al. [26] used Genetic Algorithm (GA) to solve the ELSP problem using basic period approach, although is some cases GA resulted in an unfavorable solution when compared with LB. Moon et al. [1] used the Hybrid GA to solve ELSP using time varying lot size approach. In this study we proposed a Tabu Search (TS), Simulated Annealing (SA) algorithm and two Neighborhood Search heuristics for solving the ELSP using time-varying approach, hence extending the work of Moon et al. [1]. The rest of the article is composed as follows : The ELSP model under time-varying approach is outlined in section 2. Section 3 discusses the implementation of TS and SA algorithms to ELSP and the selection of operation parameters of the algorithms using factorial Design Of Experiment (DOE). Computational experience with the algorithms is summarized in section 5, and conclusions are presented in section 6.

2   ELSP Model

In this section we outline Dobson's [10] ELSP model. The model uses following notations. The objective is to determine the optimal parameters for cycle

i = Product index.
j = Position index of the item in a production sequence $\mathbf{f}$.
m = Total number of products.
n = Total number of production runs in a cycle.
p_i = Production rate of item i, $\forall\, i = 1, 2, \ldots, m$.
d_i = Demand rate of item i, $\forall\, i = 1, 2, \ldots, m$.
h_i = Inventory holding cost ($ per unit per day), $\forall\, h_i\ i = 1, 2, 3, \ldots, m$.
A_i = Set-up cost for item i ($), $\forall\, i = 1, 2, 3, \ldots, m$.
s_i = Set up time for item i, $\forall\, i = 1, 2, \ldots, m$.
t^j = Production time for a product produced at position j in a production sequence.
u^j = Machine idle time associated with product processed at j_{th} position in a production sequence.
$\mathcal{T}$ = Length of the production cycle (in days).
J_i = Set of indexes of a schedule $\mathbf{f}$ containing product i.

length $\mathcal{T}$, a production sequence $\mathbf{f} = \{f^1, f^2, \ldots, f^n\}$, $\forall f^j \in \{1, 2, 3, \ldots, m\}$, production times $\mathbf{t} = \{t^1, t^2, \ldots, t^n\}$ and idle times $\mathbf{u} = \{u^1, u^2, \ldots, u^n\}$, such that the demand is satisfied and the total cost of setup and inventory holding is minimized. The cycle is repeated indefinitely. Let's consider the i_{th} product that is produced at j^{th} position in a production sequence $\mathbf{f}$. It's production involves a production time t^j, setup time s^j and an idle time u^j. This part will be produced again after subsequent products are produced in production sequence. The total number of parts produced in the jth position is $p^j t^j$. These parts will satisfy the demand for the product in the period $[0, v]$, where $v = \dfrac{p^j t^j}{d^j}$. The highest inventory level is $(p^j - d^j) t^j$. The total inventory holding cost for the product produced at the jth position in the sequence is $\frac{1}{2} h^j (p^j - d^j)(p^j/d^j)(t^j)^2$. Let L_k represents the set containing the products that are produced in a given sequence from k up to the position in the sequence where product k is produced again but not included in the sequence. Now, using the information provided above, an ELSP problem is written as shown below:

$$\min_{\substack{t \geq 0 \\ \inf u \geq 0 \\ \mathcal{T} \geq 0}} \frac{1}{\mathcal{T}} \left(\sum_{j=1}^{n} \frac{1}{2} h^j (p^j - d^j)(\frac{p^j}{d^j})(t^j)^2 + \sum_{j=1}^{n} A^j \right) \tag{1}$$

Subject to

$$\sum_{j \in J_i} p_i t^j = d_i \mathcal{T} \quad i = 1, 2, \ldots, m \tag{2}$$

$$\sum_{j \in L_k} (t^j + s^j + u^j) = \left(\frac{p^k}{d^k}\right) t^k \quad k = 1, 2, \ldots, n \tag{3}$$

$$\sum_{j=1}^{n} (t^j + s^j + u^j) = \mathcal{T} \tag{4}$$

Equation 2 guarantees that enough quantity of item i is produced to satisfy the demand incurring with a rate of d_i over the cycle time $\mathcal{T}$. The constraints stated in Equation 3 mention that enough quantity of an item should be produced to satisfy it's demand until it is produced again during the cycle $\mathcal{T}$. Equation 4 describes that the sum of the production, setup and idle time of the product produced in any given sequence is the cycle time $\mathcal{T}$. In this study all the model assumptions stated in Dobson [10] are kept the same.

3   Implementation of Meta-heuristics to ELSP

Both meta-heuristic techniques we described in this paper use the following procedure [1] to find the optimal solution to ELSP.

- **Step 1:** Use the LB computation procedure given in Moon et al. [1] to determine lower bound on estimates of optimal cycle times, T_i^*'s of each product.
- **Step 2:** Let z_i represent the optimal production frequency for the item i. Then z_i is determined as follows:

$$z_i = \frac{\overset{max}{\underset{i}{}} \{T_i^*\}}{T_i^*} \quad \forall\, i = 1, 2, 3, \ldots, m. \tag{5}$$

- **Step 3:** Round the production frequencies found in step 2 to the nearest integers.
- **Step 4:** Use of a meta-heuristic to determine the best production sequence using the rounded production frequencies.
- **Step 5:** Use Quick and Dirty heuristic [1], assuming a zero idle time i.e., $\mathbf{u} = 0$. For a given production sequence $\mathbf{f}$ solve Equation 3 to find $\mathbf{t}$.

3.1   Tabu Search

Fred Glover [27] introduced Tabu Search (TS) in 1989. It is an iterative heuristic for solving optimization problems. Unlike a local search which stops when no improved solution is found in the current neighborhood, TS continues the search even if the new solutions are worse than the current best solution. To prevent cycling, information pertaining to the most recently visited solutions is recorded in a list called tabu list. The tabu status of a solution is overridden when a certain criterion (*Aspiration Criterion*) is satisfied. In it's simplified form the TS algorithm requires seed solution, neighborhood generation scheme, tabu criterion and stopping criterion. The performance of the TS algorithm can be improved by the incorporation of the search *Intensification* and *Diversification* (Sait and Youssef [28]).

TS Parameter Selection Using Factorial Design. Tables 1- 2 present the DOE with three important control parameters of TS, usually called factors. It works as a good tool to improve the performance of the algorithm (Lyu et al. [29]). The DOE reveals that the TS algorithm is robust in it's performance as none of these factors significantly affects the performance. Since ELSP is usually considered an offline scheduling problem, the authors would choose a mediocre value of Candidate list size to give TS a chance to perform more aggressive search, although it may not result any improvement. For the same reason intensification and diversification is also chosen.

Table 1. Levels of Design of experiment on TS

Factor	Label	Levels		
Candidate List Size (CLS)	A	3	25	50
Tabu List Size (TLS)	B	1	7	20
Intensification and Diversification	C	No	Yes	

Table 2. Factorial design on TS response is % relative deviation form lower bound

Source	DF	Sum of Squares	F	Prob. of Larger F
A	2	1.40031	1.042	0.3552
B	2	0.00971	0.007	0.9928
C	1	0.03227	0.048	0.8268
A×B	4	0.02930	0.011	0.9998
A×C	2	0.04929	0.037	0.9640
B×C	2	0.01258	0.009	0.9907
A×B×C	4	0.01440	0.005	0.9999

- *Seed Generation*: The seed solution is also called the initial solution. A feasible production sequence $\mathbf{f}$ is generated randomly using production frequencies. Only a feasible seed is accepted in this search.
- *Neighborhood Scheme*: Several neighborhood schemes can be used to generate neighborhoods. Taillard [30] outlined a variety of schemes that can be used to generate neighborhood. In this study, we employ a simple neighborhood scheme in which two randomly selected items at distinct positions in a production sequence $\mathbf{f}$ are interchanged. A neighbor production sequence $\mathbf{f}'$ is considered unacceptable if there exists two similar items at any two adjacent positions in $\mathbf{f}'$.
- *Tabu Feature*: Tabu list (TL) is updated on a First In First Out (FIFO) basis. Two indexes of $\mathbf{f}$ that are randomly swapped are stored in TL.
- *Aspiration Criterion*: Tabu status of a move is overridden if it results a cost less than the best solution encountered in the search.

- *Candidate Selection Scheme*: Candidate List Size (CLS) is 20 and best solution among 20 candidate neighbors is selected for next iteration.
- *Diversification*: If there no improvement is observed after 500 iterations, the search is diversified with a randomly generated new seed.
- *Intensification*: Intensification is done after 250 iterations of no improvement using the procedure given in Figure 1.
- *Stopping Criterion*: Algorithm stops after 1000 iterations.

Algorithm Intensification
    (* m = Product index. *)
    (* n = Position index in the sequence. *)
    (* $\mathbf{H}_{m \times n}$ = Frequency matrix. *)
    (* $\mathbf{f}$ = Production sequence. *)
    (* z_i = Production frequency of item i. *)
Begin
    $\mathbf{f} \leftarrow \{\emptyset\}$, $l \leftarrow 1$, $j \leftarrow 1$.
    Repeat
        Find the entry in frequency matrix $\mathbf{H}$ such that $H_{ij} = \max\{\mathbf{H}\}$.
        If $(z_i \geq 1)$ **Then**
            Insert Product i at the j_{th} position in the $\mathbf{f}$.
            $z_i \leftarrow z_i - 1$.
        Endif
        Delete the i_{th} row and column j_{th} from $\mathbf{H}$.
        $l \leftarrow l + 1$
    Until $(l \leq n)$
End

Fig. 1. Intensification scheme for TS algorithm

3.2 Simulated Annealing

Simulated Annealing (SA) is another powerful tool used in optimization. Unlike TS, SA does not discriminate between perviously visited and unexplored solutions while finding new search directions. A basic SA needs seed solution, cooling schedule, equilibrium criterion and stopping criterion. More details pertinent to SA can be found in Sait and Youssef [28].

SA Parameters Selection Using Factorial Design. The DOE shows that the three control parameters (see Tables 3) do not effect the performance of the SA and hence SA is also robust in performance. ANOVA is presented in Table 4 where performance indicator is deviation from LB i.e., SA/LB.

- **Cooling Schedule:**
  - *Initial Temperature*: The initial temperature T_0 is estimated using an efficient algorithm proposed by White [31]. $T_0 = k\sigma$, where $k = -3/\ln P$ and P is acceptance ratio at T_0. In this implementation we take $P = 0.90$.

Table 3. Levels of Design of experiment on SA

Factor	Label	Levels
Acceptance ratio (P) at initial temperature	A	0.60 0.80 0.90
Metropolis loop size (M)	B	5 20 50
Temperature decrement (α)	C	0.60 0.80 0.99

Table 4. Factorial design on SA response is SA/LB

Source	DF	Sum of Squares	F	Prob. of Larger F
A	2	5.668×10^{-5}	0.021	0.9793
B	2	1.625×10^{-3}	0.599	0.5500
C	2	3.492×10^{-4}	0.129	0.8792
A×B	4	1.121×10^{-5}	0.002	1.0000
A×C	4	3.386×10^{-5}	0.006	0.9999
B×C	4	2.994×10^{-4}	0.055	0.9943
A×B×C	8	5.735×10^{-5}	0.005	1.0000

- *Temperature Decrement Rule*: $T_{k+1} = \alpha T_k$ where T_k and T_{k+1} are temperatures at k_{th} and $k_{th} + 1$ iterations respectively and α is cooling rate. Here we take $\alpha = 0.95$ to avoid any drastic decrement in temperature.
- *Final Temperature*: Final temperature is set at 0.001. Temperature decrement is stopped when the temperature in annealing process reaches that level.
- **Markov Chain Length:** Metropolis loop (see Eglese [32]) is called 20 times at each temperature level. The number is the same as CLS to give both algorithms equal search opportunity.
- **Acceptance Probability Function:** We used the statistical acceptance probability function (Sait and Youssef [28]). At a given temperature T_i, the acceptance probability function p_a of a solution $\mathbf{f}'$ can be written as:

$$p_a = \begin{cases} 1 & \text{If } \text{cost}(\mathbf{f}') < \text{cost}(\mathbf{f}) \\ \exp(\Delta/T_i) & \text{If } \text{cost}(\mathbf{f}') \geq \text{cost}(\mathbf{f}) \\ & \text{Where } \Delta = \text{cost}(\mathbf{f}') - \text{cost}(\mathbf{f}) \end{cases} \tag{6}$$

The seed solution, neighborhood scheme and stopping criterion are the same with those used in the TS algorithm given in Section 3.1.

4   Neighborhood Search Heuristics

A Neighborhood Search (NS) heuristic is greedy random search method. Randomization generates a neighborhood of the seed solution. The neighborhood solution is accepted if it is superior to the existing solutions. Hence it is unable to escape from the local minimum like other deterministic, greedy algorithms. More details on NS can be found in Sait and Youssef [28]. In this research we study two simple NS heuristics i.e., NS_a and NS_b. The NS_a is based on rounding off production frequencies to the nearest integers (Moon et al. [1]). It takes the same implementation procedure given in section 3. The only difference is that NS_a is used in step 4. The implementation of NS_b heuristic differs from NS_a only in rounding off the scheme of production frequencies. It makes use of the rounding-off algorithm given in Roundy [33] which rounds the production frequencies to the power of 2. Both the NS_a and NS_b stop after 1000 iterations.

5   Computational Results

The proposed algorithms are coded using MATLAB and tested on an Intel Pentium 4, 2.4 GHz Processor. Initial experimentation is performed with the test problem given in Mallya [34], Bomberger [6] and randomly generated problems using Set 1 shown in Table 5. In this study the deviation from the LB is estimated using ratio of cost resulted in an algorithm to the cost of LB (See Dobson [10]) and presented in Tables 8-10. Tables also show the improvements i.e., the proposed solution methods made over DH using the same idea. For the example presented by Mallya [34] both TS and SA are able to find an alternative solution resulting in the same cost $60.911 as obtained by Hybrid GA [1]. All of these algorithms round production frequencies to the nearest integer. The Dobson [10] heuristic (DH) on the other hand rounds production frequencies to the power of 2 which results in a better cost of $60.874. When production frequencies are rounded to the power of 2 in Mallya's problem (See Mallya[§] in Table 7), TS and SA outperform DH with a cost of $60.782. The NS_b is also able to converge to the same cost. We also conduct test with Bomberger's problem [6] at $K = 0.01$ i.e., 99% Utilization, clearly TS, SA and NS_a outperform existing solutions (see Table 7).   Moon et al. [1] tested Hybrid GA on 50 randomly generated problems with uniform distribution under parameters given in Set 1, Table 5. Table 8 shows the computational results with Set 1, CPU time for DH is not presented because it is constructive heuristic and can be executed for large scale ELSP in negligible time. Out of 50 randomly generated problems, the TS, SA and NS_a

Table 5. Distribution for randomly generated data for the test problems

Parameters	Set 1	Set 2	Set 3
Number of items (units)	[5, 15]	[5, 15]	[5, 15]
Production rate (units/unit time)	[2000, 20000]	[4000, 20000]	[1500, 30000]
Demand rate (units/ unit time)	[1500, 2000]	[1000, 2000]	[500, 2000]
Setup time (time/ unit)	[1, 4]	[1, 4]	[1, 8]
Setup cost ($)	[50, 100]	[50, 100]	[10, 350]
Holding cost ($)	[1/240, 6/240]	[1/240, 5/240]	[5/240000, 5/240]
Utilization		$\geq 90\%$	

Table 6. Production frequencies for Mallya and Bomberger's problem

Example	Items m	Production frequencies $\{y_1, y_2, \ldots, y_m\}$	
		Round-off to power-of 2	Round off to nearest integer
Mallya	5	$\{2, 1, 4, 2, 1\}$	$\{2, 2, 3, 3, 1\}$
Bomberger	10	$\{1, 4, 4, 8, 4, 2, 2, 16, 4, 4\}$	$\{1, 4, 4, 7, 5, 2, 1, 12, 4, 2\}$

Table 7. Comparison on Mallya and Bomberger's problem

Problem type	Lower bound	Existing solutions		Proposed solutions			
		DH	GA	SA	TS	NS_a	NS_b
Mallya	57.726	60.874	60.911	60.911	60.911	60.911	60.782
Mallya§	-	-	-	60.782	60.782	-	-
Bomberger	122.945	128.339	126.12	125.135	125.31	125.754	130.346

Table 8. Comparison of algorithms on randomly generated problems using Set 1

Parameters	Comparison with Lower Bound					Comparison with Dobson heuristic			
	$\frac{DH}{LB}$	$\frac{SA}{LB}$	$\frac{TS}{LB}$	$\frac{NS_a}{LB}$	$\frac{NS_b}{LB}$	$\frac{DH}{SA}$	$\frac{DH}{TS}$	$\frac{DH}{NS_a}$	$\frac{DH}{NS_b}$
Mean	1.0517	1.0243	1.0242	1.0250	1.0484	1.0267	1.0268	1.0260	1.0031
Min.	1.0114	1.0074	1.0074	1.0074	1.0114	0.9957	0.9956	0.9919	0.9919
Max.	1.2216	1.1098	1.1100	1.1192	1.2176	1.1056	1.1056	1.1055	1.0283
Avg. CPU time (sec.)	-	-	-	-	-	8.3641	11.0777	0.3828	0.4081
Best time (sec.)	-	-	-	-	-	0.6713	3.2662	0.0288	0.0447
Nbr. of Problems with ratio ≤ 1	0	0	0	0	0	7	8	8	27

Table 9. Comparison of algorithms on randomly generated problems using Set 2

Parameters	Comparison with Lower Bound					Comparison with Dobson heuristic			
	$\frac{DH}{LB}$	$\frac{SA}{LB}$	$\frac{TS}{LB}$	$\frac{NS_a}{LB}$	$\frac{NS_b}{LB}$	$\frac{DH}{SA}$	$\frac{DH}{TS}$	$\frac{DH}{NS_a}$	$\frac{DH}{NS_b}$
Mean	1.0503	1.0274	1.0272	1.0278	1.0491	1.0225	1.0227	1.0220	1.0011
Min.	1.0070	1.0051	1.0051	1.0051	1.0070	0.9754	0.9758	0.9722	0.9904
Max.	1.2336	1.0713	1.0708	1.0748	1.2054	1.2034	1.2034	1.1967	1.0234
Avg. CPU time (sec.)	-	-	-	-	-	8.7303	11.3887	0.4297	0.4088
Best time (sec.)	-	-	-	-	-	0.8628	4.5018	0.0528	0.0590
Nbr. of Problems with ratio ≤ 1	0	0	0	0	0	8	12	13	31

improve atleast 2.6% over DH on average, whereas GA by Moon et al. [1] improves 1.1%. More analysis is done using 50 randomly generated problem for

Table 10. Comparison of algorithms on randomly generated problems using Set 3

Parameters	Comparison with Lower Bound					Comparison with Dobson heuristic			
	$\frac{DH}{LB}$	$\frac{SA}{LB}$	$\frac{TS}{LB}$	$\frac{NS_a}{LB}$	$\frac{NS_b}{LB}$	$\frac{DH}{SA}$	$\frac{DH}{TS}$	$\frac{DH}{NS_a}$	$\frac{DH}{NS_b}$
Mean	1.2550	1.1594	1.1592	1.1745	1.2301	1.0440	1.0443	1.0311	1.0089
Min.	1.0193	1.0111	1.0107	1.0123	1.0192	0.9272	0.9304	0.9304	0.9502
Max.	8.1570	5.3715	5.3720	5.4625	7.3938	1.5186	1.5184	1.4933	1.1032
Avg. CPU time (sec.)	-	-	-	-	-	22.4303	22.1410	2.4722	5.0455
Best time (sec.)	-	-	-	-	-	11.8694	14.5094	2.2650	4.8128
Nbr. of Problems with ratio ≤ 1	0	0	0	0	0	13	12	18	21

each of Set 2 and 3. The Problems in Set 3 are known to be hard (Dobson [10]). In Set 3, the SA and TS algorithms are atleast 4.4% better than the DH on average. The NS heuristics, NS_a and NS_b produce solutions with averages of 3.11% and 0.89% better than the DH respectively. The results are summarized in Tables 8-10.

6   Conclusion

In this paper two meta-heuristics, based on Tabu Search and Simulated Annealing algorithms are proposed. Two Neighborhood Search heuristics, NS_a and NS_b are also studied. Computational studies on randomly generated problems showed that the TS and the SA algorithms outperform the best known Dobson's heuristic and Hybrid Genetic Algorithm to ELSP. It can also be inferred from the performance of NS_a and NS_b that NS_a supersedes NS_b on the majority of problems. Surprisingly, NS_a finds better solutions than Dobson's Heuristic and Hybrid GA to most of the problems solved in this study. In most cases both TS and SA result the best solutions to the problem and they are found most consistent in their performance.

Acknowledgements. Authors are thankful to Dr. Danielle Morin for her help with statistical analysis. Authors are also grateful to anonymous referees.

References

1. Moon, I., Silver, E., Choi, S.: Hybrid genetic algorithm for the economic lot-scheduling problem. International Journal of Production Research **20** (2002) 809–824
2. Eilon, S.: Economic batch-size determination for multi-product scheduling. Operations Research **10** (1959) 217–227
3. Rogers, J.: A computational approach to the economic lot scheduling problem. Management Science **4** (1958) 264–291

4. Maxwell, W.L.: The scheduling of economic lot sizes. Naval Research Logisitcs Quarterly **11** (1964) 89–124
5. Hanssmann, F.: Operations Research in Production Planning and Control. John Wiley, New York (1962)
6. Bomberger, E.E.: A dynamic programming approach to a lot size scheduling problem. Management Science **12** (1966) 778–784
7. Elmaghraby, S.: The economic lot scheduling problem (ELSP): Review and extension. Management Science **24** (1978) 587–598
8. Delporte, C., Thomas, L.: Lot sizing and sequencing for N products on one facility. Management Science **23** (1978) 1070–1079
9. Hsu, W.: On the general feasibility test for scheduling lot sizes for several products on one machine. Management Science **29** (1983) 93–105
10. Dobson, G.: The economic lot-scheduling problem: Achieving feasibility using time-varying lot sizes. Operations Research **35** (1987) 764–771
11. Gallego, G., Shaw, X.: Complexity of the ELSP with general cyclic schedules. IIE Transactions **29** (1997) 109–113
12. Dobson, G.: The cyclic lot scheduling problem with sequence-dependent setups. Operations Research **40** (1992) 736–749
13. Allen, S.J.: Production rate planning for two products sharing a single process facility: A real world case study. Production and Inventory Management **31** (1990) 24–29
14. Silver, E.: Deliberately slowing down output in a family production context. International Journal of Production Research **28** (1990) 17–27
15. Moon, I., Gallego, G., Simchi-Levi, D.: Controllable production rates in a family production context. IIE Transaction **30** (1991) 2459–2470
16. Gallego, G.: Reduced production rates in the economic lot scheduling problem. International Journal of Production Research **31** (1993) 1035–1046
17. Khouja, M.: The economic lot scheduling problem under volume flexibility. International Journal of Production Research **48** (1997) 73–86
18. Moon, D., Christy, D.: Determination of optimal priduction rates on a single facility with dependent mold lifespan. International Journal of Production Economics **54** (1998) 29–40
19. Silver, E.: Dealing with shelf life constraint in cyclic scheduling by adjusting both cycle time and production rate. International Journal of Production Research **33** (1995) 623–629
20. Viswanathan, S., Goyal, S.K.: Optimal cycle time and production rate in a family production context with shelf life considerations. International Journal of Production Research **35** (1997) 1703–1711
21. Gellego, G., Moon, I.: The effect of externalizing setups in the economic lot scheduling problem. Operations Research **40** (1992) 614–619
22. Hwang, H., Kim, D., Kim, Y.: Multiproduct economic lot size models with investment costs for set-up reduction and quality improvement. International Journal of Production Research **31** (1993) 691–703
23. Moon, I.: Multiproduct economic lot size models with investment costs for setup reduction and quality improvements: Reviews and extensions. International Journal of Production Research **32** (1994) 2795–2801
24. Moon, I., Hahm, J., Lee, C.: The effect of the stabilization period on the economic lot scheduling problem. IIE Transactions **30** (1998) 1009–1017
25. Silver, E.A.: An overview of heuristic solution methods. Journal of Operational Research Society **55** (2004) 936–956

26. Khouja, M., Michalewicz, Z., Wilmot, M.: The use of genetic algoritms to solve the economic lot size scheduling problem. European Journal of Operational Research **110** (1998) 509–524
27. Glover, F.: Tabu Search- Part I. OSAR Journal on Computing **1** (1989) 190–206
28. Sait, S.M., Youssef, H.: Iterative Computer Algorithms with Applications in Engineering. IEEE Computer Society (1999)
29. Lyu, J., Gunasekaran, A., Ding, J.H.: Simulated annealing algorithm for solving the single machine early/tardy problem. International Journal of Systems Science **27** (1996) 605–610
30. Taillard, E.: Some efficient heuristic methods for the flow shop sequencing problem. European Journal of Operational Research **47** (1990) 65–74
31. White, S.R.: Concept of scale in simulated annealing, IEEE, IEEE International Conference of Computer Design (1984) 646–651
32. Eglese, R.W.: Simulated annealing: A tool for operational research. European Journal of Operational Research **46** (1990) 271–281
33. Roundy, R.: Rounding off to powers of two in continuous relaxation of capcitated lot sizing problems. Management Science **35** (1989) 1433–1442
34. Mallya, R.: Multi-product scheduling on a single machine: A case study. OMEGA: International Journal of Management Science **20** (1992) 529–534

Making the Edge-Set Encoding Fly by Controlling the Bias of Its Crossover Operator

Franz Rothlauf and Carsten Tzschoppe

Department of Business Administration and Information Systems,
University of Mannheim,
68131 Mannheim/Germany
rothlauf@uni-mannheim.de, carsten.tzschoppe@gmx.de

Abstract. Edge-sets encode spanning trees directly by listing their edges. Evolutionary operators for edge-sets may be heuristic, considering the weights of edges they include in offspring, or naive, including edges without regard to their weights. Crossover operators that heuristically prefer shorter edges are strongly biased towards minimum spanning trees (MST); EAs that apply heuristic crossover generally perform poorly on spanning tree problems whose optimum solutions are not very similar to MSTs. For the edge-set encoding, a modified heuristic crossover called γ-TX implements variable bias towards low-weight edges and thus towards MSTs. The bias can be set arbitrarily between the strong bias of the heuristic crossover operator, or being unbiased. An investigation into the performance of EAs using the γ-TX for randomly created OCST problems of different types and OCST test instances from the literature present good results when setting the crossover-specific parameter γ properly. The presented results suggest that the original heuristic crossover operator of the edge-sets should be substituted by the modified γ-TX operator that allows us to control the bias towards the MST.

1 Introduction

A spanning tree T of an undirected graph $G(V, E)$ is a subgraph that connects all vertices of G and contains no cycles. Relevant constrained minimum spanning tree (MST) problems are, for example, the optimal communication spanning tree (OCST) problem [1], or the degree-constrained minimum spanning tree problem [2,3]. When using evolutionary algorithms (EAs) for tree problems it is necessary to encode a tree such that the evolutionary search operators like crossover or mutation can be applied. There are two different possibilities for doing this. Indirect representations usually encode a tree (phenotype) as a list of strings (genotypes) and apply standard search operators to the genotypes. The phenotypes are constructed by an appropriate genotype-phenotype mapping (representation). In contrast, direct representations encode a tree as a set of edges and apply search operators directly to the set of edges. Therefore, no representation is necessary. Instead, tree-specific search operators must be developed, as standard search operators can not be used any more. Examples for direct encodings

G.R. Raidl and J. Gottlieb (Eds.): EvoCOP 2005, LNCS 3448, pp. 202–212, 2005.

are the edge-set encoding [3], or the NetDir encoding [4–sec. 7.2]. Raidl and Julstrom [3] proposed two different variants of search operators for the edge-set encoding: Heuristic variants where the operators consider the weights of the edges, and non-heuristic versions. Results for the degree-constrained MST problem and the traveling salesman problem indicated a good performance of the heuristic variants [5, 3].

Representations and search operators can have a bias towards some solutions. A representation is biased if it is redundant and some phenotypes are over-represented. A search operator is biased if its iterative application results in a biased population that means not all possible phenotypes are represented with the same probability by the population. Consequently, heuristic search is biased if it pushes a population of solutions towards some solution even if no selection operator is used. Biased representations and operators do change the performance of evolutionary search. If the optimal solutions are similar to the solution towards which the bias points, EA performance increases [4–section 3.1]. However, if there is a bias towards a solution that is not similar to the optimal solution, EA performance is low. Tzschoppe et al. [6] examined the bias of the edge-set encoding and found a strong bias of the heuristic crossover operator of the edge-sets towards the MST. Therefore, problems where the optimal solution is the MST can be easily solved. However, as the bias of the heuristic crossover towards the MST is strong, EAs fail if the optimal solution is only slightly different from the MST.

This paper proposes a modified version of the heuristic crossover operator of the edge-set encoding that allows us to control the strength of the bias towards the MST. Therefore, the problems arising from the oversized bias of the heuristic crossover operator can be overcome and its bias can be adjusted according to the properties of the problem at hand. If it is known a-priori that optimal solutions of a problem are similar to the MST, a modest bias towards the MST allows EAs to solve the problem more efficiently. Experiments on the performance of the modified crossover operator are performed for the optimal communication spanning tree (OCST) problem. Results for random problems and problem instances from the literature show that by controlling the strength of the heuristic bias EA performance increases.

The paper is structured as follows. The following section describes the functionality of the edge-set encoding with and without heuristics and introduces the modified crossover operator. Section 3 investigates the bias of the crossover operators of the edge-set encoding and shows that the bias can be controlled when using the modified crossover operator. Its influence on EAs when solving OCST problems is examined in section 4. The paper ends with concluding remarks.

2   The Edge-Set Encoding

The edge-set encoding [3] is a direct representation for trees. Therefore, the search operators are applied directly to sets of edges. There are two different variants of search and initialization operators of the edge-set encoding: either with or without heuristics. When using operators with heuristics the weights of the edges are considered for the construction of the offspring. In the following paragraphs we briefly review the functionality of the initialization method and the crossover operator. We do not consider the mutation operator as Raidl and Julstrom [3] already proposed a version of the mutation operator that allows us to control its bias [7].

2.1   The Edge-Set Encoding Without Heuristics

Initialization. In order to create feasible solutions for the initial population, the edge-set encoding uses the Kruskal random spanning tree (RST) algorithm, a slightly modified version of the algorithm from Kruskal. In contrast to Kruskals' algorithm, KruskalRST chooses edges (i, j) not according to their weight w_{ij} but randomly. Raidl and Julstrom [3] have shown that this algorithm for creating random spanning trees, KruskalRST, has a small bias towards star-like trees.

procedure KruskalRST(V, E): $//E$: set of edges; V: set of vertices
$T \leftarrow \emptyset, \ A \leftarrow E$; $//T$: to be constructed spanning tree
while $|T| < |V| - 1$ **do**
   choose an edge $\{(u, v)\} \in A$ at random;
   $A \leftarrow A - \{(u, v)\}$;
   if u and v are not yet connected in T **then**
      $T \leftarrow T \cup \{(u, v)\}$;
return T.

Recombination. The non-heuristic KruskalRST* crossover operator [3] includes in a first step all edges that are common to both parents T_1 and T_2 in the offspring T_{off}. Then, in a second step, KruskalRST is applied to $G_{cr} = (V, T_1 \cup T_2)$. KruskalRST* has high heritability as in the absence of constraints, only parental edges are used to create the offspring. Crossover becomes more complicated for constrained MST problems as it is possible that the RST algorithm can create no feasible tree from $G_{cr} = (V, T_1 \cup T_2)$. Then, additional edges have to be chosen randomly to complete an offspring.

2.2   Heuristic Recombination Operators for the Edge-Set Encoding

The heuristic crossover operator [3] is a modified version of KruskalRST* crossover. In a first step, the operator transfers all edges $T_1 \cap T_2$ that exist in both parents to the offspring. Then, the remaining edges are chosen randomly from $E' = (T_1 \cup T_2) \setminus (T_1 \cap T_2)$ using a tournament with replacement of size two. If the underlying optimization problem is constrained, it is possible that the offspring has to be completed using edges not in E'. This version of the heuristic crossover operator is denoted as 2-tournament-crossover (TX). Two other variants of the heuristic crossover operator were proposed [5]. They differ in the strategy of completing the offspring with the edges available in E':

- **Greedy crossover:** When using this strategy, the edge with the smallest weight is chosen from E'.

- **Inverse-weight-proportional crossover:** This strategy selects each edge from E' according to probabilities inversely proportional to the edges' weights.

Julstrom and Raidl [5] examined the performance of the different crossover variants for the traveling-salesperson problem and the degree-constrained MST problem. The results indicated that Greedy crossover shows good performance for simple and easy problem instances. For large problems TX crossover resulted in the best performance.

All three crossover strategies have a strong bias towards the MST. The bias of the TX operator is already so strong that EAs are only able to find optimal solutions if they are very similar to the MST [7]. Problems where the optimal solutions are slightly different from the MST could no longer be solved by using the TX operator. The bias of the Greedy crossover is higher than the bias of the TX crossover. Therefore, Greedy crossover also results in low EA performance if the optimal solution is not the MST. The inverse-weight-proportional crossover introduces a bias to the MST similar to the TX crossover. However, the bias can not be controlled in a systematic way but depends on the specific weights of the edges.

We want to propose a modified version of the heuristic TX operator. The modification is only small but allows us to control the bias towards the MST. In the new crossover variant (denoted as γ-TX crossover) the tournament of size two that chooses one edge from E' is not always performed but only with the probability γ. Therefore, for $\gamma = 0$ an edge is randomly chosen from E' and we see the same behavior as KruskalRST*. For $\gamma = 1$ all edges are chosen by a tournament of size 2 and we get the same behavior as TX crossover. The bias of γ-TX towards the MST can be set arbitrarily small with $\gamma \to 0$.

3 Bias of the Crossover Operators for Edge-Sets

We investigate the bias of the TX and γ-TX operator for randomly created trees with $n = 10$ and $n = 16$ nodes. To every edge (i, j) a non-negative weight w_{ij} is associated. We want to consider two different possibilities for the weights w_{ij}:

- **Random weights:** The real-valued weights w_{ij} are generated randomly and are uniformly distributed in $]0, 100]$.
- **Euclidean weights:** The nodes are randomly placed on a 1000x1000 grid. The weights w_{ij} between nodes i and j are the Euclidean distances between nodes i and j.

As the weights w_{ij} are randomly created and $w_{ij} \neq w_{kl}$, $\forall i \neq k, j \neq l$, there is a unique MST for every problem instance. T is the MST if $c(T) \leq c(T')$ for all other spanning trees T', where $c(T) = \sum_{(i,j) \in T} w_{ij}$. The similarity between two spanning trees T_i and T_j can be measured using the distance $d_{ij} \in \{0, 1, \ldots, n - 1\}$ as $d_{ij} = \frac{1}{2} \sum_{u,v \in V, \, u < v} |l_{uv}^i - l_{uv}^j|$, where l_{uv}^i is 1 if an edge from u to v exists in T_i and 0 if it does not exist in T_i.

For the experiments we randomly generate an initial population of 500 individuals using the non-heuristic KruskalRST initialization and apply the crossover operators iteratively. As no selection operator is used, no selection pressure pushes the population to high-quality solutions. An operator is unbiased if the statistical properties of the population do not change by applying crossover alone. In the experiments we measure in each generation the average distance $d_{mst-pop} = 1/N \sum_{i=1}^{n} d_{i,MST}$ of the individuals T_i in the population to the MST. If $d_{mst-pop}$ decreases, the crossover operator is biased towards the MST. If $d_{mst-pop}$ remains constant, the crossover operator is unbiased and no MST-like solutions are overrepresented.

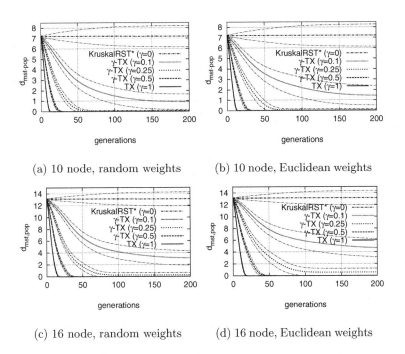

(a) 10 node, random weights   (b) 10 node, Euclidean weights

(c) 16 node, random weights   (d) 16 node, Euclidean weights

Fig. 1. The plots show the mean and the standard deviation of the distance $d_{mst-pop}$ between a population of 500 randomly generated individuals towards the MST over the number of generations when only using crossover (no selection pressure). The results show that the bias of the γ-TX crossover can be controlled and lies between the strong bias of TX crossover ($\gamma = 1$) and the no-bias of the KruskalRST* crossover ($\gamma = 0$)

We performed this experiment on 500 randomly generated 10 and 16 node problem instances with random, and Euclidean weights w_{ij}. For every problem instance we performed 50 runs. In each run, the crossover operator was applied 200 generations. Fig. 1 shows the mean and the standard deviation of the distance $d_{mst-pop}$ over the number of generations. The plots compare the non-heuristic KruskalRST* crossover with the heuristic TX and γ-TX operator (no selection is used).

The results show that the non-heuristic KruskalRST* operator is unbiased. In contrast, the heuristic TX operator shows a strong bias towards the MST and a population converges to the MST after a few generations. When using the γ-TX crossover the bias towards the MST can be controlled. With lower γ the bias gets smaller and for $\gamma = 0$ we get the same results as for KruskalRST*.

4   Performance of the γ-TX Crossover for OCST Problems

This section investigates how the performance of different crossover variants of the edge-set encoding depends on the properties of the optimal solutions. We perform the experiments for the optimal communication spanning tree (OCST) problem as all trees are feasible solutions and there are no additional constraints.

4.1   Optimal Solutions for Randomly Created OCST Problems

The OCST problem was first introduced by Hu [1] and is $\mathcal{MAX}\ \mathcal{NP}$-hard [8]. The problem seeks a spanning tree that connects all given nodes and satisfies their communication requirements for a minimum total cost. The problem can be defined as follows: Let $G = (V, E)$ be a complete undirected graph with $n = |V|$ nodes and $m = |E|$ edges. To every pair of nodes (i, j) a non-negative weight w_{ij} and a non-negative communication requirement r_{ij} is associated. The communication cost $c(T)$ of a spanning tree T is defined as

$$c(T) = \sum_{i,j\in V,\ i<j} r_{ij} \cdot w(p_{i,j}^T),$$

where $w(p_{i,j}^T)$ denotes the weight of the unique path from node i to node j in the spanning tree T. The OCST problem seeks the spanning tree with minimal costs among all other spanning trees. The OCST problem becomes the MST problem if there are no communication requirements r_{ij} and $c(T) = \sum_{(i,j)\in T} w_{ij}$.

It was shown [9] that on average optimal solutions for OCST problems are similar to the MST, that means the average distance $d_{opt,MST}$ between the optimal solution and the MST is significantly lower than the average distance $d_{rand,MST}$ between a randomly created tree and the MST. Therefore, as the optimal solutions of OCST problems are biased towards the MST, representations as well as operators that are biased to the MST are expected to solve the OCST problem efficiently.

To investigate how the performance of EAs using different crossover variants of edge-sets depend on the structure of the optimal solution, an optimal or near-optimal solution for the OCST problem must be determined. We identified optimal (or near-optimal) solutions for the OCST problem by an EA whose population size N is doubled in every iteration until the same solutions are found in subsequent iterations. Details of the experimental setting for finding optimal solutions for OCST problems can be found in Rothlauf et al. [9].

4.2   Edge-Set Crossover for Randomly Created OCST Problems

This section investigates for randomly created OCST problems how the performance of EAs using different variants of the crossover operator depends on the distance $d_{opt,MST}$ between the optimal solution and the MST.

We randomly generated 500 problem instances with 10 and 16 nodes using either random or Euclidean distance weights. The demands r_{ij} are chosen randomly and are uniformly distributed in $]0,\ldots,100]$. Then, we determine the optimal solutions using the experimental setting described in [9]. For comparing the performance of the different crossover variants (KruskalRST*, TX, and γ-TX) we use a simple generational EA with no mutation and tournament selection without replacement of size two. The population size N is chosen with respect to the performance of KruskalRST*. The aim is to find the optimal solution with a probability of about 25-75 %. Therefore, we choose for the 10 node problems a population size $N = 100$ and for the 16 node problems $N = 250$. Each run is stopped after the population is fully converged or the number of generations exceeds 200. 50 runs are performed for each of the 500 problem instances.

The results of our experiments are presented in Fig. 2. It shows the percentage of EA runs that find the optimal solutions (left) and the gap $\frac{c(T_{found})-c(T_{opt})}{c(T_{opt})}$ between the cost of the optimal solution T_{opt} and the cost of the best found solution T_{found} (right) at the end of a run over the distance $d_{opt,MST}$ between the optimal solution and the MST. Results are plotted for KruskalRST*, different variants of γ-TX, and TX. The initial population was generated using the non-heuristic initialization from section 2.1. We only show results for those $d_{opt,MST}$ with more than 10 problem instances (out of 500).

The results reveal that with increasing $d_{opt,MST}$ the performance of EAs is reduced. The decrease in performance is emphasized with larger γ. When using a crossover operator with a strong bias like TX or γ-TX with $\gamma = 0.5$ EA performance is high if and only if $d_{opt,MST} \approx 0$; with larger $d_{opt,MST}$ EA performance drops rapidly. The strong bias pushes the population towards the MST and makes it difficult to find the optimal solution. In contrast, when using the γ-TX operator with a low γ ($\gamma = 0.05$ or $\gamma = 0.2$) the bias towards the MST is small and reasonable and EAs perform better or equal than when using the non-heuristic version. These results are confirmed when examining the gap $\frac{c(T_{found})-c(T_{opt})}{c(T_{opt})}$. With increasing bias and increasing $d_{opt,MST}$, the quality of the found solutions decreases.

In summary, using a strong bias towards the MST results in high EA performance for $d_{opt,MST} \approx 0$ but low performance elsewhere. With lower γ, problems with larger $d_{opt,MST}$ can be solved. EAs using the γ-TX operator with a low bias towards the MST ($\gamma \approx 0.05 - 0.2$ for 10 nodes and $\gamma \approx 0.05$ for 16 node problems) outperform the non-heuristic KruskalRST* crossover for low $d_{opt,MST}$ and also show good results for larger $d_{opt,MST}$.

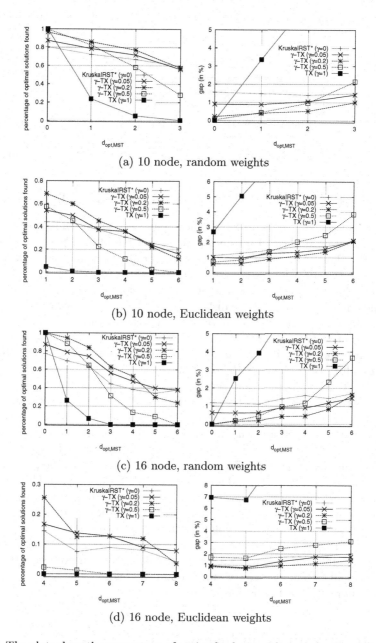

(a) 10 node, random weights

(b) 10 node, Euclidean weights

(c) 16 node, random weights

(d) 16 node, Euclidean weights

Fig. 2. The plots show the percentage of optimal solutions that can be found (left) and the gap between the cost of the best found solution and the optimal solution (right) over $d_{opt,MST}$ for different crossover operators and different types of OCST problems. The plots show that with increasing $d_{opt,MST}$, EAs using the TX operators fail due to their strong bias. When using the γ-TX operator with a low bias, EA performance is high. With increasing γ the bias towards the MST becomes stronger and problems with larger $d_{opt,MST}$ can no longer be solved

Table 1. Performance of EA using different crossover operators for OCST test problems from the literature

problem instance	n	$d_{opt,MST}$	opt solution $c(T_{opt})$	N	KruskalRST* μ	σ	γ-TX (γ = 0.05) μ	σ	γ-TX (γ = 0.2) μ	σ	TX (γ = 1) μ	σ
palmer6	6	1	693,180	50	698,200	8,447	696,301	5,833	698,217	8,524	706,784	6,438
palmer12	12	7	3,428,509	60	3,589,154	92,885	3,582,433	91,529	3,534,935	71,125	3,707,947	56,691
palmer24	24	12	1,086,656	800	1,088,231	915	1,088,007	665	1,088,615	695	1,873,835	36,453
raidl10	10	3	53,674	60	57,046	5,266	55,275	3,481	55,077	2,261	57,200	927
raidl20	20	4	157,570	400	159,714	4,038	159,943	3,984	157,922	1,426	164,811	2,671
berry6	6	0	534	50	539	13	534	3	534	0	534	0
berry35u	35	-	16,273	800	16,621	173	16,622	180	16,604	190	16,577	187
berry35	35	0	16,915	800	17,263	381	16,975	138	16,915	0	16,915	0

4.3   Edge-Set Crossover for Test Instances from the Literature

Several test instances for the OCST problem have been proposed in the literature [10, 11, 12]. Details of the test instances and an analysis of their properties can be found in Rothlauf et al. [9]. The following paragraphs examine the performance of EAs using the different crossover variants for these test instances.

Table 1 lists the properties of the optimal solutions for the test instances. It shows the number of nodes n, the distance $d_{opt,MST}$, and the cost $c(T_{opt})$ of the optimal solution. In the instance berry35u, all distances are uniform ($w_{ij} = 1$), so all spanning trees are minimal. For all test instances, $d_{opt,MST}$ is smaller than the average distance of a randomly created solution towards the MST [9]. Therefore, all test problems are biased towards the MSTs.

For the experiments the same generational EA with population size N as in the previous section is used. The table presents the mean μ and the standard deviation σ of the cost of the best solution found at the end of the runs averaged over 50 runs for each problem instance. The results show that the heuristic TX crossover ($\gamma = 1$) is only able to find the optimal solution if $d_{opt,MST} = 0$. Otherwise, the performance of EAs using it is low. In contrast, the performance of EAs using the non-heuristic and unbiased KrukalRST* is high and high quality solutions can be found. When using the modified γ-TX operator with a low γ ($\gamma = 0.05$ or $\gamma = 0.2$) the performance of EAs can be increased in comparison to KruskalRST* as the optimal solutions for all problem instances are biased towards the MST. EAs using the γ-TX with low γ (e. g. $\gamma = 0.05$) always find solutions of similar or higher quality. Only for berry35u can no better solutions be found as all spanning trees are minimal and therefore a low bias towards the MST does not increase EA performance.

5   Summary and Conclusions

This work proposed a new variant (γ-TX) of the heuristic crossover (TX) operator of the edge-set encoding. When using the standard TX operator an offspring tree is created from two parents by inserting all edges that are common in both parents into the offspring. The offspring is completed by parental edges chosen by a tournament of size two. Edges with lower weight are preferred. In contrast to the standard TX operator, the γ-TX operator only performs a tournament with probability γ and otherwise inserts a random edge from one of the parents.

The TX operator shows a strong bias towards the minimum spanning tree (MST). Using it for the optimal communication spanning tree (OCST) problem allows EAs only to solve the problem if the optimal solution is the MST. If the optimal solution is slightly different from the MST, EAs fail. The γ-TX operator allows us to control the bias which can be set arbitrarily (according to γ) between the strong bias of the TX operator ($\gamma = 1$) and no-bias ($\gamma = 0$). Therefore, the problems of the TX operator with the strong bias towards the MST can be overcome while still allowing the crossover operator to be slightly biased towards the MST. The experimental results for random OCST problem instances and problem instances from the literature show that EAs using the γ-TX operator with a proper setting of γ show good performance.

The problems of the TX operator of the edge-set encoding emphasize the difficulties of a proper design of representations and operators. If it is known a priori that the optimal solutions for a problem are biased towards some solutions this bias can be exploited by developing representations and operators that are biased in a proper way. Then, problems can be solved more efficiently than when using non-biased encodings. However if the bias is too great, EAs fail if the optimal solutions are only slightly different from the solutions the representation and operator are biased to. Therefore, biased representations should be used with great care and only if there is a priori knowledge about the problem available. Otherwise, either non-biased representations should be used or appropriate mechanisms should be developed that allow to identify problem-specific knowledge during the EA run. For the edge-set encoding, the parameter γ can be, for example, incorporated into the genotype and evolved on-the-fly by the EA.

References

1. Hu, T.C.: Optimum communication spanning trees. SIAM Journal on Computing **3** (1974) 188–195
2. Narula, S.C., Ho, C.A.: Degree-constrained minimum spanning trees. Computers and Operations Research **7** (1980) 239–249
3. Raidl, G.R., Julstrom, B.A.: Edge-sets: An effective evolutionary coding of spanning trees. IEEE Transactions on Evolutionary Computation **7** (2003) 225–239
4. Rothlauf, F.: Representations for Genetic and Evolutionary Algorithms. 1 edn. Number 104 in Studies on Fuzziness and Soft Computing. Springer, Heidelberg (2002)
5. Julstrom, B., Raidl, G.: Weight-biased edge-crossover in evolutionary algorithms for two graph problems. In et al., G.L., ed.: Proceedings of the 16th ACM Symposium on Applied Computing, ACM Press (2001) 321–326
6. Tzschoppe, C., Rothlauf, F., Pesch, H.J.: The edge-set encoding revisited: On the bias of a direct representation for trees. In Deb, Kalyanmoy et al., ed.: Proceedings of the Genetic and Evolutionary Computation Conference 2004, Heidelberg, Springer (2004) 1174–1185

7. Rothlauf, F., Tzschoppe, C.: On the bias and performance of the edge-set encoding. Technical Report 2004/11, Department of Information Systems, University of Mannheim (2004)

8. Papadimitriou, C.H., Yannakakis, M.: Optimization, approximation, and complexity classes. J. Comput. System Sci. **43** (1991) 425–440

9. Rothlauf, F., Gerstacker, J., Heinzl, A.: On the optimal communication spanning tree problem. Technical Report 15/2003, Department of Information Systems, University of Mannheim (2003)

10. Palmer, C.C.: An approach to a problem in network design using genetic algorithms. unpublished PhD thesis, Polytechnic University, Troy, NY (1994)

11. Berry, L.T.M., Murtagh, B.A., McMahon, G.: Applications of a genetic-based algorithm for optimal design of tree-structured communication networks. In: Proceedings of the Regional Teletraffic Engineering Conference of the International Teletraffic Congress, Pretoria, South Africa (1995) 361–370

12. Raidl, G.R.: Various instances of optimal communication spanning tree problems. personal communciation (2001)

Ant Algorithm for the Graph Matching Problem

Olfa Sammoud[1], Christine Solnon[2], and Khaled Ghédira[1]

[1] SOIE, Institut Supérieur de Gestion de Tunis,
41 rue de la Liberté, Cité Bouchoucha, 2000 Le Bardo, Tunis
{olfa.sammoud,khaled.ghedira}@isg.rnu.tn
[2] LIRIS, CNRS UMR 5205, bât. Nautibus, University of Lyon I
43 Bd du 11 novembre, 69622 Villeurbanne cedex, France
christine.solnon@liris.cnrs.fr

Abstract. This paper describes a new Ant Colony Optimization (ACO) algorithm for solving Graph Matching Problems, the goal of which is to find the best matching between vertices of multi-labeled graphs. This new ACO algorithm is experimentally compared with greedy and reactive tabu approaches on subgraph isomorphism problems and on multivalent graph matching problems.

1 Introduction

Numerous applications require to measure the similarity of objects. For instance, Case-Based Reasoning (CBR) relies on the hypothesis that similar problems have similar solutions, so that CBR systems solve new problems by retrieving similar ones, for which solutions are known and can be adapted [1]. Also, information retrieval systems must be able to measure the similarity of documents and images in order to retrieve relevant documents from a database.

In many of these applications, objects are described by graphs, so that measuring objects similarity turns into determining graphs similarity, *i.e.,* matching graph vertices to identify their common features [6, 8, 10]. This may be done by looking for an exact graph or subgraph isomorphism in order to show graph equivalence or inclusion. However, the objects to be compared are usually not identical and the assumption of the existence of an isomorphism between the corresponding graphs is usually too strong. As a consequence, error-tolerant graph matchings such as maximum common subgraph and graph edit distance have been proposed [7, 10]. Such matchings drop the condition that all vertices and edges must be preserved: the goal is to find a *"best"* matching, *i.e.,* one which preserves a maximum number of vertices and edges.

Most recently, three different papers proposed to go one step further by introducing multivalent matchings, where a vertex in one graph may be matched with a set of vertices of the other graph:

- In [4], graph matching is used for model-based pattern recognition of brain images. In this case, the assumption of a bijection between regions of models and images is too strong: models have schematic aspects easy to segment

G.R. Raidl and J. Gottlieb (Eds.): EvoCOP 2005, LNCS 3448, pp. 213–223, 2005.

while images are noised and usually over-segmented. Therefore, scene recognition is better expressed as a multivalent matching problem where a set of vertices of the scene may be linked to a same vertex of the model.

- Guided by very similar motivations, [2] proposes a new graph edit distance that introduces two new edit operations —vertex splitting and merging— in order to handle the fact that images may be over- or under- segmented.
- In [9], graphs are used to model design objects in a computer-aided design application. In this context, vertices are used to represent object components and one single component of an object may play the same role than a set of components of another object, depending of the granularity of object description. Therefore, the authors introduce a similarity measure based on multivalent matchings so that one vertex in a graph may be associated with a set of vertices of the other graph.

The graph similarity measure of [9] is generic and is parameterized by functions that allow one to express domain dependent knowledge. Hence, [16] shows that the matchings introduced in [4] and [2] are special cases of the graph similarity measure of [9].

In this paper, we address the problem of computing this graph similarity measure. Indeed, [9] has proposed a first greedy algorithm that incrementally builds multivalent matchings. These matchings are quickly computed but are usually far from optimality. Hence, we propose to improve the quality of the constructed matchings by using the Ant Colony Optimization (ACO) meta-heuristic [11]: the idea is to use pheromone trails to keep track of the best components of matchings built by the greedy algorithm.

Section 2 briefly describes the generic graph similarity measure and the associated greedy algorithm introduced in [9]. Section 3 describes a new ACO algorithm —Ant Graph Matching (ANT-GM)— for computing this measure. Section 4 presents experimental results on different graph matching problems, and compares ANT-GM with the greedy algorithm of [9] and a reactive tabu search algorithm introduced in [16].

2   A Generic Similarity Measure for Multi-labeled Graphs

2.1   Definition of Multi-labeled Graphs

A directed graph is defined by a couple $G = (V, E)$, where V is a finite set of vertices and $E \subseteq V \times V$ is a set of directed edges. Vertices and edges may be associated with labels that describe their properties. Given a set L_V of vertex labels and a set L_E of edge labels, a multi-labeled graph is defined by a triple $G = \langle V, r_V, r_E \rangle$ such that:

- V is a finite set of vertices,
- $r_V \subseteq V \times L_V$ is a relation associating labels to vertices, *i.e.*, r_V is the set of couples (v_i, l) such that vertex v_i is labeled by l,

− $r_E \subseteq V \times V \times L_E$ is a relation associating labels to edges, *i.e.*, r_E is the set of triples (v_i, v_j, l) such that edge (v_i, v_j) is labeled by l. Note that the set E of edges of the graph can be defined by $E = \{(v_i, v_j) | \exists l, (v_i, v_j, l) \in r_E\}$.

We shall call the tuples of r_V and r_E the vertex and edge features of G. The set $descr(G) = r_V \cup r_E$ of all vertex and edge features of a graph G completely describes the graph G.

2.2 Similarity Measure

We now briefly describe the graph similarity measure introduced in [9], we refer the reader to [9] for more details. This similarity measure is defined for two multi-labeled graphs $G = \langle V, r_V, r_E \rangle$ and $G' = \langle V', r_{V'}, r_{E'} \rangle$, defined over the same sets of vertex and edge labels L_V and L_E, and such that $V \cap V' = \emptyset$.

The first step for measuring graph similarity is to match vertices. The matching function considered here is multivalent, *i.e.*, each vertex of one graph is matched with a possibly empty set of vertices of the other graph. More formally, a multivalent matching of two graphs G and G' is a set $m \subseteq V \times V'$ which contains every couple $(v, v') \in V \times V'$ such that vertex v is matched with vertex v'.

Once a multivalent matching is defined, the next step is to identify the set of features that are common to the two graphs with respect to this matching. This set contains all the features from both G and G' whose vertices (resp. edges) are matched by m to at least one vertex (resp. edge) that has the same feature. More formally, the set of common features $descr(G) \sqcap_m descr(G')$, with respect to a matching m, is defined as follows:

$$
\begin{aligned}
descr(G) \sqcap_m descr(G') \doteq{}& \{(v, l) \in r_V | \exists (v, v') \in m, (v', l) \in r_{V'}\} \\
& \cup \{(v', l) \in r_{V'} | \exists (v, v') \in m(v), (v, l) \in r_V\} \\
& \cup \{(v_i, v_j, l) \in r_E | \exists (v_i, v'_i) \in m, \exists (v_j, v'_j) \in m\, (v'_i, v'_j, l) \in r_{E'}\} \\
& \cup \{(v'_i, v'_j, l) \in r_{E'} | \exists (v_i, v'_i) \in m, \exists (v_j, v'_j) \in m\, (v_i, v_j, l) \in r_E\}
\end{aligned}
$$

Given a multivalent matching m, we also have to identify the set of split vertices, *i.e.*, the set of vertices that are matched to more than one vertex, each split vertex v being associated with the set s_v of its matched vertices:

$$
\begin{aligned}
splits(m) = {}& \{(v, s_v) \mid v \in V, \ s_v = \{v' \in V' | (v, v') \in m\}, |s_v| \geq 2\} \\
& \cup \{(v', s_{v'}) \mid v' \in V', \ s_{v'} = \{v \in V | (v, v') \in m\}, \ |s_{v'}| \geq 2\}
\end{aligned}
$$

The similarity of G and G' with respect to a matching m is then defined by:

$$
sim_m(G, G') = \frac{f(descr(G) \sqcap_m descr(G')) - g(splits(m))}{f(descr(G) \cup descr(G'))} \tag{1}
$$

where f and g are two functions that are defined to weight features and splits, depending on the considered application. For example, if f is the cardinality function and g is the null function, then the similarity is proportional to the number of common features with respect to the total number of features. If g is the cardinality function, instead of the null function, then the similarity is decreased proportionally to the number of split vertices.

Finally, the maximal similarity $sim(G, G')$ of two graphs G and G' is the greatest similarity with respect to all possible matchings, *i.e.*,

$$sim(G, G') = \max_{m \subseteq V \times V'} sim_m(G, G')$$

Note that the denominator in the definition of formula (1) does not depend on the matching m —this denominator is introduced to normalize the similarity value between zero and one. Hence, it will be sufficient to find the matching that maximizes the *score* function below:

$$score(m) = f(descr(G) \sqcap_m descr(G')) - g(splits(m))$$

2.3   Greedy Algorithm

A greedy algorithm for approximating $sim(G, G')$ is introduced in [9]. We briefly describe it as it is used as a starting point of our ACO algorithm.

The algorithm starts from the empty matching $m = \emptyset$, and iteratively adds to this matching couples of vertices that are chosen within the set $cand = V \times V' - m$ in a greedy way: at each step, the algorithm first selects the set of couples $(u, u') \in cand$ that most increase the *score* function. This set of best scored couples often contains more than one couple. To break ties between them, the potentiality of each candidate (u, u') is looked ahead by taking into account the features that are shared by edges starting from (resp. ending to) both u and u' and that are not already in $descr(G) \sqcap_{m \cup \{(u,u')\}} descr(G')$. More formally, one defines the set $look\_ahead_m(u, u')$ of potential common edge features by:

$$\{(u, v, l) \in r_E \mid \exists v' \in V', (u', v', l) \in r_{E'}\} \cup \{(u', v', l) \in r_{E'} \mid \exists v \in V, (u, v, l) \in r_E\}$$
$$\cup \{(v, u, l) \in r_E \mid \exists v' \in V', (v', u', l) \in r_{E'}\} \cup \{(v', u', l) \in r_{E'} \mid \exists v \in V, (v, u, l) \in r_E\}$$
$$- descr(G) \sqcap_{m \cup \{(u,u')\}} descr(G')$$

The next couple to enter the matching is randomly selected within the set of couples (u, u') that most increase the *score* function and that maximize $f(look\_ahead(u, u'))$.

This greedy algorithm stops iterating when every couple neither directly increases the score function nor has looked-ahead common edge features.

3   Description of ANT-GM

The greedy algorithm may be run several times, in order to compute different matchings. We propose to combine such iterated greedy constructions with the Ant Colony Optimization (ACO) meta-heuristic, in order to take benefit of the previously computed matchings when building new ones.

The ACO meta-heuristic is a bio-inspired approach that has been used to solve different hard combinatorial optimization problems [14, 12]. The main idea is to model the problem to solve as the search for a minimum cost path in a graph —called construction graph— and to use artificial ants to search for good paths. The behavior of artificial ants is inspired from real ants: they lay pheromone trails on graph components and they choose their path with respect to probabilities that depend on pheromone trails that have been previously laid, these pheromone trails progressively decrease by evaporation. Intuitively, this indirect stigmergic communication means aims at giving information about the quality of path components in order to attract ants, in the following iterations, towards the corresponding areas of the search space.

The proposed ACO algorithm for computing graph similarity follows the classical ACO algorithmic scheme for static combinatorial optimization problems [11]. At each cycle, each ant constructs a complete matching in a randomized greedy way, and then pheromone trails are updated. The algorithm stops iterating either when an ant has found an optimal matching, or when a maximum number of cycles has been performed.

3.1   Construction Graph

The construction graph is the graph on which artificial ants lay pheromone trails. Vertices of this graph are solution components that are selected by ants to generate solutions. In our graph matching application, ants build matchings by iteratively selecting couples of vertices to be matched. Hence, given two attributed graphs $G = (V, r_V, r_E)$ and $G' = (V', r_{V'}, r_{E'})$, the construction graph is the complete non-directed graph that associates a vertex to each couple $(u, u') \in V \times V'$.

3.2   Pheromone Trails

Ants communicate by laying pheromone trails on edges of the construction graph[1]. The amount of pheromone on an edge $< (u, u'), (v, v') >$ is noted $\tau_{<(u,u'),(v,v')>}$ and represents the learnt desirability of matching together u with u' and v with v'. Hence, to reward a matching m_i, ants lay pheromone trails between every pair of matched vertices $((u, u'), (v, v')) \in m_i^2$. Then, when constructing a new matching m_k, vertices that are matched in m_i will be more likely to be matched in m_k *if m_k already contains some matched vertices of m_i*. More precisely, the more m_k will contain matched vertices of m_i, the more the other matched vertices of m_i will be attractive.

3.3   Construction of a Matching by an Ant

At each cycle, each ant constructs a matching, starting from the empty matching $m = \emptyset$, by iteratively adding couples of vertices that are chosen within the set $cand = \{(u, u') \in V \times V' - m\}$. As usually in ACO algorithm, the choice of the next couple to be added to m is done with respect to a probability that depends

---

[1] We have defined another ACO algorithm where pheromone trails are laid on vertices of the construction graph (instead of edges). However, experiments showed us that this algorithm obtains much worse results than when pheromone is laid on edges.

on pheromone and heuristic factors. More formally, given a matching m and a set of candidates $cand$, the probability $p_m(u, u')$ of selecting $(u, u') \in cand$ is:

$$\frac{[\tau_m(u, u')]^\alpha \cdot [h1_m(u, u')]^{\beta_1} \cdot [h2_m(u, u')]^{\beta_2}}{\sum_{(v,v') \in cand} [\tau_m(v, v')]^\alpha \cdot [h1_m(v, v')]^{\beta_1} \cdot [h2_m(v, v')]^{\beta_2}} \qquad (2)$$

where

- $\tau_m(u, u')$ is the pheromone factor and is defined by the sum of all pheromone trails laying between the candidate (u, u') and every couple (v, v') already selected in m, i.e.,

$$\tau_m(u, u') = \sum_{(v,v') \in m} \tau_{<(u,u'),(v,v')>}$$

  When $m = \emptyset$, i.e., when choosing the first couple, $\tau_m(u, u') = 1$ so that the probability only depends on heuristic factors).
- $h1_m(u, u')$ is a first heuristic factor that aims at favoring couples that most increase the score function, i.e.,

$$h1_m(u, u') = score(m \cup \{(u, u')\}) - score(m)$$

- $h2_m(u, u')$ is a second heuristic factor that aims at favoring couples that have many looked-ahead features (as defined for the greedy algorithm described in Section 2), i.e.,

$$h2_m(u, u') = f(look\_ahead_m(u, u'))$$

- α, β_1, and β_2 are three parameters that determine the relative importance of the three factors.

Ants stop adding new couples to the matching m when every couple neither directly increases the score function nor has looked-ahead common edge features, or when the score function has not been increased since the last three iterations.

3.4   Pheromone Updating Step

Once every ant has constructed a matching, pheromone trails are updated according to the ACO meta-heuristic. First, evaporation is simulated by multiplying every pheromone trail $\tau_{<(u,u'),(v,v')>}$ by $(1 - \rho)$, where ρ is the pheromone evaporation rate such that $0 \le \rho \le 1$.

Then, the best ant of the cycle deposits pheromone. More precisely, let m_k be the best matching (with respect to the score function) built during the cycle (if there are several best matchings, ties are randomly broken), and m_{best} be the best matching built since the beginning of the run (including the current cycle). The quantity of pheromone laid is inversely proportional to the gap of score between m_k and m_{best}, i.e. it is equal to $1/(1 + score(m_{best}) - score(m_k))$. This quantity of pheromone is deposited on every edge $((u, u'), (v, v'))$ connecting two different couples (u, u') and (v, v') of m_k.

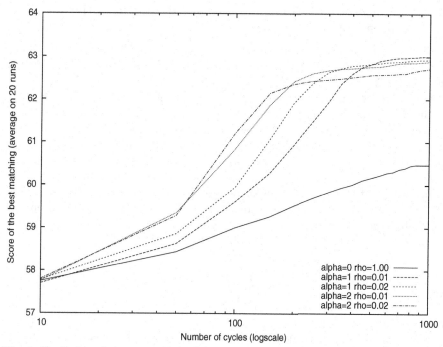

Fig. 1. Evolution of the score of the best found matching w.r.t. the number of cycles, for different settings of α and ρ (with 10 ants, $\beta_1 = 8$ and $\beta_2 = 3$)

4   Experimental Study of ANT-GM

4.1   Influence of Pheromone on the Solution Quality

As usually in ACO algorithms, the behavior of ANT-GM depends on its parameters, and more particularly on α, the pheromone factor weight, and ρ, the evaporation rate. Diversification can be emphasized both by decreasing α, so that ants become less sensitive to pheromone trails, and ρ, so that pheromone evaporates more slowly. When increasing the exploratory ability of ants in this way, better solutions are found, but as a counterpart it takes more longer time.

This is illustrated in Figure 1 on the $si2r001s80$ UNINA instance [13]. On this figure, one can remark that when α or ρ increase, ants converge quicker towards a matching: convergence occurs around cycle 500 when $\alpha=1$ and $\rho=0.01$, around cycle 350 when $\alpha=1$ and $\rho=0.02$, and around cycle 200 when $\alpha=2$ and $\rho=0.02$. As a counterpart, ants find better matchings, at the end of the solution process, when α and ρ are set to lower values such as $\alpha=1$ and $\rho=0.01$.

Note also that when $\alpha=0$ and $\rho=1$, *i.e.*, when pheromone is totaly ignored, so that the solution process is a pure randomized greedy one, the constructed matchings have a much lower score and hardly reach 60.5 at the end of the solution process, instead of more than 62.5 when pheromone is used. This shows that pheromone improves the solution quality.

4.2 Comparison of ANT-GM with Greedy and Reactive Approaches

Considered algorithms. We compare our ACO algorithm (ANT-GM) with the Greedy Search algorithm (GS) of [9] described in Section 2 and a Reactive Tabu Search algorithm (RTS) described in [16].

RTS improves a matching built by the greedy search algorithm of [9] by performing local search: the idea is to iteratively move from a matching to one of its neighbours (obtained by either adding or removing one couple of vertices) until the optimal solution is found or until a maximum number of moves have been performed. At each step, the search moves towards the best neighbour of the current matching (with respect to the same criteria than for the greedy algorithm). To avoid being trapped in locally optimal matchings, a Tabu list is used that memorizes the last moves in order to forbid backward moves. As proposed in [3], the length of this Tabu list is dynamically adapted during the search, depending on the need for diversification/intensification.

Experimental Setup. ANT-GM, GS, and RTS have been implemented in C++, and run on a 1.8Ghz pentium M with 512Mo RAM.

For ANT-GM, we have set α to 1, ρ to 0.01, β_1 to 8, β_2 to 3, the maximum number of cycles $MaxCycle$ to 1000 and the number of ants $nbAnts$ to 10, so that each run builds 10,000 matchings. Parameters of RTS have been set as recommended in [16].

To compare algorithms independently from implementation issues, all runs on a given instance are limited to a same number of moves, where a move is defined by the addition or removal of one couple of vertices to a matching. This limit on the number of moves depends on the considered instance. Indeed, one run of ANT-GM builds 10,000 matchings, but the size of these matchings, and therefore the number of moves performed by ANT-GM, depends on the considered instance. Hence, let x be the average size of the matchings built by ANT-GM for a given instance, the number of moves performed by ANT-GM is $x * 10,000$ so that the maximum number of moves for this instance is set to $x * 10,000$.

Each algorithm has been run 20 times on each instance of each benchmark.

Results on subgraph isomorphism problems. We first consider 11 benchmarks of subgraph isomorphism problems, for non labeled graphs, issued from a UNINA benchmark [13] and available at http://amalfi.dis.unina.it.graph. For each of these 11 benchmarks, we have considered the 30 first instances.

Each instance is composed of two graphs $G=(V,E)$ and $G'=(V',E')$ such that $|V| \leq |V'|$, and the goal is to find an injective function $\phi: V \to V'$ such that $(v_1, v_2) \in E \Rightarrow (\phi(v_1), \phi(v_2)) \in E'$.

To solve subgraph isomorphism problems with the generic similarity measure of formula (1), we define function g as the cardinality function and function f as a weighted sum where the weight of the features of G (resp. G') is 1 (resp. 0). In this case, $sim(G, G')=1$ if and only if there exists a mapping m such that $descr(G) \subseteq descr(G) \sqcap_m descr(G')$ (as $f(descr(G) \cup descr(G'))=|descr(G)|$)

Table 1. Results on 11 benchmark sets of subgraph isomorphism problems. For each benchmark set, the table first reports its name and the number of vertices of the two graphs to be matched. Then, for each algorithm, it reports the global success rate (GSR), *i.e.*, the percentage of successful runs over all runs for all instances of the benchmark, the instance success rate (ISR), *i.e.*, the percentage of instances that have been solved at least once over the twenty runs, and the number of moves (Mv) and the CPU time (T) spent to find the solution (average on successful runs only)

Benchmark	ANT–GM				GS				RTS			
Name (nb vertices)	GSR	ISR	Mv	T	GSR	ISR	Mv	T	GSR	ISR	Mv	T
si2r001s100 (20/100)	76.8	86.7	40492	33.2	33.3	33.3	89	0.2	67.5	100.0	9758	6.6
si2r001s80 (16/80)	93.3	100.0	42240	10.1	33.3	33.3	37	0.0	90.0	100.0	5585	2.4
si2r001s60 (12/60)	99.7	100.0	22164	2.8	46.7	46.7	15	0.0	99.2	100.0	1590	0.4
si4r001s80 (32/80)	81.3	90.0	110818	44.1	23.3	23.3	507	0.5	85.7	100.0	8292	7.5
si4r001s60 (24/60)	99.2	100.0	44539	9.0	40.0	40.0	39	0.1	93.2	100.0	5066	2.5
si4r001s40 (16/40)	100.0	100.0	8634	0.7	53.3	53.3	41	0.0	99.7	100.0	1759	0.4
si4r001s20 (8/20)	100.0	100.0	166	0.0	83.3	83.3	9	0.0	100.0	100.0	219	0.0
si4r005s40 (16/40)	89.7	96.7	34976	4.4	6.7	6.7	67	0.0	88.0	96.7	4647	1.0
si6r001s60 (36/60)	99.7	100.0	79738	21.0	63.3	63.3	110	0.1	94.5	100.0	6964	5.2
si6r001s40 (24/40)	100.0	100.0	16547	1.9	86.7	86.7	44	0.0	98.3	100.0	3101	1.0
si6r001s20 (12/20)	100.0	100.0	352	0.0	93.3	93.3	24	0.0	100.0	100.0	266	0.0
Average		94.5	97.6	36424	11.6	51.2	51.2	89	0.1	92.4	99.7	4295	2.45

and $splits(m) = \emptyset$, *i.e.*, $sim(G, G') = 1$ if and only if there exists a subgraph isomorphism.G and G'.

Table 1 reports results obtained on these subgraph isomorphism problems. These results first show that GS is much less successfull than both ANT–GM and RTS, being able to solve nearly twice as less instances. Moreover, global and instance success rates of GS are always equal and, when a solution is found, the number of moves performed to find it is always very low. Indeed, the search is not much diversified in GS: random choices are performed only to break ties between candidates that have equally highest scores. As a consequence, GS always computes very similar matchings and, given an instance, either it very quickly finds a solution, or it never finds it.

When comparing ANT–GM with RTS, one can note that the global success rate of ANT–GM is nearly always greater or equal to the one of RTS: 94.5% of the $20 * 30 * 11$ runs of ANT–GM have succeeded instead of 92.4% for RTS. However, the instance rate of ANT–GM is always smaller or equal to the instance success rate of RTS: 97.6% of the $30 * 11$ considered instances have been solved at least once over the 20 runs of ANT–GM instead of 99.7% for RTS. Actually, given an instance, the result of an execution of ANT–GM is nearly always the same (*i.e.*, either it nearly always fail or it nearly always succeed), whereas the result of an execution of RTS is more variable and highly depends on the starting point of the local search.

Table 1 also shows that ANT–GM performs 8.5 times as more moves as RTS to find a solution. However, as one move of ANT–GM is performed twice as fast, RTS is 4.7 times as fast as ANT–GM.

Table 2. Results on 5 multivalent matching problems. For each problem and for each algorithm, the table displays the average similarity (Sim), the average number of moves (Mv) and the average CPU time in seconds (T) needed to find the best solution

Problem name	ANT-GM			RTS		
	Sim	Mv	T	Sim	Mv	T
hom-v20-e60	0.795	303167	30.9	0.798	17747	2.2
hom-v30-e90	0.863	512746	155.0	0.865	14187	4.4
hom-v40-e120	0.885	685155	477.9	0.895	24801	13.7
hom-v45-e135	0.895	717767	709.5	0.904	60085	40.5
hom-v50-e150	0.804	847699	1075.6	0.913	53922	47.9
Average		0.848	613307	489.8	0.875	34149	21.7

Experimental comparison on multivalent matching problems. We have also compared ANT-GM and RTS on 5 multivalent graph matching problems that have been randomly generated. Each problem named hom-vN-eM is composed of a couple of non labeled graphs such that the first graph has N vertices and M edges (randomly generated) and the second graph is obtained by randomly removing 6 vertices and their incident egdes of the first graph, and then randomly splitting 5 vertices and their incident edges.

Table 2 shows the results obtained by ANT-GM and RTS on these multivalent matching problems. On this table, one can note that similarities computed by ANT-GM are slightly worse than those computed by RTS. Moreover, when graph sizes increase, this difference in quality becomes more important. Also, ANT-GM needs more moves to converge towards its best solution, and therefore it is more time consuming.

5   Conclusion

We have introduced in this paper ANT-GM, a new ACO algorithm for solving multivalent graph matching problems. First experiments on benchmarks of subgraph isomorphism problems showed us that ANT-GM is able to solve to optimality a wide majority of these problems. A key point of the multivalent graph matching problem is that each vertex may be mapped to a set of vertices, so that it can be used to evaluate the similarity of two graphs, and not only their equivalence, inclusion or intersection. As there does not yet exist benchmarks dedicated to this problem, we have generated random instances. Experiments showed us that, on these problems, ANT-GM is outperformed by a Reactive Tabu Search approach.

Further work will mainly concern the integration within ANT-GM of some local search technics such as the one used by RTS. Indeed, experiments showed us that results obtained by ANT-GM and RTS are rather complementary, each algorithm being able to solve instances that the other one cannot solve. Actually, the best performing ACO algorithms for many combinatorial problems are hybrid algorithms that combine probabilistic solution construction by a colony of ants with local search [12, 15].

References

1. A. Aamodt and E. Plaza. Case-Based Reasoning: Foundational Issues, Methodological Variations, and System Approaches. AI Communications, IOS Press, Amsterdam (NL), 7(1):39-59, 1994.
2. R. Ambauen, S. Fischer, and H. Bunke. Graph Edit Distance with Node Splitting and Merging, and Its Application to Diatom Identification. IAPR-TC15 Wksp on Graph-based Representation in Pattern Recognition, LNCS, Springer Verlag, 95-106, 2003.
3. R. Battiti and M. Protasi. Reactive Local Search for the Maximum Clique Problem. Algorithmica, Springer-Verlag, (29), 610-637, 2001.
4. M. Boeres, C. Ribeiro, and I. Bloch. A Randomized Heuristic for Scene Recognition by Graph Matching. WEA 2004, 100-113, 2004.
5. H. Bunke. Error-tolerant Graph Matching: A Formal Framework and Algorithms. Lecture Notes in Computer Science. Springer, Berlin, 1998.
6. H. Bunke and B.T. Messmer. Recent advances in graph matching. International Journal of Pattern Recognition and Artificial Intelligence, (11):169-203, 1997.
7. H. Bunke and K. Shearer. A graph distance metric based on maximal common subgraph. Pattern recognition letters, (19):255-259, 1998.
8. H. Bunke and X. Jiang. Graph matching and similarity. Volume Teodorescu, H-N, Mlynek, D. Kandel, A. Zimmermann, H-J. (ds.): Intelligent Systems and Interfaces, chapter 1, 2000.
9. P. Champin and C. Solnon. Measuring the similarity of labeled graphs. 5th International Conference on Case-Based Reasoning (ICCBR). Lecture Notes in Computer Science - Springer Verlag, 2003.
10. D. Conte, P. Foggia, C. Sansone, and M. Vento. Thirty years of graph matching in pattern recognition. International Journal of Pattern Recognition and Artificial Intelligence, 18(3):265-298, 2004.
11. M. Dorigo and G. Di Caro. The Ant Colony Optimization Meta-heuristic. In D. Corne, M. Dorigo, and F. Glover, editors, New Ideas in Optimization. McGraw Hill, London, UK, pages 11-32, 1999.
12. M. Dorigo and L. Gambardella. Ant Colony System: A cooperative learning approach to traveling salesman problem. IEEE transactions on evolutionary computation, 1(1):53-66, 1997.
13. P. Foggia, C. Sansone, and M. Vento. A database of graphs for isomorphism and sub-graph isomorphism benchmarking. In 3rd IAPR-TC15 Workshop on Graph-based Representations in Pattern recognition, pages 176-187, 2001.
14. V. Maniezzo and A.Colorni. The Ant System Applied to the Quadratic Assignement Problem. IEEE Transactions on Data and Knowledge Engineering, 11(5):769-778, 1999.
15. T. Stützle and H.H. Hoos. $\mathcal{MAX} - \mathcal{MIN}$ Ant System. Journal of Future Generation Computer Systems, 16:889-914,2000.
16. S. Sorlin and C. Solnon. Reactive Tabu Search for Measuring Graph Similarity. to appear in 5th IAPR Workshop on Graph-based Representations in Pattern Recognition (GbR 2005), LNCS, Springer Verlag, 2005.

An Adaptive Genetic Algorithm for the Minimal Switching Graph Problem

Maolin Tang

School of Software Engineering and Data Communications,
Queensland University of Technology,
2 George Street, Brisbane, Australia
m.tang@qut.edu.au

Abstract. Minimal Switching Graph (MSG) is a graph-theoretic representation of the constrained via minimization problem — a combinatorial optimization problem in integrated circuit design automation. From a computational point of view, the problem is NP-complete. Hence, a genetic algorithm (GA) was proposed to tackle the problem, and the experiments showed that the GA was efficient for solving large-scale via minimization problems. However, it is observed that the GA is sensitive to the permutation of the genes in the encoding scheme. For an MSG problem, if different permutations of the genes are used the performances of the GA are quite different. In this paper, we present a new GA for MSG problem. Different from the original GA, this new GA has a self-adaptive encoding mechanism that can adapt the permutation of the genes in the encoding scheme to the underlying MSG problem. Experimental results show that this adaptive GA outperforms the original GA.

1 Introduction

Minimal Switching Graph (MSG) [1] is a graph-theoretic representation of the constrained via minimization problem — a combinatorial optimization problem in integrated circuit design automation. From a computational point of view, the problem is NP-complete [2]. Hence, a genetic algorithm (GA) was proposed to tackle the problem [3], and the experiments showed that the GA was good for solving large-scale via minimization problems. However, it is observed that the GA is sensitive to the permutation of the genes in the encoding scheme. For an MSG problem, if different permutations of the genes are used the performances the GA are quite different.

The power of GAs comes from their ability to preserve good pieces, or building blocks, from parents and to combine them to produce highly fit children. However, when the genes of building blocks are spread across the chromosome of a GA representation, the building blocks are most likely destroyed by the crossover operator, and therefore it is less likely for the GA to obtain an optimal solution. The problem of building block disruption is often referred to as the *linkage problem* [4]. The unscalability of the GA was caused by the linkage problem.

G.R. Raidl and J. Gottlieb (Eds.): EvoCOP 2005, LNCS 3448, pp. 224–233, 2005.

The techniques for handling the linkage problem are categorized into two classes. The first class of techniques is based on changing the representation of solutions in the algorithm or evolving the recombination operators [4, 5, 6, 7]. The second class of techniques is based on extracting some information from the entire set of promising solutions in order to generate new solutions [8, 9, 10]. In this paper, we will focus on the first class of techniques.

This paper presents a new GA for the MSG problem. Different from conventional GAs, this GA uses a knowledge-based self-adaptive encoding mechanism to optimize the permutation of the genes before it actually starts generating the initial population and evolving the population. This GA has been implemented and experimental results show that this GA outperforms the conventional GA.

The remaining paper is organized as follows. In Section 2 the MSG problem is formalized. Then a GA encoding scheme for the MSG problem is presented in Section 3. Section 4 details the self-adaptive encoding mechanism, and Section 5 shows the experimental results for the adaptive GA. Finally, this research work is concluded in Section 6.

2 Minimal Switching Graph Problem

A *directed bigraph* is a directed graph whose vertices can be partitioned into two disjoint sets such that no vertices within the same set are adjacent. A directed bigraph can be denoted as $G = (V_1 \cup V_2, E)$, where V_1 and V_2 are two disjoint vertex sets, and E is a set of directed edges, or arcs. Figure 1 shows a directed bigraph, where $V_1 = \{v_1, v_2, v_3, v_4, v_5, v_6, v_7\}$, $V_2 = \{v_8, v_9, v_{10}, v_{11}, v_{12}, v_{13}\}$, and $E = \{< v_1, v_8 >, < v_1, v_{10} >, < v_{12}, v_2 >, < v_{10}, v_3 >, < v_{11}, v_3 >, < v_3, v_{12} >, < v_{13}, v_3 >, < v_9, v_4 >, < v_4, v_{11} >, < v_4, v_{12} >, < v_5, v_9 >, < v_{13}, v_6 >, < v_7, v_8 >\}$.

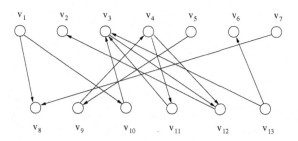

Fig. 1. A directed bigraph

A directed bigraph can be represented in a matrix $M_{|V_1| \times |V_2|} = [m_{ij}]$. The rows of the matrix present the vertices in V_1 and the columns of the matrix present the vertices in V_2. If $< v_i, v_j > \in E$, $v_i \in V_1$, and $v_j \in V_2$, then $m_{ij} = 1$, if $< v_i, v_j > \in E$, $v_i \in V_2$, and $v_j \in V_1$, then $m_{ij} = -1$, otherwise, $m_{ij} = 0$, where

m_{ij} is the ith row and j^{th} column element in $M_{|V_1| \times |V_2|}$. For example, the directed bigraph shown in Figure 1 is represented by the following matrix:

$$
\begin{bmatrix}
1 & 0 & 1 & 0 & 0 & 0 \\
0 & 0 & 0 & 0 & -1 & 0 \\
0 & 0 & -1 & -1 & 1 & -1 \\
0 & -1 & 0 & 1 & 1 & 0 \\
0 & 1 & 0 & 0 & 0 & 0 \\
0 & 0 & 0 & 0 & 0 & -1 \\
1 & 0 & 0 & 0 & 0 & 0
\end{bmatrix}
$$

In the above matrix, the rows from the top to the bottom represent v_1, v_2, $\cdots$, and v_7 respectively, and the columns from the left to the right represent v_8, v_9, $\cdots$, and v_{13} respectively.

Given a directed bigraph G, a *switching graph* of G, denoted as $G(S)$, is defined as the directed graph obtained by reversing the direction of all the arcs being incident to the vertices in $S \subseteq V_1$. Figure 2 displays a switching graph of G, $G(\{v_3, v_6\})$.

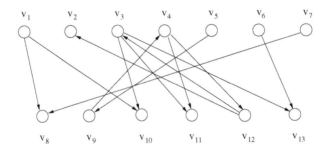

Fig. 2. A switching graph

It is pointed out that a directed bigraph G itself is a special switching graph of G in which $S = \phi$. A directed bigraph $G = (V_1 \cup V_2, E)$ has $2^{|V_1|}$ switching graphs.

The vertices in V_2 are categorized into two types: *desirable vertices* and *undesirable vertices*. Desirable vertices are those vertices whose either in-degree or out-degree is zero, and undesirable vertices are those vertices whose both in-degree and out-degree are not zero.

Given a directed bigraph G, the objective is to find a switching graph $G(S)$ such that the number of desirable vertices is maximal. This is so-called MSG problem.

3 GA Encoding Scheme for the MSG Problem

GA encoding transforms a potential solution to an underlying problem into a set of parameters, known as *genes*. These genes are structured to form a string of values, often referred to as a *chromosome*.

A *schema H* is a similarity template describing a subset of strings with similarities at certain string positions [5]. For example, $0*1***1$ is a schema representing a subset of 7-bit binary strings in which the first bit is 0, the third and seventh bits are 1. The $*$ symbol means *don't care*. The *defining length* of a schema H, denoted as $\delta(H)$, is the distance between the first and last positions at which values are fixed. For example, $\delta(0*1***1) = 6$. A string of n binary bits contains 2^n schemata. Those highly fit, meaningful schemata are *building blocks*. In a GA encoding scheme, building blocks are expected to be short in defining length; otherwise, they may be destroyed by the crossover operator.

Suppose $G = (V_1 \cup V_2, E)$ is a directed bigraph, and $G(S)$ is a switching graph of G. Let $n_1 = |V_1|$ and $n_2 = |V_2|$. Then, $G(S)$ can be represented in a binary string of n_1 bits, $b_1 b_2 \cdots b_{n_1}$, where

$$b_i = \begin{cases} 1 & \text{if } v_i \in S \\ 0 & \text{otherwise ,} \end{cases}$$

and $1 \le i \le n_1$. For example, the switching graph $G(\{v_3, v_6\})$ shown in Figure 2 is encoded as 0010010, and $G = G(\phi)$ is encoded 0000000. In this GA, each individual is a switching graph encoded in a binary string.

This encoding scheme maps the underlying problem to a binary string GA representation naturally. It uses minimal alphabets, and the crossover and mutation operators never produce any invalid individuals under this representation because all the binary strings of length n_1 represent a valid switching graph of the MSG problem.

4 Self-Adaptive Encoding

There are $n_1!$ different permutations of the genes $(b_1, b_2, \cdots, b_{n_1})$, and the linkages of building blocks in different permutations are different. In one permutation of the genes the defining length of a building block may be longer, while in another permutation of the genes the defining length of the building block may be shorter. Hence, it is desirable to find a permutation of the genes in which all building blocks have a shorter defining length.

Building blocks and their linkages cannot be known beforehand by the GA. Hence, given an MSG problem the GA needs to find a good permutation at runtime before it actually starts evolving the individuals in the initial population. To find a good permutation of the genes for the encoding scheme, we need to identify building blocks and have a strategy to discover a good permutation of the genes. In the following we address the two issues.

4.1   Identifying Building Blocks

An *atomic building block* for the MSG problem is a schema which a vertex is a desirable vertex in the corresponding switching graph. For instance, $1******1$ is an atomic building block for the MSG problem shown in Figure 1 as it represents a collection of switching graphs on which v_8 is a desirable vertex. The following lists all the atomic building blocks for that MSG problem:

$$
\begin{aligned}
s_1 &= \begin{bmatrix} 1 * * * * * 1 \end{bmatrix} \\
s_2 &= \begin{bmatrix} 0 * * * * * 0 \end{bmatrix} \\
s_3 &= \begin{bmatrix} * * * 1 0 * * \end{bmatrix} \\
s_4 &= \begin{bmatrix} * * * 0 1 * * \end{bmatrix} \\
s_5 &= \begin{bmatrix} 1 * 0 * * * * \end{bmatrix} \\
s_6 &= \begin{bmatrix} 0 * 1 * * * * \end{bmatrix} \\
s_7 &= \begin{bmatrix} * * 0 1 * * * \end{bmatrix} \\
s_8 &= \begin{bmatrix} * * 1 0 * * * \end{bmatrix} \\
s_9 &= \begin{bmatrix} * 1 0 0 * * * \end{bmatrix} \\
s_{10} &= \begin{bmatrix} * 0 1 1 * * * \end{bmatrix} \\
s_{11} &= \begin{bmatrix} * * 0 * * 0 * \end{bmatrix} \\
s_{12} &= \begin{bmatrix} * * 1 * * 1 * \end{bmatrix}
\end{aligned}
$$

Among these atomic building blocks, s_1 and s_2 represent the class of switching graphs on which v_9 is a desirable vertex; s_3 and s_4 stand for the set of switching graphs on which v_{10} is a desirable vertex; etc. It is expected that these building blocks are preserved in the population of the GA and are combined to form highly fit individuals. For example, by combining an individual containing s_1 and an individual containing s_3 it is expected to produce a new individual containing the schema $1**10*1$, which represents both v_9 and v_{10} as desirable vertices on the corresponding switching graph. The schema $1**10*1$ is a *compound building block*. It is pointed out that some building blocks are exclusive from each other. For example, it is impossible for an individual to contain both s_7 and s_9 due to the conflict at bits 3 and 4.

When identifying building blocks, we only identify atomic building blocks and do not identify compound building blocks because of the following reasons: firstly, the atomic building blocks are essential to the GA to generate an optimal solution, but not for the compound building blocks because the compound building blocks can be built during the evolution of the GA as long as the atomic building blocks are preserved in the population; secondly, it is not practical to identify all the compound building blocks. In fact, to identify all the compound building blocks may be as complex as the original MSG problem. Hence, the GA identifies atomic building blocks only.

There are two atomic building blocks associated to a vertex in V_2. Hence, for a directed bigraph $G = (V_1 \cup V_2, E)$, there are $2 \times |V_2|$ atomic building blocks. It takes $O(|V_1|)$ time to identify two atomic building blocks associated to a vertex in V_2. Therefore, the computational complexity for identifying all atomic building blocks is $O(|V_1| \times |V_2|)$.

4.2   Self-Adaptive Encoding Algorithm

Given an MSG problem represented by a matrix, the GA uses the encoding scheme to transform the matrix into the binary string genetic representation. The permutation of the genes in this initial encoding is in the order as in the matrix. Then, the GA identifies all atomic building blocks. Once the atomic building blocks have been identified, the GA uses a so called *self-adaptive encoding* mechanism to optimize the permutation of the genes in the encoding scheme online, and then generate the initial generation and evolves it until the termination condition is satisfied.

It is observed that the permutation of the genes does not have to be perfect for the GA to get a satisfactory solution, and that the longest defining length of the atomic building blocks directly affects the performance of the GA although the total defining length of all atomic building blocks is also important. Therefore, we design a computationally efficient hill-climbing algorithm to gradually minimize the longest defining length of the atomic building blocks without increasing the total defining length of the atomic building blocks. Below is the heuristic algorithm:

1. Calculate the defining length of the atomic building blocks and find the atomic building block that has the longest defining length;
2. Extract the genes associated with the building block that has the longest defining length;
3. Generate a new permutation of the genes by moving the right-most gene to the left of the second right-most gene of the longest defining length building block;
4. Update the atomic building blocks according to the new permutation of the genes;
5. Calculate the defining length of the atomic building blocks and find the atomic building block that has the longest defining length;
6. If the longest defining length under the new permutation is shorter than that under the current permutation and the total defining length of the atomic building blocks under the new permutation is not longer than that under the current permutation, then replace the current permutation with the new permutation and go to 1; otherwise, stop.

Denote the permutation of the genes in the initial GA representation for the MSG problem shown in Figure 1 as $b_1 b_2 b_3 b_4 b_5 b_6 b_7$. The new encoding obtained by the self-adaptive encoding algorithm would be $b_7 b_1 b_6 b_3 b_4 b_2 b_5$. In this new encoding, the building blocks become:

$$s_1 = \begin{bmatrix} 1 & 1 & * & * & * & * & * \end{bmatrix}$$
$$s_2 = \begin{bmatrix} 0 & 0 & * & * & * & * & * \end{bmatrix}$$
$$s_3 = \begin{bmatrix} * & * & * & * & 1 & * & 0 \end{bmatrix}$$
$$s_4 = \begin{bmatrix} * & * & * & * & 0 & * & 1 \end{bmatrix}$$
$$s_5 = \begin{bmatrix} * & 1 & * & 0 & * & * & * \end{bmatrix}$$
$$s_6 = \begin{bmatrix} * & 0 & * & 1 & * & * & * \end{bmatrix}$$

$$s_7 = \begin{bmatrix} * * * 0\,1 * * \end{bmatrix}$$
$$s_8 = \begin{bmatrix} * * * 1\,0 * * \end{bmatrix}$$
$$s_9 = \begin{bmatrix} * * * 0\,0\,1 * \end{bmatrix}$$
$$s_{10} = \begin{bmatrix} * * * 1\,1\,0 * \end{bmatrix}$$
$$s_{11} = \begin{bmatrix} * * 0\,0 * * * \end{bmatrix}$$
$$s_{12} = \begin{bmatrix} * * 1\,1 * * * \end{bmatrix}$$

In the initial encoding, the longest defining length and the total defining length of the atomic building blocks were 6 and 30 respectively; in the new encoding the total defining length has been dropped to 2 and the average defining length has been reduced to 18.

5   Experiment

5.1   A Benchmark MSG Problem

It is desirable to use large-scale and complex benchmarks with known optimal solutions to evaluate the performance of the adaptive GA. Unfortunately, there are no such benchmarks available. Hence, we construct such a large MSG problem.

The large MSG problem consists of 16 components of the directed bigraph graph shown in Figure 3. The size of the component MSG problem is 3, and individuals 100 and 011 represent two optimal solutions in which all the three vertices in V_2 are desirable vertices (the fitness of the solutions is 3). The maximal defining length of the atomic building blocks for the component MSG problem is 2.

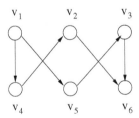

Fig. 3. A component directed bigraph

The size of the benchmark MSG problem is 48, thereby the search space is 2^{48}. A combination of 16 3-bit binary strings 100 or 001 constitutes an optimal solution to the large problem, in which all 48 vertices are desirable ones. The fitness of an optimal solution is 48.

5.2   Experimental Results

This experiment focuses on investigating the effect of the adaptive encoding on the performance of the GA. Therefore, we implement a GA for the MSG problem and develop two versions of the GA: one using the self-adaptive encoding and one not using the self-adaptive encoding, and compare the performance of the two versions of the GA.

To make the experiment simple and to exclude factors that may influence the performance and behavior of the GA, we implement the GA as a conventional one with a normal 1-point crossover operator and a normal mutation operator. The selection strategy used by the GA is the roulette selection. In order to get unbiased experimental results, we randomly generate the initial permutation of the genes, use the same initial population for the two versions of the GA. The parameters of the GA are fixed at the following values: the probability of crossover is 0.95, the probability of mutation is 0.05, and the size of population is $|V_1| * |V_2|/2$.

We run the two versions of the GA 10 times each. Each time the permutation of the genes is randomly generated. we keep the GA running until an optimal solution is found. Table 1 shows the experimental results, which include the characteristics of the randomly generated encoding (the longest defining length and total defining length of the atomic building blocks), and the computation time for each run.

Table 1. Comparisons of GA without self-adaptive encoding (GA without SAE) and GA with self-adaptive encoding (GA with SAE)

	GA without SAE			GA with SAE		
Runs	Longest δ	Total δ	Time (sec)	Longest δ	Total δ	Time (sec)
1	43	698	27.74	2	62	9.51
2	44	805	26.88	6	99	20.01
3	44	763	30.68	8	178	11.95
4	37	666	24.78	11	166	18.56
5	32	611	31.03	8	123	19.60
6	42	658	23.55	4	90	5.69
7	41	563	30.00	5	99	17.55
8	40	773	34.26	3	74	14.12
9	45	753	29.29	6	118	22.38
10	45	575	29.67	4	79	16.14

It can be seen from Table 1 that the GA with the self-adaptive encoding spends significantly less time than the GA without the self-adaptive encoding to find an optimal solution. This indicates that the GA with the self-adaptive encoding is more efficient than the GA without the self-adaptive encoding.

The GA with the self-adaptive encoding uses encodings in which the longest defining length and total defining length of the atomic building blocks are relatively short. Therefore, the building blocks in the GA with the self-adaptive encoding have more chances to be preserved in the crossover operation, and therefore better quality offsprings can be produced by the crossover operator. As a result, the total fitness value of the population in the GA with the self-adaptive encoding increases quicker than in the GA without the self-adaptive encoding, and thus an optimal solution can be built quicker in the GA with self-adaptive encoding. This was observed when we monitored the dynamics of the two versions of the GA.

6   Conclusions

The MSG problem is a combinatorial optimization problem in integrated circuit design automation, and GAs are suitable for solving the problem. This paper has presented an adaptive GA for the MSG problem.

This adaptive GA tackles the potential linkage problem using a knowledge-based self-adaptive encoding mechanism, which can adapt the permutation of the genes in the encoding scheme to the MSG problem online. This GA has been implemented and experimental results have shown that this adaptive GA outperforms the original GA.

This GA is similar to messy GAs in that it uses a preprocessing to adapt the encoding to the underlying problem before actually starting generating the initial population and evolving the population. However, this GA is funda-mentally different from messy GAs in that it treats the encoding scheme as a white-box and directly uses the domain-specific knowledge to improve the en-coding while messy GAs treat the encoding scheme as a black-box and do not use any domain-specific knowledge. Generally speaking, a GA that exploits the domain-specific knowledge is more efficient and effective than a GA that does not use any domain-specific knowledge. Since the adaptive GA uses domain-specific knowledge, however, the adaptive encoding mechanism cannot be directly em-ployed by other GAs algorithm although the basic idea behind the mechanism can be used.

This paper has presented some preliminary research on an adaptive GA for the MSG problem. This research focuses on how to improve the performance of the GA by optimizing the permutation of the genes. Further research on the adaptive GA can be extended to adaptive GA parameter setting. In addition, the performance of the adaptive GA could be further improved if a sophisticated GA is used. For example, a steady-state GA with higher mutation rates, no duplicates allowed, and always deleting the worst population member may speed up the evolution of the GA. Finally, it is interesting to compare this adaptive GA with a GA using a uniform crossover operator, a messy GA, and a LLGA.

Acknowledgement

The author would like to thank the anonymous reviewers for their valuable comments and suggestions on the manuscript of this paper.

References

1. Tang, M., Eshraghian, K., and Cheung, H.N.: An Efficient Approach to Constrained Via Minimization for Two-Layer VLSI Routing. Proc. IEEE Asia and South Pacific Design Automation Conference. Hong Kong (1999) 149–152
2. Naclerio, N.J., Masuda, S., and Nakajima, K.: The Via Minimization problem is NP-complete. IEEE Transactions on Computers. **38** (1989) 1604–1608
3. Tang, M., Eshraghian, K., and Cheung, H.N.: A Genetic Algorithm for Constrained Via Minimization. Proc. IEEE International Conference on Neural Information Processing. Perth (1999) 435–440
4. Harik, G.R. and Goldberg, D.E.: Linkage Learning. IlliGAL Technical Report 96006 (1996)
5. Goldberg, D.E.: Genetic Algorithms in Search, Optimization, and Machine Learning. Addison-Wesley, Reading (1989)
6. Goldberg, D.E., Korb, B.,and Deb, K.: Messy genetic algorithms: Motivation, analysis, and first results. Complex Systems **3** (1989) 493–530
7. Harik, G.R.: Lingake Learning Via Probabilistic Modeling. IlliGAL Technical Report 99010 (1999)
8. Mühlenbein, H. and Paaß, G.: From Recombination of Genes to the Estimation of Distributions I. Binary Parameters. In: Eiben, A., Bäck, T., Shoenauer, M. and Schwefel, H.-P. (eds): Parallel Problem Solving from Nature - PPSN IV, Springer-Verlag, Berlin (1996) 178–187
9. Harik, G. R., Lobo, F.G. and Goldberg, D.E.: The Compact Genetic Algorithm. International Conference on Evolutionary Computation, IEEE, New Jersey (1998) 523–528.
10. Pelikan, M., Goldberg D.E. and Cantu-Paz, K.: Linkage Problem, Distribution Estimation, and Bayesian Networks. Evolutionary Computation. **8** 311-340
11. Sastry, K.: Analysis of Mixing in Genetic Algorithms: A survey. IlliGAL Report No. 2002012 (2002)
12. Sastry, K. and Goldberg D.E.: How Well Does A Single-Point Crossover Mix Building Blocks with Tight Linkage?. IlliGAL Report No. 2002013 (2002)
13. Michalewicz, Z.: Genetic Algorithms + Data Structures = Evolution Programs. 3rd edn. Springer-Verlag, Berlin Heidelberg New York (1996)
14. Prügel-Bennett, A.: Modeling Cross-Induced Linkage in Genetic Algorithms. IEEE Trans. on Evolutionary Computation **5** (2001) 376–387
15. Thierens, D. and Goldberg, D.E.: Mixing in Genetic Algorithms. Proc. the 5th Int. Conf. on Genetic Algorithms. (1993) 38–45

An Improved Simulated Annealing Method for the Combinatorial Sub-problem of the Profit-Based Unit Commitment Problem

T. Aruldoss Albert Victoire[1] and A. Ebenezer Jeyakumar[2]

[1] Department of Electrical and Electronics Engineering,
Karunya Institute of Technology, Coimbatore, India
aruldoss@karunya.edu
[2] Department of Electrical and Electronics Engineering,
Government College of Technology, Coimbatore, India
ebyjaya@rediffmail.com

Abstract. Here is presented an improved simulated annealing (SA) method for solving the combinatorial sub-problem of profit-based unit commitment (UC) problem in electric power and energy systems. The UC problem is divided into a combinatorial sub-problem in unit status variables and a non-linear programming sub-problem in unit power output variables. The simulated annealing method with an improved random perturbation of current solution scheme is proposed to solve the combinatorial sub-problem. A simple scheme for generating initial feasible commitment schedule for the SA method to solve the combinatorial problem is also proposed. The non-linear programming sub-problem is solved using the sequential quadratic programming (SQP) technique. Several example systems are solved to validate the robustness and effectiveness of the proposed technique for the profit-based UC problem.

1 Introduction

Unit commitment is the problem of determining the optimal set of generating units within a power system to be used during the next one to seven days [1]. The UC problem to minimize production costs (mainly fuel cost) and transition costs (start-up/shut-down costs) is traditionally referred as cost-based unit commitment (CBUC) problem [2-7]. In a CBUC problem, utilities has to produce power to satisfy their customers with the minimum production cost fulfilling the condition that all power demand and reserve must be met.

In the recent past, due to deregulation of electricity industry, power producers are identified as GENCOs (power GENerating COmpanies), have more freedom to utilize their generation unit capabilities to enhance their benefits in the power market [8-10]. Here GENCOs sells power in the spot market, and sell spinning reserves in the reserve markets [8]. The GENCOs can consider a UC schedule that produces less than the predicted power demand, with its ultimate aim to maximize its own profit, regardless of the system wide profit [9-10]. For this UC problem the objective is the profit

G.R. Raidl and J. Gottlieb (Eds.): EvoCOP 2005, LNCS 3448, pp. 234–245, 2005.

which is not the negative of production cost, rather it is defined as the revenue minus production cost [9-10]. This updated UC problem with a different objective is referred as price-based UC (PBUC) problem [9]. Deregulation of electricity industry, also brought challenges is to maintain efficient use of generation resources in market-based environment, thereby maximizing the profit of GENCOs [3,10].

In General, a UC problem is referred as a mixed combinatorial and non-linear optimization problem [3]. It is very complex to solve because of its enormous dimension, non-linear objective function, and coupling constraints [2]. Thus a need for optimality exists for this highly nonlinear and computationally difficult power system problem. Researchers studied this complex problem for decades and many conventional and Meta-heuristic techniques have been developed [3].

Solving this complex problem using several conventional solution techniques involves a number of simplifying assumptions, and consumes a large amount of computation time. Unlike strict conventional techniques, meta-heuristics has the apparent ability to adapt to non-linearities and discontinuities commonly found in power system problems. All these techniques have the attractive feature of assured convergence under appropriate assumptions. The main difficulty in heuristics is their sensitivity to the choice of parameters and long computation time [3]. Many advanced operators are proposed for reducing search time to acceptable values, but these sophisticated operators are problem specific. To overcome these difficulties, hybrid techniques combining different optimization techniques were proposed [3,7]. These techniques accommodate more constraints and produce better solutions [7]. Despite the extensive work carried out, building a consistent technique to solve UC problem is still evolving.

This article emphasizes the effectiveness of heuristic techniques for the UC problem, by proposing an improved SA technique. The SA [6,11] method has proved itself as a powerful technique for solving combinatorial optimization problems. It has the ability of escaping local minima by incorporating a probability function in accepting or rejecting new solutions. Another advantage of SA method is it does not need large computer memory [6,7,11].

SQP [12] seems to be the best nonlinear programming methods for constrained optimization. It outperforms every other nonlinear programming method in terms of efficiency, accuracy, and percentage of successful solutions, over a large number of test problems. In this article the combinatorial part of the UC problem is solved using the SA method. A modified approach for generating the trial schedule as the neighbourhood of the current feasible schedule and a simple procedure to generate the initial feasible schedule for the SA method is proposed. The non-linear optimization part (Economic Dispatch Problem (EDP)) is solved using SQP technique.

2 Nomenclature

$FC_i(P_{it})$	: Fuel Cost function of unit i (\$)
ST_i	: Start up of cost of unit i (\$)
PF	: Profit of the GENCO (\$)
D_t	: Power demand at time t (MW)

SR_t	: Power reserve at time t (MW)
P_{it}	: Power produced by unit i at time t (MW)
R_{it}	: Reserve generation of unit i at time t (MW)
X_{it}	: ON/OFF status of unit i at time t
N	: Number of generating units.
H	: Number of hours in the scheduling horizon.
$P_{i\min}/P_{i\max}$	: Minimum/Maximum generation capability of unit i (MW)
$Toff_i/Ton_i$	: OFF/ON time of unit i (hr)
$Tup_i/Tdown_i$	: Minimum up/down time of unit i (hr)
S_{hi}/S_{ci}	: Start-up costs incurred for a hot/cold start for unit i ($)
$t_{icold\ start}$	: Number of hours that it takes for the boiler to cool down. (hr)
SP_t/RP_t	: Forecasted spot/reserve price at hour t ($)
ρ	: Probability that the reserve is called and generated.
$rand(0,1)$	: Random number between 0 and 1.

3 Problem Formulation

A PBUC problem is formulated mathematically by the following equations [10]:
The objective function

$$Min \quad TC - RV \quad or \quad Max \quad PF = RV - TC \ (\$) \tag{1}$$

Where, TC is the total cost for the GENCO ($) to commit N units during the given time. The total cost, is minimized by economically dispatching the units and this is a non-linear programming sub-problem of the UC problem, commonly referred as EDP. RV is the revenue earned by the GENCO ($) by selling the generated power.

$$RV = \sum_{i=1}^{N}\sum_{t=1}^{H}(P_{it}.SP_t).X_{it} + \sum_{i=1}^{N}\sum_{t=1}^{H}\rho.RP_t.R_{it}.X_{it} \tag{2}$$

$$TC = (1-\rho)\sum_{i=1}^{N}\sum_{t=1}^{H}FC_i(P_{it}).X_{it}$$
$$+ \rho\sum_{i=1}^{N}\sum_{t=1}^{H}FC_i(P_{it}+R_{it}).X_{it} + ST_i.X_{it}(1-X_{i(t-1)}) \tag{3}$$

The classic EDP minimizes the following incremental fuel cost function associated to dispatchable units,

$$FC_i(P_{it}) = a_i P_{it}^2 + b_i P_{it} + c_i \qquad \text{i=1,2, ..., N and t=1,2,...T} \qquad (4)$$

where, a_i, b_i, c_i are cost coefficients of the ith generating unit.

Start-up cost:

$$ST_i = \begin{cases} S_{ci} & if \quad Toff_i \geq t_{icoldstart} \\ \\ S_{hi} \end{cases} \qquad (5)$$

The start-up cost is the cost spent for committing a unit which was de-committed in previous hours. Hot start represents the starting of the generating unit before the boiler parameters reaches below the specified values. Cold start indicates the start of the unit when the boiler parameters reach below the specified values.

Subject to the following constraints:
Power demand is to be met to by the committed units at a given time:

$$\sum_{i=1}^{N} P_{it} X_{it} \leq D_t, \quad t = 1,...,H. \qquad (6)$$

Reserve need to be maintained at a given time:

$$\sum_{i=1}^{N} R_{it} X_{it} \leq SR_t, \quad t = 1,...,H. \qquad (7)$$

Minimum uptime and minimum downtime:
$$Toff_i \geq Tdown_i$$
$$Ton_i \geq Tup_i \qquad (8)$$

A generating unit cannot be committed for $Tdown_i$ hours once it is de-committed and similarly it cannot be de-committed for Tup_i hours once it is committed.

Combining constraints given in Eqn (6) and (7), we may write,

Power and Reserve Limits:

$$\sum_{i=1}^{N} X_{it} P_{i\max} \leq D_t + SR_t \quad t = 1,......,H \qquad (9)$$

In a PBUC problem or even a cost based unit commitment problem of the regulated system, the predicted demand, reserve, and the prices are important parameters. It is assumed that all data are readily available, as was assumed in [21]. But the predicted load demand is only known through short-term load forecasting, errors always exist in the forecasted system loads. Moreover, the spinning reserve constraint practically is based on the probability of abnormal conditions that might result in insufficient generation capacity to cover the load demand; hence this constraint could be a soft, not a hard, limit constraint. Consequently, it is advisable to formulate the problem within the uncertainty frame [2].

Fuzzy logic plays a very successful role in dealing with the uncertainties in the system. Thus the power demand and reserve requirements are treated as the fuzzy variables and represented in membership functions based on error statistics. A penalty factor based on both power demand and reserve fuzzy membership functions is then determined to guide the solution. Thus the final schedule and the optimum solution will take into consideration the uncertainties in the constraints of the unit commitment problem. This way of solving the UC problem would be more appropriate and exact though this article does not consider the uncertainties in the unit commitment problem for the sake of the comparison of the final results of various techniques for combinatorial problems.

4 Simulated Annealing Algorithm

SA models the process of annealing in solids. Kirkpatrick et al. [11] adopted the idea of SA for solving difficult combinatorial problems. Essentially, the SA method generates a sequence of solutions, which are successively modified until a stopping criterion is satisfied. A temperature parameter is used to control the acceptance of modifications. Initially, the temperature is set to a high value and is decreased over iterations. If the modified solution has better fitness value than the current solution, it replaces the current solution. If the modified solution is less fit, it is still retained as current solution but with a probability condition. As the algorithm proceeds, the temperature becomes cooler, and it is then less likely to accept deteriorated solutions.

In each iteration, the process of generating and testing a new trial solution is repeated for a specified number of trials, to establish the 'thermal equilibrium'. The last of the accepted solution becomes the initial solution for the next iteration, after the temperature is reduced, according to the 'annealing schedule'. Thus the main features of the SA process are: the transition mechanism; and the cooling scheme. The transition mechanism consists of three components [7]:

(a) Generation of candidate solution by perturbing the current solution according to a probabilistic distribution function.
(b) Acceptance test for the solution based on better objective values or a probability of acceptance in case of higher values
(c) Iterative procedure.

In the last component, the first and the second components are used to produce a chain of tested candidate solutions. The last accepted solution becomes the initial solution of the next iteration. The way by which the control temperature parameter is reduced is called the cooling schedule. Optimal choice of this parameter is very critical to the success of the SA method [7].

5 Proposed SA Based UC Solution Methodology

The following steps enumerate the algorithm for the SA based UC problem. As discussed previously, the EDP sub-problem will be solved using the SQP [12] technique.

Step 1: Get the system data

Step 2: Initialise the temperature $T_k = T^o$, and set the iteration counter k=0.

Where, T^o is the initial temperature.

Step 3: Randomly generate an initial feasible commitment schedule, and evaluate (solve the economic dispatch problem using SQP) its cost function (Eqn.(1)), Set this schedule as current schedule UC_C, with cost value FS_C .

Step 4: Find the neighbourhood trial feasible schedule UC_P by randomly perturbing the current schedule, and evaluate its cost function, and assign it as FS_P.

Step 5: Perform the acceptance test:

(a) If $FS_P < FS_C$, corresponding commitment schedule is accepted and set $UC_C = UC_P$ and also $FS_C = FS_P$

(b) Else if $\exp[(FS_C - FS_P)/T_k] \geq rand(0,1)$ the commitment schedule is accepted and set $UC_C = UC_P$ and $FS_C = FS_P$

(c) Otherwise reject the trial schedule UC_P

Step 6: If the equilibrium condition (the specified number of iterations at T_k is completed) is satisfied, go to Step 7. Otherwise go to Step 4.

Step 7: If T_k reaches below the final temperature, exit. Otherwise Set k = k +1, up date temperature $T_k = \tau^k T^o$ and go to Step 3, where $0 < \tau < 1$.

5.1 Improved Random Perturbation (IRP) of the Current Schedule

The algorithm begins with a randomly generated feasible commitment schedule UC_C (a commitment schedule denotes the ON/OFF status of the N units over the H hours) as the current schedule. Then a trial feasible schedule is generated by random perturbation of the current schedule. A feasible schedule is the one, which satisfies the constraints given by Eqns. (5-9). The random perturbation scheme proposed in this article is a modification of the one proposed by Mantawy et. al. [6].

Mantawy et.al, perturbed the current schedule as follows. Given a current feasible schedule, randomly select a generating unit i, and an hour t. The current schedule is examined to find the state of generating unit i, at hour t. If the selected unit is OFF during t and if the OFF interval (the length of this interval is found by moving forward and backward around t), $Toff_i$ is equal to $Tdown_i$, then change the status of that unit to ON over the entire $Toff_i$ interval. If $Toff_i > Tdown_i$, then there is a flexibility in determining the number of hours, Tch for which the status shall be changed. In [7], Purushothama et.al, suggested an option to select, changing the status at either the beginning or at the end of the interval $Toff_i$.

In this article, the modification proposed is as follows.

Let, $Tch = Tr + Tdown_i$ (10),

Where, $Tr = UD(0, Toff_i - Tdown_i)$ (11)

Instead of selecting a single Tch (17) corresponding to a single Tr (11) using, it is suggested to select all possible $Tr = 0,1,2,...Toff_i - Tdown_i$. From which the best feasible schedule (commitment schedule) is chosen as the next feasible neighbourhood trial schedule. The following steps enumerate this for finalizing a schedule.

Step I : Identify all feasible commitment schedules from all $Tr = 0,1,2,...Toff_i - Tdown_i$.

Step II : Find the corresponding cost value by evaluating the cost function of all the feasible schedules obtained in Step I.

Step III: Find the best commitment schedule UC_p corresponding to the best cost value from all the cost values in Step II.

The best commitment schedule UC_p of Step III above is the neighbourhood trial schedule of the current schedule UC_C.

A CBUC problem with 3-units, expected to meet a 12 hours power demand pattern is used to demonstrate the performance of the proposed IRP scheme over the other two perturbation schemes proposed in [6,7]. The data for this test system is adopted from [1]. Table 1(a) shows a feasible commitment schedule for this system and taken as the current schedule UC_C and a cost value FS_C of $75282. Random perturbation of this solution starts randomly with unit 3 at the 9th hour. Here $Tdown_3 = 3$ also $Toff_i = (12 - 8 + 1) = 5$ therefore, Tch_i can either take 3 or 4 or 5. Using the random perturbation (RP) scheme of [6] let the random value of $Tr = 1$. For this value, Table 1(b) shows the commitment schedule generated using the RP scheme proposed in [6] and cost value FS_C is $73460. This commitment schedule is also valid for the RP scheme proposed in [7], since the option of changing the status at the end of the interval will result an infeasible commitment schedule. For this schedule Now using the IRP scheme, instead of choosing randomly a single Tr , all values of Tr are checked before concluding the next trial schedule. Thereby for $Tr = 2$ the cost value FS_C is $72992, which is better, compared to the cost value of RP scheme and the schedule is also feasible. Table 1(c) shows the final commitment schedule using the proposed random perturbation (IRP) scheme. This schedule is finalized as the next trial feasible schedule for the SA method. This commitment schedule would have been possibly generated by the RP schemes proposed in [6 or 7], but the success probability is only 33.33%. Thus the proposed IRP scheme, much more effectively explores the solution space around the current solution compared with that was possible with the other two RP schemes.

If the randomly selected unit is ON at t, replace $Tdown$ to Tup & $Toff$ to Ton in the above procedure and may be proceeded.

Table 1. Commitment schedules of the 3 unit system to illustrate the performance of the proposed perturbation scheme

(a)

Hour	1	2	3	4	5	6	7	8	9	10	11	12
Unit 1	0	100	100	120	300	450	500	400	250	100	100	150
Unit 2	100	150	300	400	400	400	400	400	400	230	300	400
Unit 3	70	0	0	0	0	200	200	0	0	0	0	0

(b)

Hour	1	2	3	4	5	6	7	8	9	10	11	12
Unit 1	0	100	100	120	300	450	500	200	100	100	100	150
Unit 2	100	150	300	400	400	400	400	400	350	100	100	400
Unit 3	70	0	0	0	0	200	200	200	200	130	200	0

(c)

Hour	1	2	3	4	5	6	7	8	9	10	11	12
Unit 1	0	100	100	120	300	450	500	200	100	100	100	100
Unit 2	100	150	300	400	400	400	400	400	350	100	100	250
Unit 3	70	0	0	0	0	200	200	200	200	130	200	200

5.2 Generating the Initial Feasible Commitment Schedule

A simple procedure for generating initial feasible schedule for the SA method is as follows,

Step a: Identify the must-not-run units from the initial status of the units, prior to scheduling. If there are no must-not-run units, go to Step b: else go to Step c:

Step b: If $PC_{max} \geq D_{max}$ & $PC_{min} : D_{min}$, switch ON all the N units for all H hours and exit.

Step c: If $PC_{max}^{m} \geq D_{max}$ & $PC_{min}^{m} \leq D_{min}$, switch ON all the N units (excluding the must-not-run units) for all H hours and exit.

Where, $PC_{\max} = \sum_{i=1}^{N} P_{i\max}, \quad PC_{\min} = \sum_{i=1}^{N} P_{i\min},$

$$PC_{\max}^m = \sum_{\substack{i=1 \\ i \neq NRU}}^{N} P_{i\max}, \quad PC_{\min}^m = \sum_{\substack{i=1 \\ i \neq NRU}}^{N} P_{i\min}$$

$D_{\max} / D_{\min}$ is the maximum/minimum power demand in the scheduling horizon.

NRU is the index of must-not-run units at hour t, so that at hour t, the maximum number of the units are kept must-not-run.

This approach is validated from the experiments conducted to generate initial feasible schedule on several UC problems available in ref of [3]. If this way of generating initial feasible schedule is not possible for any UC problem, then the technique proposed by Mantawy et al [6], is suggested. The above procedure is for generating initial feasible commitment schedule for CBUC problem. For generating initial feasible commitment schedule for PBUC problem, following modifications are made on the above procedure. Identify the must-run units from the initial status of the units, prior to scheduling. Switch ON those must-run units and switch OFF all other units.

6 Numerical Results

The proposed technique has been implemented in MATLAB on a 933 MHz Pentium PC. The performance of the algorithm has been evaluated through simulation. Simulation studies have been carried out on a example system, over a scheduling time horizon of 24 hours. The example system has 10 units and is taken from [10].

Hereinafter, SA-RP based technique is referred to the technique proposed in [6] and SA-IRP based technique is referred to the proposed technique of this article. After performing several trial experiments using the proposed technique for the UC problems of this article, the following control parameters are found to be most fit for the SA method. Initial temperature: 700, Final temperature: .001, equilibrium condition: 30 and $\tau = 0.97$. These values are also used for the SA-RP based technique for comparison purpose. Since the prime emphasis of this article is to compare the performance and solutions obtained using the proposed technique (SA-IRP based technique) with that of the SA-RP based technique, the proposed UC problem formulation does not include the ramp rate limit and security constraints.

To validate the robustness of the solution procedure, 30 different trial runs were conducted. In each run, the control parameters remained the same. To compare the performance of the SA method with IRP (SA-IRP) scheme with that of the RP scheme of [6], this approach is also coded in MATLAB and tested for 30 different trial runs. All simulation results represent the average of 30 trial runs.

In addition the GA approach reported in [4] is also coded in MATLAB and used to solve the PBUC problem for 30 different trial runs. The simulation parameters for GA are population size 100, Probability of crossover 0.8, probability of mutation 0.001 and Maximum number of generations 500.

Based on market's forecasted information, the proposed technique considers both power and reserve generation at the same time. The system data is available in [10] and had previously been solved using the EP-LR method [10]. The final results show that the profit using the proposed technique is $113134 compared to $112819 of EP-LR method, $110633 of SA-RP based technique and $111035 of GA method. The final commitment schedule using the proposed technique is shown in Table 2.

As can be seen from the above final solutions the proposed IRP scheme enhances the performance of the SA method. Also the minimum operators of SA compared to the GA make it easy to implement. In GA the initialization of initial population are very difficult and tedious, also repairing mechanisms leads it further cumbersome of the simulation. Thus the SA approach in total is robust and simple compared to the GA method for the Unit commitment problems.

Table 2. Final commitment schedule for example sytsem using the proposed technique

Hr	Unit power output (MW)										Unit reserve output (MW)									
	1	2	3	4	5	6	7	8	9	10	1	2	3	4	5	6	7	8	9	10
1	455	245	0	0	0	0	0	0	0	0	0	70	0	0	0	0	0	0	0	0
2	455	295	0	0	0	0	0	0	0	0	0	75	0	0	0	0	0	0	0	0
3	455	395	0	0	0	0	0	0	0	0	0	60	0	0	0	0	0	0	0	0
4	455	455	0	0	0	0	0	0	0	0	0	0	0	0	0	0	0	0	0	0
5	455	455	0	0	62	0	0	0	0	0	0	0	0	0	100	0	0	0	0	0
6	455	455	0	130	52	0	0	0	0	0	0	0	0	0	110	0	0	0	0	0
7	455	455	0	130	47	0	0	0	0	0	0	0	0	0	115	0	0	0	0	0
8	455	455	0	130	42	0	0	0	0	0	0	0	0	0	120	0	0	0	0	0
9	455	455	130	130	32	0	0	0	0	0	0	0	0	0	130	0	0	0	0	0
10	455	455	130	130	162	64	0	0	0	0	0	0	0	0	0	16	0	0	0	0
11	455	455	130	130	162	80	0	0	0	0	0	0	0	0	0	0	0	0	0	0
12	455	455	130	130	162	80	0	0	0	0	0	0	0	0	0	0	0	0	0	0
13	455	455	130	130	25	0	0	0	0	0	0	0	0	0	137	0	0	0	0	0
14	455	455	130	130	32	0	0	0	0	0	0	0	0	0	130	0	0	0	0	0
15	455	455	130	130	30	0	0	0	0	0	0	0	0	0	120	0	0	0	0	0
16	455	455	0	0	57	0	0	0	0	0	0	0	0	0	105	0	0	0	0	0
17	455	455	0	0	62	0	0	0	0	0	0	0	0	0	100	0	0	0	0	0
18	455	455	0	0	52	0	0	0	0	0	0	0	0	0	110	0	0	0	0	0
19	455	455	0	0	42	0	0	0	0	0	0	0	0	0	120	0	0	0	0	0
20	455	455	0	0	25	0	0	0	0	0	0	0	0	0	137	0	0	0	0	0
21	455	455	0	0	32	0	0	0	0	0	0	0	0	0	130	0	0	0	0	0
22	455	455	0	0	52	0	0	0	0	0	0	0	0	0	110	0	0	0	0	0
23	455	445	0	0	0	0	0	0	0	0	0	10	0	0	0	0	0	0	0	0
24	455	345	0	0	0	0	0	0	0	0	0	80	0	0	0	0	0	0	0	0

To investigate the scalability of the proposed technique, simulations on 10-, 20-, 40-, 60-, 80- and 100- unit systems had been carried out. The larger systems are obtained by an appropriate scaling of the 10-unit systems. To obtain the 20-unit test system, each of the 10 units was duplicated, and the power demand at each hour was doubled. Although such duplication does not create the true diversity of a typical large system, it is believed that this approach does demonstrate scalability of the technique [7], which is not influenced by any pair of units being identical to one another. The simulation parameters are kept as same for all these simulations. For each case, 30 trial runs are taken, and observed that, even the mean cost of SA-IRP based method is better compared to the best cost of SA-RP based method and GA based method.

Here are some comments on the SA approach of proposed by Mantawy et.al, and the proposed SA approach. The proposed approach needs less transactions per temperature compared to the Mantawy et.al, approach. This further leads to quick convergence of the solution procedure.

As the execution times are platform dependent, the simulation times are not directly compared. However, due to the increased rate of convergence at each temperature of the SA algorithm with an IRP scheme, fewer function evaluation (transitions) are required to reach the thermal equilibrium, and there is a significant reduction in time overhead. The proposed technique require more simulation time when compared with the LR method, for all example systems, due to the nature of the basic SA technique, but this is acceptable, as there is an improvement in the quality of the solution produced. In the two SA techniques, the simulation taken by the SA-IRP technique is less. Further reduction in simulation time can be achieved by parallel coding of the approach. Research for inclusion of uncertainties in the unit commitment problem using the fuzzy logic approach and inclusion of other practical constraints is underway and the simulation and analysis results for the unit commitment problem will be presented in another article.

7 Conclusion

An improved SA technique for solving the UC problem is presented. An IRP scheme and a simple procedure for generating initial feasible commitment schedule are proposed for the SA method. Proposed improved random perturbation for the SA method scheme effectively generates feasible schedules for the UC problems. The nonlinear programming sub-problem is solved using the SQP technique. A PBUC problem is solved to demonstrate the validity and effectiveness of the proposed technique. Comprehensive numerical results show the applicability of the proposed technique for combinatorial part of UC problem in the real-time operation of electric power systems.

References

1. A.J. Wood, B.F. Wollenberg, Power Generation, Operation, & Control, Wiley, NY, 1984.
2. G.B. Sheble, T.T. Maifeld, Unit commitment by genetic algorithm and expert system', Electr. Power Syst. Res. 30 (1994) 115-121.

3. H.Y. Yamin, Review on methods of generation scheduling in electric power systems, Electr. Power Syst. Res. 69 (2004) 227-248.

4. S.O. Orero, M.R Irving, A genetic algorithm for generator scheduling in power systems, Electr. Power Energy Syst. 18 (1) (1996) 19–26.

5. K.A. Juste, H. Kita, E. Tanaka, J. Hasegawa, An evolutionary programming solution to the unit commitment problem, IEEE Trans. Power Syst. 14 (1999) 1452–1459.

6. A.H. Mantawy, Y.L. Abdel-Magid, S.Z. Selim, A simulated annealing algorithm for unit commitment, IEEE Trans. Power Systems, 9 (1) (1998) 197–204.

7. G.K. Purushothama, U.A. Narendranath, L. Jenkins, Unit commitment using a stochastic extended neighbour hood search, IEE Proc., Gener. Transm. Distrib., 150, (2003) 67-72.

8. E.H. Allen, M.D. Ilic, Reserve markets for power systems reliability, IEEE Trans. Power Syst., 15 (1) (2000) 228–233.

9. S.M. Shahidehpour, H.Y. Yamin, Z. Li, Market Operations in Electric Power Systems, Wiley, February 2002.

10. A. Pathom, H. Kita, E. Tanaka, J. Hasegawa, A Hybrid LR–EP for New Profit-Based UC Problem Under Competitive Environment, IEEE Trans. Power Syst., 18 (2003) 229–237.

11. S. Kirkpatrick, C.D. Gelatt, M.P. Vecchi, Optimization by simulated annealing. Science, 1983, 220 (4598), 671-680.

12. P.T. Boggs, J.W. Tolle, Sequential Quadratic Programming, Acta Numerica, (1995) 1–52.

A New Hybrid GA/SA Algorithm for the Job Shop Scheduling Problem

Chaoyong Zhang, Peigen Li, Yunqing Rao, and Shuxia Li

School of Mechanical Science & Engineering, Huazhong University of
Science & Technology, Wuhan, 430074, P.R. China
zcyhust@sohu.com

Abstract. Among the modern heuristic methods, simulated annealing (SA) and genetic algorithms (GA) represent powerful combinatorial optimization methods with complementary strengths and weaknesses. Borrowing from the respective advantages of the two paradigms, an effective combination of GA and SA, called Genetic Simulated Algorithm (GASA), is developed to solve the job shop scheduling problem (JSP). This new algorithm incorporates metropolis acceptance criterion into crossover operator, which could maintain the good characteristics of the previous generation and reduce the disruptive effects of genetic operators. Furthermore, we present two novel features for this algorithm to solve JSP. Firstly, a new full active schedule (FAS) based on the operation-based representation is presented to construct schedule, which can further reduce the search space. Secondly, we propose a new crossover operator, named Precedence Operation Crossover (POX), for the operation-based representation. The approach is tested on a set of standard instances and compared with other approaches. The Simulation results validate the effectiveness of the proposed algorithm.

Keywords: Genetic Algorithm, Simulated Annealing, Crossover, Local Search.

1 Introduction

The job shop scheduling problem (JSP) is a well-known NP-hard problem of combinatorial optimization and has a very wide engineering background. The general JSP can be described as follows: given n jobs, each composed of several operations that must be processed on m machines. Each operation uses one of the m machines for a fixed duration. Each machine can process at most one operation at a time and once an operation initiates processing on a given machine it must complete processing on that machine without interruption. The operations of a given job have to be processed in a given order. The problem consists in finding a schedule of the operations on the machines, taking into account the precedence constraints, which minimizes the makespan, i.e. the finish time of the last operation completed in the schedule.

JSP is also known to be a very difficult combinatorial optimization problem such that some test problems of moderate size are still unsolved. During the last three

G.R. Raidl and J. Gottlieb (Eds.): EvoCOP 2005, LNCS 3448, pp. 246–259, 2005.
© Springer-Verlag Berlin Heidelberg 2005

decades, the approaches proposed to solve the scheduling problem can be divided into two categories: exact methods and approximation algorithms. Exact methods, such as Branch and Bound, Linear Programming, Lagrangian Relaxation and Decomposition methods, have been successful in solving small instances, including the notorious 10×10 instance of Fisher and Thompson proposed in 1963 and only solved twenty years later. But for the big instances there is a need for approximation algorithms that include priority dispatch, shifting bottleneck approach, heuristic methods and so on. In recent years, hybridizing the modern heuristic methods, such as Genetic Algorithm (GA) [1-2-3], Simulated Annealing (SA) [4-5], Tabu Search (TS) [6-7], has been captured many researchers attention, because they are able to attain high-quality solutions within reasonable computational times. A comprehensive survey of job shop scheduling techniques can be found in [8-9].

Among the modern heuristic methods, GA and SA represent powerful combinatorial optimization methods with complementary strengths and weaknesses. GA exhibits implicit parallelism and can retain useful redundant information about what is learned from previous searches by its representation in individuals in the population. But GA is also prone to loss of solutions and theirs substructures due to the disruptive effects of genetic operators and suffers from poor convergence properties [10]. By contrast to GA, SA is a naturally serial algorithm and currently possesses a formal proof of convergence, but its behavior can be controlled by the cooling schedule. Since SA maintains only one solution at a time, whenever it accepts a new structure, it must discard the old one, so there is no redundancy and no history of past structures. The end result is that good structures and substructures can be discarded [11]. Moreover, it has been demonstrated that SA and GA, especially SA, are very sensitive to parameters and their performances are largely dependent on fine-tuning of the parameters [12]. How to set optimal parameters and efficiently find global optima are still open problems. In recent years, by reasonably combining the respective advantages of the two paradigms, researchers have been investigating hybrid algorithms that attempt to mix GA and SA techniques and presented several hybrid frameworks [11-12-13-14]. Those previous SA/GA hybrid frameworks benefit from closer resemblance to SA than to the GA. In this paper, we develop a new hybrid genetic algorithm GASA, which most closely resembles GA, and investigate its potential on solving JSP.

The organization of remain contents is as follows. In Section 2, the framework of GA and SA is presented. In Section 3, a new type of schedule, full active schedule, is proposed. In Section 4, we present our approach to solve the job shop scheduling problem: the encoding, decoding scheme, schedule generation procedure and new genetic operators. Section 5 reports the computational results and simulation. The conclusions are made in Section 6.

2 Hybrid GASA Methods

Genetic algorithms and simulated annealing are similar, natural motivated, general purpose optimization procedures [15]. The GA is based on the survival of the fittest and involves some selection, crossover and mutation. A GA exhibits parallelism, contains certain redundancy and historical information of past solutions, and is suitable for implementation on massively parallel architecture. Critical components of past good solutions can be captured, which can be combined together via crossover to form high quality solutions. Unfortunately, GA is also prone to loss of solutions and theirs substructures due to the disruptive effects of genetic operators. This is because new solutions produced by the genetic operators are always accepted, even if they are significantly inferior to older solutions. This characteristic can lead to disruption, where good solutions are lost or damaged, so that a pure GA may easily produce premature convergence and poor results. By contrast to GA, SA possesses good convergence property and the ability to probabilistically escape from local optima; it accepts or rejects the newly generated candidate solution probabilistically by the *Metropolis acceptance criterion*, where inferior candidate can be accepted some of the time. By exploiting the efficient parallelization of the GA and the convergence control of SA, the new hybrid GA/SA algorithm (GASA), which incorporates metropolis acceptance criterion into crossover operator, is proposed with the framework illustrated in Fig.1.

In contrast to a general genetic algorithm, the GASA has superior features. First, this new algorithm incorporates metropolis acceptance criterion into crossover operator, which motivated by both empirical and analytical evidence (Goldberg [15]) that the distribution of population members over time is nearly Boltzmann. If the fitness f_{temp1}, f_{temp2} of the individual *temp, temp2* which is generated by two parent satisfy $f_{tempi} < f_{avg}$ or probability $exp(f_{tempi} - f_{avg})/T$ *(i=1,2)*, then the solution *temp_i* is accepted. Otherwise crossover continues until the two children are updated. The two unequal best individuals in these children are selected to next generation. This approach could maintain the good characteristics of the previous generation and reduce the disruptive effects of genetic operators. Secondly, a novel approach of crossover rate is proposed and replaces the general crossover rate. If the fitness f_{P1}, f_{P2} of individuals *P1, P2* which select from the P_{old} is unequal, then implement the crossover operation. Otherwise implement the mutation operation. This new method of crossover rate can vary dynamically, adaptively in response to the state of premature convergence. For example, at the beginning of the evolution period, the mutation rate is small; whereas at the end of the convergence period, the mutation rate increases accordingly, so GA can avoid premature convergence better, as well as to avoid the difficulty of choosing the crossover rate. Our experimental results shown this new method could obtain better results and reduce the search times. This novel method could also be used for the general genetic algorithm. Thirdly, the schedule generation procedure is used for constructing schedule, i.e., after a schedule is obtained a local search heuristic is applied to improve the solution. In addition, Temperature ($f_{avg}/120$) is adjusted to control the number of the crossover times, at a high temperature, GA performs a course search with the low number of the crossover times; while at a low temperature, SA performs a fine search with the high the number of the crossover times. Moreover, such hybrid framework can convert to traditional GA by omitting the SA unit.

Such hybrid algorithm retains the generality of GA, SA and can easily be implemented and applied to any combinatorial optimization problems. For different problems, the representation scheme, optimization operators and parameters should be designed suitably. In the next sections, we will explore the potential of such hybrid GASA optimization strategy to JSP.

Begin

   Initialize population randomly with P_{size} individual;

   Evaluate the initial population with the Schedule Generation Procedure;

While stopping criteria not satisfied Repeat

{

   Reproduce the 10% elite individuals from P_{old} to P_{new};

   Select a pair of individuals $P1$, $P2$ from the P_{old}, which the fitness is f_{P1}, f_{P2} respectively.

   $UpdateChild_i$ = False;($i = 1,2$)

   If (f_{P1} != f_{P2})

   {

      Do {

            Implement crossover operation and generate two children individual $temp1$, $temp2$, then calculate fitness f_{temp1}, f_{temp2} with the Schedule Generation Procedure. f_{avg} is the average fitness of this generation.

            Set the temperature T is f_{avg} /120;

            if (ftempi<f_{avg} || $exp(f_{tempi}$ -f_{avg})/T > random[0,1])   i=1,2;

            { $UpdateChild_i$ = True;   }

            Select two unequal best individuals in these children to next generation.

         }Until($UpdateChild_1$ || $UpdateChild_2$)

   }

   else

   {   Mutation the chosen pair of individuals by the probability of P_m .   }

}

End

Fig. 1. The outline of the hybrid GASA

3   Types of Schedules

In general, schedule can be classified into three types: semi-active schedule, active schedule and non-delay schedule. The set of active schedules is a subset of semi-active schedules, which can be obtained by shifting the operation of a semi-active schedule to the left without delaying other jobs. A schedule with no more permissible left shifts is called an active schedule. An optimal schedule must be an active schedule, so it is safe and efficient to limit the search space to the set of all active schedules.

Table 1. A sample 2×2 problem

Job	Operations routing (Processing machine)		Processing time	
1	2	1	2	4
2	2	1	2	3

Although repairing a semi-active schedule to the active one improves the makespan, the set of active schedules is usually very large and contain many schedules with relatively large delay times. The simple 2×2 problem described in Table 1 is taken for example. Fig.2 (a) shows an active solution of this problem. Fig.2 (b) is attained by shifting the operation (J1, 1) to the right of the operation (J2, 1) in M2 (machine2). It can be seen from Fig.2 that sometimes there are obvious improvements that can be attained by right shift the operation of the active schedule.

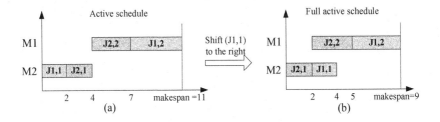

Fig. 2. Permissible right shift for an active schedule

Fig. 3. Full active schedules

This paper proposes a new type of schedule—Full Active Schedule (FAS). A FAS can be obtained by shifting operations of an active schedule to the right without delaying other jobs (The example can be seen from Fig. 2). So the full active schedule can be defined as a schedule with no more permissible left and right shifts. The set of full active schedules is a subset of active schedules. Fig. 3 illustrates where the set of full active schedules is, A full active schedule generator procedure based the operation-based representation is proposed in section 4.1. Using the FAS set, we can further reduce the search space and get better optimal or near optimal schedule.

4 Hybrids GASA for JSP

4.1 Schedule Generator Procedure

4.1.1 Representation
The genetic algorithm described in this paper uses an operation-based representation that uses an unpartitioned permutation with m-repetitions of job numbers [2-16-17]. In this representation, each job number occurs m times in the chromosome. By scanning the chromosome from left to right the k-th occurrence of a job number refers to the k-th operation in the technological sequence of this job. The important feature of the operation-based representation is that any permutation of the chromosome can be decoded to a feasible schedule.

Table 2. Example of 3×3 problem

Job	Operations routing(Processing time)		
j1	1 (3)	2 (1)	3 (2)
j2	3 (1)	1 (5)	2 (3)
j3	2 (3)	3 (2)	1 (3)

Table 3. A 3×3 feasible solution

Machine	Job sequence		
m1	1	2	3
m2	3	1	2
m3	2	3	1

Consider the 3-job and 3-machine problem given in Table 2. Suppose a chromosome is given as [2 1 1 2 2 3 3 1 3]. Because each job consists of three operations, it occurs exactly three times in the chromosome. The fourth gene of the chromosome in this example is 2. Here, 2 represents the second operation of job 2 because number 2 has been repeated twice. A schedule is decoded from a chromosome with the following procedure: the first operation in the list is scheduled firstly, then the second operation, and so on. Each operation under treatment is allocated in the best available processing time for the corresponding machine the operation requires. The process is repeated until all operations are scheduled. A schedule generated by the procedure can be guaranteed to be an active schedule [18]. Then, using the active-decoding process, we can get the corresponding feasible solution shown in Table 2 and the active chromosome [2 3 1 1 2 2 3 1 3]. The active chromosome and the feasible solution can be converted into each other, however two or more different chromosomes can be decoded to an identical solution.

4.1.2 Schedule Generation Procedure

The objective of the schedule generation procedure is to improve the chromosome and compute the corresponding makespan, which makes use of a full active schedule generator procedure and a local search procedure and feedback of the makespan and the corresponding full active chromosome.

The algorithm described in section 4.1.1 can generate an active schedule. Using the same algorithm to this active schedule, a full active schedule can be attained with only small modifications described as follows. By reversing the chromosome based on the operation-based representation and all of the technological sequences, a given schedule can be converted to another schedule. The new schedule is equivalent to the original one with the same makespan (the same critical path) and the reversed chromosome (reversing the job processing sequences in same machine). Through shifting left all operations of the new schedule, we can obtain the makespan of the full active schedule. Therefore the makespan and chromosome of a full active schedule can be obtained by using the algorithms (in Section 4.1) to the reversed chromosome and the reversed technological sequences of an active schedule.

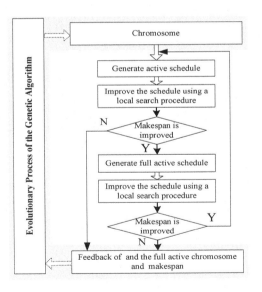

Fig. 4. The schedule generation procedure

Fig. 4 illustrates the steps of the schedule generation procedure applied to each chromosome generated by the hybrid GASA. The schedule generation procedure generates the active schedule and it's full active schedule, and then applies the local search to each of them to improve them. If the local search improves the makespan, the schedule generation procedure continuously generates the full active schedule and applies the local search to improve it. Otherwise, Improvement of the chromosome (or schedule) ends, and we obtain the makespan and the corresponding full active chromosome for the evolutionary process. A local search procedure is introduced in the next section.

4.1.3 Local Search Procedure

A neighborhood, $N(x)$ is a function which defines a simple transition from a solution x to another solution by inducing a change that typically may be viewed as a small perturbation [19]. The objective of these strategies is to progressively perturb the current configuration through a succession of neighbors in order to direct the search to an improved solution.

For the JSP, a key component of a solution is the critical path, which is the longest route from source to sink in the disjunctive graph. It is possible to decompose the critical path into a number of blocks where a block is a maximal sequence of adjacent critical operations that require the same machine. Because the permutation of non-critical operation cannot improve the objective function and may create an infeasible solution, an efficient method can be obtained by introducing a transition operator that exchanges a pair of consecutive operations only on the critical path and forms a neighborhood.

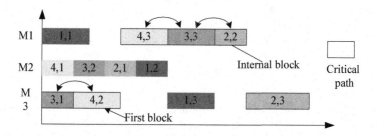

Fig. 5. Permutation of operations on a critical path

In this paper, we focus particularly on the neighborhood of Nowicki and Smutnicki (1996), which is noted for proposing and implementing the most restrictive neighborhood in the literature. In the approach only a single arbitrary critical path is generated, but our approach generates all critical paths. The critical path thus gives rise to the following neighborhood of moves (see Fig. 5). The local search used in this paper is the standard hill climbing. If the swap improves the makespan, it is accepted. Otherwise, the swap is undone. Once a swap is accepted, the critical path is changed and a new critical path must be identified. If no swap of the first or the last operations in any block of critical path improves the makespan, the local search ends [20].

4.2 Crossover Operation

Crossover can be regarded as the backbone of the genetic algorithm. It intends to inherit the information of two parent solutions to one or more offspring solutions. To apply crossover operation successfully to the JSP, the following criteria should be satisfied: completeness, feasibility, non-redundancy and characteristics- preserving-ness [21]. In this paper, a new crossover operator named Precedence Operation Cross-over (POX) is proposed for the operation-based representation, which can satisfy the characteristics-preservingness, completeness and the feasibility properly between parents and their children.

The effective crossover operator proposed in this paper is described as follows. Given chromosome, parent1 and parent2, POX generates the children, child1 and child2 by the following procedure:

(1) Randomly choose the set of job numbers, {1, 2... n}, into one nonempty ex-clusive subset J1.

(2) Copy those numbers in J1 from parent1 to child1 and from parent2 to child2, preserving their locus.

(3) Copy those numbers in J1, which are not copied at step 2, from parent2 to child1 and from parent1 to child2, preserving their order.

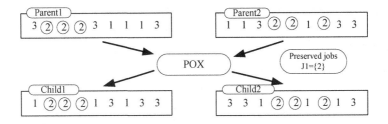

Fig. 6. POX crossover

Fig. 6 shows an example of the precedence operation crossover (POX) of the 3×3 problem; the locus of job {2} is preserved. The crossover of parent1 and parent2 generates two children chromosomes, child1 {1 2 2 2 1 3 1 3 3} and child2 {3 3 1 2 2 1 2 1 3}. It can be seen that child1 preserves the locus and order of job {2} in parent1 and the order of job {1, 3} in parent2 respectively, and child2 preserves the locus and order of job {2} in parent2 and the order of job {1, 3} in parent1 respectively. There-fore POX is excellent in the characteristics-preservingness.

4.3 Mutation Operation

Mutation is just used to produce perturbations on chromosomes in order to maintain the diversity of population. In this paper, an insertion mutation is introduced, which is used not only to produce small perturbations but also to perform intensive search. The insertion mutation is described as follows: select two positions randomly within the chromosome and insert the back one before the front one, implements the procedure the number of jobs times and chooses the best one (except the parent chromosome) to the next generation.

4.4 Selection and Reproduction

In the GA application, it is necessary to define an evaluation function, which determines the probability of survival of an individual to the next generation. In this paper, an individual chromosome x is assigned to the fitness value:

$$Fitness(x) = (worst\_fitness - currentfitness(x)) + samll\_int;$$

Here, worst_fitness is the fitness of the schedule produced from the worst chromosome in the current generation. Currentfitness(x) is the fitness of the schedule generated from the chromosome x by the schedule generation procedure, and small_int means the lowest fitness and it is a fixed parameter based on the instance size.

Roulette wheel selection incorporating elitist strategy is used to generate a new population for the next generation. In this paper, the elitist strategy is accomplished by copying the top 10% from the previous population chromosomes to the next generation. The advantage of an elitist strategy over the traditional probabilistic reproduction is that the best solution is monotonically improving from one generation to the next. The potential problem in this way is population convergence to a local minimum. However, this can be overcome by the new crossover rate proposed in our paper.

5 Computational Results and Simulation

To illustrate the effectiveness and the performance of the algorithm described in this paper, we consider instances from three classes of standard JSP test problems: Fischer and Thompson (1963) instances FT06, FT10, FT20, Lawrence (1984) instances LA01-LA40 and Adams et al (1988) instances ABZ7- ABZ9. Here, we regard FT06, LA01, LA06, LA11 and so on as easy problems, because they can be easily solved by many methods. So we don't consider those problems. All test instances were downloaded from Beasley's OR-Library, http://mscmga.ms.ic.ac.uk.

The proposed algorithm is compared with the following algorithms:

Hybrid SA/GA (PRSA)(SAGen)	M. Kolonko(1999)[22]
Tabu Search (TSAB)	Nowicki and Smutnicki (1996)[7]
Tabu Search method guided by shifting bottleneck (TSSB)	Ferdinando Pezzella and Emanuela Merelli(2000)[23]

The Hybrid GASA Algorithm for JSP mentioned above was implemented in Visual C++ and the tests were run on a computer with Pentium IV1.6G and 256M RAM. In our experiments, if the number of operations in the problem is less than or equal to 100, population size is 200. Otherwise population size is 300. The top 10% elite solutions from the old population chromosomes are copied to the next generation; the mutation rate (P_m in Fig.1) is 0.8. The algorithm was terminated when an optimal solution was found or after 60-100 generations of the algorithm.

Table 4. Results of the fifteen tough problems

Prob	Size	LB	GASA Best	GASA Average	GASA(s) AVGTime	SAGen	TSAB	TSSB
FT10	10×10	930	930*	930	18.7	930	930	930
FT20	20×5	1165	1165*	1166.3	29.8	1165	1165	1165
LA21	15×10	1046	1046*	1052.6	259.8	1047	1047	1046
LA24	15×10	935	935*	941.5	278.4	938	939	938
LA25	15×10	977	977*	980.8	203.8	977	977	979
LA27	20×10	1235	1235*	1254.2	553.2	1236	1236	1235
LA29	20×10	1157	1164	1176.1	557.4	1167	1160	1168
LA36	15×15	1268	1274	1282	397.9	1268	1268	1268
LA37	15×15	1397	1397*	1402.3	406.8	1401	1407	1411
LA38	15×15	1196	1196*	1202.6	530.8	1201	1196	1201
LA39	15×15	1233	1238	1239.8	514.6	–	1233	1240
LA40	15×15	1222	1224	1233.2	463.3	1226	1229	1233
ABZ7	20×15	656	666	673.9	1195.6	658	670	666
ABZ 8	20×15	665(645)	675	684.5	1461.7	670	682	678
ABZ 9	20×15	679(661)	681	687.1	1281.4	–	695	693

*The optimal solutions were found by our algorithm.

The GASA finds the optimal solutions for the ft10 problems almost every time in less than half minute on average. Table 4 shows the makespan performance statistics of the hybrid GASA for the fifteen difficult benchmark problems. It lists problem name, problem dimension (number of jobs × number of operations), the best-known lower bound (**LB**), and the solution obtained by our algorithm (GASA) including best solution, the average value and the average time and the solution by each of the other algorithms. In the table, the columns named best, average and average time values are obtained over 10 runs respectively. Optimal solutions were found for eight out of the fifteen difficult problems. The GASA compared with the other famous algorithm (SAGen, TSAB and TASB). The computation results validate the effectiveness and robustness of the proposed algorithm. Fig. 7 shows the optimal solution of the La27 20×10 problem that is one of the hardest problems.

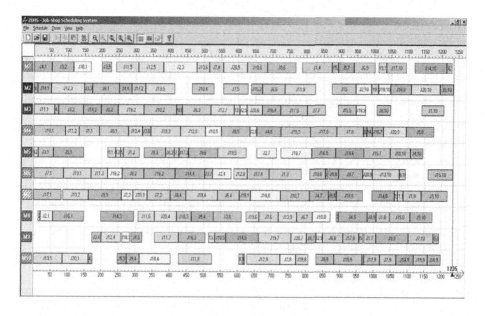

Fig. 7. The optimal solution of the La27 20×10 problem

6 Conclusions

This paper describes a hybrid algorithm (GASA) combining simulated annealing and local search with genetic algorithm for the JSP. By reasonably combining the advantage aspects of these global probabilistic search algorithms, the hybrid framework of GA and SA can achieve more efficient optimization results and relax the parameter dependence to some extent. The chromosome representation of Our GA is based on the operation-based representation. A new type of schedule, full active schedule (FAS), is presented to construct schedule. After a schedule is obtained, a local search heuristic is applied to improve the solution. To preserve the characteristics properly, a new crossover operator named Precedence Operation Crossover (POX) is proposed for the operation-based representation. The approach is tested on a set of standard instances and compared with the other approaches. The computational results show the effectiveness and robustness of our algorithm. Further research is necessary to reduce the computational time, which can be improved by integrating TS into the algorithms.

Acknowledgements

This paper is supported by the National Natural Science Foundation, China (Grant No. 50105006, 50305008). The authors would like to thank the referees for their helpful comments and suggestions.

References

1. Davis, L.: Job shop scheduling with genetic algorithms, in: Proceedings of the International Conference on Genetic Algorithms and their Applications. Hillsdale, Lawrence Erlbaum (1985) 136–149
2. Bierwirth, C.: A generalized permutation approach to job shop scheduling with genetic algorithms. OR Spektrum. 17 (1995) 87–92
3. Croce, F.D., Tadei, R., Volta, G.: A genetic algorithm for the job shop problem. Computers & Operations Research. 22(1) (1995) 15-24
4. Laarhoven, P.V., Aarts, E., Lenstra, J.K.: Job shop scheduling by simulated annealing. Operations Research. 40 (1) (1992) 113–125
5. Aarts, E.H.L., van Laarhoven, P.J.M., Lenstra, J.K., Ulder, N.L.J.: A computational study of local search algorithms for Job Shop Scheduling. ORSA Journal on Computing. 6 (1994) 118-125
6. Taillard, E.D.: Parallel taboo search techniques for the job-shop scheduling problem. ORSA J. on Comput. 6(2) (1994) 108–117
7. Nowicki, E., Smutnicki, C.: A fast taboo search algorithm for the job shop problem. Management Science. 42(6) (1996) 797–813
8. Jain, A.S., Meeran, S.: Deterministic job-shop scheduling: Past, present and future. European Journal of Operational Research .113 (1999) 390-434
9. Blazewicz, J., Domschke, W., Pesch, E.: The job shop scheduling problem: Conventional and new solution techniques. European Journal of Operational Research. 93 (1996) 1-33
10. DeJong, K.A.: An analysis of the behavior of a class of genetic adaptive systems. Dissertation Abstracts International 36(10) (1975),5140B(University Microfilms No.76-9381), Ph.D. Thesis, University of Michigan, Ann Arbor
11. Mahfoud, S.W., Goldberg, D.E.: Parallel Recombinative Simulated Annealing:A Genetic Algorithm. Parallel Computing. 21(1995) 1-28
12. Ingber, L., Rosen, B.: Genetic algorithms and very fast simulated reannealing: a comparison. Mathematical Computer Modeling. 16(11) (1992) 87-100
13. Brown, D.E., Huntley, C.L., Spillane, A.R.: A Parallel Genetic Heuristic for the Quadratic Assignment Problem. in: Proceedings of the Third International Conference on Genetic Algorithms. Fairfax, VA (1989) 406-415
14. Lin, F.T., Kao, C.Y., Hsu, C.C.: Incorporating Genetic Algorithms into Simulated Annealing. in: Proceeding of the Fourth International Symposium on Artificial Intelligence. (1991) 290-297
15. Goldberg, D.E.: A note on Boltzmann tournament selection for genetic algorithms and population-oriented simulated annealing. Complex Systems. 4 (1990) 445-460
16. Gen, M., Tsujimura Y., Kubota, E.: Solving Job-Shop Scheduling Problems by Genetic Algorithm. in: Proceedings of the 1995 IEEE International Conference on Systems, Man, and Cybernetics. Institute of Electrical and Electronics Engineers, Vancouver (1995) 1577-1582
17. Shi, G. Y., IIMA, H., Sannomiya, N.: A new encoding scheme for Job Shop problems by Genetic Algorithm. in: Proceedings of the 35th Conference on Decision and Control. Kobe, Japan (1996) 4395-4400
18. Cheng, R., Gen, M., Tsujimura, Y.: A tutorial survey of job-shop scheduling problems using genetic algorithms ⅠI. Representation. Computers and Industrial Engineering. 30 (9) (1996) 83-97
19. F. Glover, and M. Laguna, Tabu Search. Kluwer Academic Publishers, Norwell, MA (1997)

20. Gonçalves J. F.: A Hybrid Genetic Algorithm for the Job Shop Scheduling Problem. AT&T Labs Research Technical Report TD-5EAL6J, September 2002
21. Kobayashi, S, Ono, I., Yamamura, M: An Efficient Genetic Algorithm for Job Shop Scheduling Problems. in: Proceedings of the 6th International Conference on Genetic Algorithms. (1995) 506-511
22. Kolonko, M.: Some new results on simulated annealing applied to the job shop scheduling problem. European Journal of Operational Research. 113 (1999) 123-136
23. Pezzella F., Merelli, E.: A tabu search method guided by shifting bottleneck for the job shop scheduling problem. European Journal of Operational Research. 120 (2000) 297-310

An Agent Model for Binary Constraint Satisfaction Problems

Weicai Zhong, Jing Liu, and Licheng Jiao

Institute of Intelligent Information Processing (224#),
Xidian University, Xi'an 710071, China
neouma@163.com

Abstract. With the intrinsic properties of constraint satisfaction problems (CSPs) in mind, several behaviors are designed for agents by making use of the ability of agents to sense and act on the environment. These behaviors are controlled by means of evolution, so that multiagent evolutionary algorithm for constraint satisfaction problems (MAEA-CSPs) results. To overcome the disadvantages of the general encoding methods, the minimum conflict encoding is also proposed. The experiments use 250 benchmark CSPs to test the performance of MAEA-CSPs, and compare it with four well-defined algorithms. The results show that MAEA-CSPs outperforms the other methods. In addition, the effect of the parameters is analyzed systematically.

1 Introduction

A large number of problems coming from artificial intelligence as well as other areas of computer science and engineering can be stated as Constraint Satisfaction Problems (CSPs). Historically, CSPs have been approached from many angles by Evolutionary Algorithms (EAs). Among the available methods, some ones put the emphasis on the usage of heuristics, such as Glass-Box [1] and H-GA [2], while some others handle the constraints by fitness function adaptation, such as SAW [3].

Agent-based computation has been studied for several years in the field of distributed artificial intelligence. In this paper, with the intrinsic properties of CSPs in mind, several behaviors are designed for agents. Furthermore, all such behaviors are controlled by means of evolution, so that a new algorithm, multiagent evolutionary algorithm for constraint satisfaction problems (MAEA-CSPs), results. In MAEA-CSPs, all agents live in a latticelike environment. Making use of the designed behaviors, MAEA-CSPs realizes the ability of agents to sense and act on the environment that they live in. During the process of interacting with the environment and the other agents, each agent increases its energy as much as possible, so that MAEA-CSPs can find solutions. Experimental results show that MAEA-CSPs has a good performance.

2 Constraint Satisfaction Agents

The meaning of an agent is very comprehensive, and what an agent represents depends on problems. In general, four elements should be defined when multiagent

G.R. Raidl and J. Gottlieb (Eds.): EvoCOP 2005, LNCS 3448, pp. 260–269, 2005.

systems are used to solve problems. The first is the meaning and the purpose of each agent. The second is the environment where all agents live. Since each agent has only local perceptivity, so the third is the definition of the local environment. The last is the behaviors that each agent can take to achieve its purpose [6].

2.1 CSPs

A CSP has three components [7]:

(1) A finite set of variables, $x=\{x_1, x_2, ..., x_n\}$;
(2) A domain set D, containing a finite and discrete domain for each variable:

$$D=\{D_1, D_2, ..., D_n\}, \quad x_i \in D_i = \{d_1, d_2, \cdots, d_{|D_i|}\}, \quad i=1, 2, \cdots, n \qquad (1)$$

   where $|\cdot|$ stands for the number of elements in the set;
(3) A constraint set, $C=\{C_1(x_1), C_2(x_2), ..., C_m(x_m)\}$, where x_i, $i=1, 2, ..., m$ is a subset of x, and $C_i(x_i)$ denotes the values that the variables in x_i cannot take simultaneously.

Thus, the search space of a CSP, S, is a Cartesian product of the n sets of finite domains, namely, $S=D_1 \times D_2 \times ... \times D_n$. A solution for a CSP, $s=<x_1, x_2, ..., x_n> \in S$, is an assignment to all variables so that the values satisfy all constraints. Here is a example:
Example 1: A CSP is described as follows:

$$\begin{cases} x = \{x_1, x_2, x_3\} \\ D = \{D_1, D_2, D_3\}, \quad D_i = \{1, 2, 3\}, \quad i = 1,2,3 \\ C = \{C_1(\{x_1,x_2\}) = \langle 1,3 \rangle, \; C_2(\{x_1,x_2\}) = \langle 3,3 \rangle, \; C_3(\{x_1,x_3\}) = \langle 2,1 \rangle, \; C_4(\{x_1,x_3\}) = \langle 2,3 \rangle, \\ \quad\quad C_5(\{x_1,x_3\}) = \langle 3,1 \rangle, \; C_6(\{x_1,x_3\}) = \langle 3,3 \rangle, \; C_7(\{x_2,x_3\}) = \langle 1,1 \rangle, \; C_8(\{x_2,x_3\}) = \langle 1,2 \rangle, \\ \quad\quad C_9(\{x_2,x_3\}) = \langle 1,3 \rangle, \; C_{10}(\{x_2,x_3\}) = \langle 2,1 \rangle, \; C_{11}(\{x_2,x_3\}) = \langle 3,1 \rangle \} \end{cases} \qquad (2)$$

All solutions for this CSP are <1, 2, 2>, <1, 2, 3>, <2, 2, 2>, <2, 3, 2>, and <3, 2, 2>.
2.2 Definition of constraint satisfaction agents
A constraint satisfaction agent (CSAgent) is defined as follows:
Definition 1: A constraint satisfaction agent, a, represents an element in the search space, and its energy is equal to

$$\forall a \in S, \quad Energy(a) = -\sum_{i=1}^{m} \chi(a, C_i) \qquad (3)$$

where $\chi(a, C_i) = \begin{cases} 1 & a \text{ violates } C_i \\ 0 & \text{otherwise} \end{cases}$. The goal of each CSAgent is to increase its

energy as much as possible, and it has some behaviors to achieve its goal.

When one uses EAs to solve problems, the search space must be encoded such that individuals can be represented as a uniform form. For CSPs, an effective coding method is a permutation coding with a corresponding decoder. For example, in [3], each individual is represented as a permutation of the problem variables, and the permutation is transformed to a partial instantiation by a simple decoder that considers the variables in the order they occur in the permutation and assigns the first possible domain value to that variable. If no value is possible without introducing a constraint violation, the variable is left uninstantiated. In what follows, we label this decoder as Decoder1 and the set of permutations of the problem variables as $\mathbf{S}_x^P$.

Because Decoder1 uses a greedy algorithm, there exists a serious problem. That is, for some CSPs, Decoder1 cannot decode any permutation to a solution, so the algorithms based on Decoder1 cannot find solutions at all. Here is a simple example:

Example 2: The CSP is given in Example 1 and its $\mathbf{S}_x^P$ is

$$\mathbf{S}_x^P = \{\langle x_1, x_2, x_3 \rangle, \ \langle x_1, x_3, x_2 \rangle, \ \langle x_2, x_1, x_3 \rangle, \ \langle x_2, x_3, x_1 \rangle, \ \langle x_3, x_1, x_2 \rangle, \ \langle x_3, x_2, x_1 \rangle\} \quad (4)$$

According to Decoder1, each element in $\mathbf{S}_x^P$ can be transformed to a partial instantiation of the variables, namely,

$$
\begin{aligned}
\langle x_1, x_2, x_3 \rangle \to \langle 1, 1, * \rangle \quad \langle x_1, x_3, x_2 \rangle \to \langle 1, 1, * \rangle \quad \langle x_2, x_1, x_3 \rangle \to \langle 1, 1, * \rangle \\
\langle x_2, x_3, x_1 \rangle \to \langle 1, *, 1 \rangle \quad \langle x_3, x_1, x_2 \rangle \to \langle 1, 1, * \rangle \quad \langle x_3, x_2, x_1 \rangle \to \langle 1, *, 1 \rangle
\end{aligned}
\quad (5)
$$

where the " $*$ " represents that the corresponding variable is left uninstantiated. As can be seen, no element in $\mathbf{S}_x^P$ can be transformed to a solution by Decoder1.

On the basis of Decoder1, we propose the minimum conflict encoding (MCE). In MCE, each CSAgent is not only represented as an element in $\mathbf{S}_x^P$, but also an element in $\mathbf{S}$, so that we can design some behaviors to deal with the values of the variables directly. Therefore, each CSAgent must record some information, and it is represented by the following structure:

```
CSAgent=Record
    P:  P∈S_x^P;  V:  V∈S;  E:  E=Energy(V);
    SL: If SL is True, the self-learning behavior can be
        performed on the CSAgent, otherwise, it cannot;
End.
```

CSAgent($\cdot$) is used to represent the corresponding component in the above structure. When computing the energy of a CSAgent, *P* must be first decoded to *V*. In MCE, it is decoded by the minimum conflict decoding:

Algorithm 1 Minimum conflict decoding (MCD)
```
Input: CSAgent: the CSAgent needs to be decoded;
           Pos: the position to start decoding;
Output: CSAgent(V);
Let CSAgent(P) = ⟨x_{P₁}, x_{P₂}, ⋯, x_{Pₙ}⟩ ,  CSAgent(V) = ⟨v₁, v₂, ⋯, vₙ⟩ ,  and
```

$$Conflicts(v_i) = \sum_{j=1}^m \chi(v_i, C_j) , \quad \text{where} \quad \chi(v_i, C_j) = \begin{cases} 1 & v_i \text{ violates } C_j \\ 0 & \text{otherwise} \end{cases} \quad (6)$$

Conflicts(v_i) only considers the variables assigned values.

Step 1: If $(Pos=1)$, $v_{P_i} \leftarrow 1$, and $i \leftarrow 2$; otherwise $i \leftarrow Pos$;

Step 2: If $i > n$, stop; otherwise, $v_{P_i} \leftarrow 1$, $Min_C \leftarrow Conflicts(v_{P_i})$, $Min_V \leftarrow 1$; and $j \leftarrow 2$;

Step 3: $v_{P_i} \leftarrow j$; if $Conflicts(v_{P_i}) < Min_C$, then $Min_C \leftarrow Conflicts(v_{P_i})$ and $Min_V \leftarrow j$;

Step 4: Let $j \leftarrow j+1$; if $j \leq |D_{P_i}|$, then go to Step 3; otherwise, go to Step 5;

Step 5: $v_{P_i} \leftarrow Min_V$, $i \leftarrow i+1$, and go to Step 2.

The behaviors performing on P (P-behaviors) and the behaviors performing on V (V-behaviors) are designed for CSAgents. When V-behaviors perform on a CSAgent, its energy can be updated directly. But when P-behaviors perform on a CSAgent, it must be decoded by MCD before updating its energy. If MCD starts to decode from the first variable in the permutation for any CSAgent, the information generated by V-behaviors would lost. Therefore, we set a parameter, Pos, for MCD, which is the first position that has been changed by P-behaviors. Thus, the value of the variables before Pos is left untouched such that some information generated by V-behaviors can be reserved. For a completely new CSAgent, Pos is set to 1.

2.2 Environment of Constraint Satisfaction Agents

In order to realize the local perceptivity of agents, the environment is organized as a latticelike structure, which is defined as follows:

Definition 2: All CSAgents live in a latticelike toroidal environment, L, which is called an agent lattice. The size of L is $L_{size} \times L_{size}$, where L_{size} is an integer. Each CSAgent is fixed on a lattice-point and it can only interact with its neighbors. Suppose that the CSAgent located at (i, j) is represented as $L_{i,j}$, $i, j=1,2,\ldots,L_{size}$, then the neighbors of $L_{i,j}$, $Neighbors_{i,j}$, are defined as $Neighbors_{i,j} = \{L_{i-1,j}, L_{i,j-1}, L_{i+1,j}, L_{i,j+1}\}$.

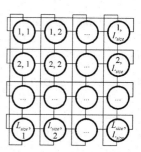

Fig. 1. The agent lattice

Therefore, the agent lattice can be represented as the one in Fig.1. Each circle represents a CSAgent, the data represent the position in the lattice, and two CSAgents can interact with each other if and only if there is a line connecting them.

2.3 Behaviors of Constraint Satisfaction Agents

Three behaviors are designed for CSAgents to realize their purposes, that is, the competition behavior, the self-learning behavior, and the mutation behavior. The former two belong to P-behaviors, while the last one belongs to V-behaviors.

Competition Behavior: The energy of a CSAgent is compared with those of its neighbors. If its energy is greater than that of any CSAgent in its neighbors, then it can survive; otherwise, it must die, and its lattice-point will be taken up by the child of the CSAgent whose energy is maximum in its neighbors. Suppose that the competition behavior is performed on the CSAgent located at (i, j), $L_{i,j}$, and $Max_{i,j}$ is the CSAgent with maximum energy among the neighbors of $L_{i,j}$, namely, $Max_{i,j} \in Neighbors_{i,j}$ and $\forall \ CSAgent \in Neighbors_{i,j}$, then $CSAgent(E) \le Max_{i,j}(E)$. If $L_{i,j}(E) \le Max_{i,j}(E)$, then $Max_{i,j}$ generates a child CSAgent, $Child_{i,j}$, to replace $L_{i,j}$, and the method is shown in Algorithm 2; otherwise, $L_{i,j}$ is left untouched.

Algorithm 2 Competition behavior
Input: $Max_{i,j}$: $Max_{i,j}(P) = \langle x_{m_1}, x_{m_2}, \cdots, x_{m_n} \rangle$;
          p_c: A predefined parameter, and it is a real number
             between 0 and 1;
Output: $Child_{i,j}$: $Child_{i,j}(P) = \langle x_{c_1}, x_{c_2}, \cdots, x_{c_n} \rangle$;
$Swap(x, y)$ exchanges the values of x and y. $U(0,1)$ is a uniform random number between 0 and 1. $Random(n, i)$ is a random integer among $1, 2, ..., n$ and is not equal to i. $Min(i, j)$ is the smaller one between i and j.
Step 1: $Child_{i,j}(P) \leftarrow Max_{i,j}(P)$, $i \leftarrow 1$, and $Pos \leftarrow n+1$;
Step 2: If $U(0, 1) < p_c$, then $l \leftarrow Random(n, i)$;
Step 3: Perform $Swap(x_{c_i}, x_{c_l})$, and go to Step 4;
Step 4: If $Min(l, i) < Pos$, then $Pos \leftarrow Min(l, i)$;
Step 5: Let $i \leftarrow i+1$; if $i \le n$, then go to Step 2;
Step 6: Perform MCD($Child_{i,j}$, Pos); $Child_{i,j}(SL) \leftarrow True$.

Self-Learning Behavior: A CSAgent increases its energy by using the knowledge about CSPs. Suppose that the self-learning behavior is performed on $L_{i,j}$. Then, $L_{i,j}$ increases its energy by the method shown in Algorithm 3.

Algorithm 3 Self-learning behavior
Input: $L_{i,j}$: $L_{i,j}(P) = \langle x_{m_1}, x_{m_2}, \cdots, x_{m_n} \rangle$, $L_{i,j}(V) = \langle v_1, v_2, \cdots, v_n \rangle$;
Output: $L_{i,j}$;
Step 1: $Repeat \leftarrow False$, $k \leftarrow 1$, and $Iteration \leftarrow 1$;
Step 2: If $(Conflicts(v_{m_k}) = 0)$, then go to Step 7;
Step 3: $Energy_{old} \leftarrow L_{i,j}(E)$, $l \leftarrow Random(n, k)$;
Step 4: Perform $Swap(x_{m_k}, x_{m_l})$, MCD($L_{i,j}$, Min(k, l)); $Energy_{new} \leftarrow L_{i,j}(E)$;
Step 5: If $Energy_{new} > Energy_{old}$, then $Repeat \leftarrow True$; otherwise,
           $Swap(x_{m_k}, x_{m_l})$, and perform MCD($L_{i,j}$, Min(k, l));
Step 6: If $Iteration < n-1$, then $Iteration \leftarrow Iteration+1$, go to Step 2;
           otherwise, $Iteration \leftarrow 1$, go to Step 7;
Step 7: Let $k \leftarrow k+1$; If $k \le n$, then go to Step 2;
Step 8: If $Repeat = True$, then go to Step1; otherwise, $L_{i,j}(SL)$
           $\leftarrow False$, and stop.

Mutation Behavior: This behavior is similar to the mutation operator used in traditional EAs. Its function is to assist the above behaviors. It can enlarge the search area so as to make up for the disadvantage of the decoding method. Suppose that the mutation behavior is performed on $L_{i,j}$, and $L_{i,j}(V) = \langle v_1, v_2, \cdots, v_n \rangle$. Then, the following operation is performed on $L_{i,j}$.

$$\text{If } U_k(0, 1) < p_m, \text{ then } v_k \leftarrow Random(|D_k|) \tag{7}$$

Where $k=1, 2, \ldots, n$, and $U_k(0, 1)$ represents the random number is anew for each k. p_m is a predefined parameter, and it is a real number between 0 and 1. $Random(|D_k|)$ is a random integer among $1, 2, \ldots, |D_k|$.

3   Multiagent Evolutionary Algorithm for CSPs

To solve CSPs, all CSAgents must orderly adopt the three behaviors aforementioned. Here these behaviors are controlled by means of evolution. The details are described in Algorithm 4.

Algorithm 4 Multiagent evolutionary algorithm for CSPs

Input: $Evaluation_{Max}$: The maximum number of evaluations;
         L_{size}: The scale of the agent lattice;
         p_c: The parameter used in the competition behavior;
         p_m: The parameter used in the mutation behavior;
Output: s: A solution or an approximate solution for the CSP
         under consideration, and $s \in \mathbf{S}$;
L^t represents the agent lattice in the tth generation. $CSAgent_{Best}^t$ is the best CSAgent in $L^0, L^1, \ldots, L^t$, and $CSAgent_{tBest}^t$ is the best CSAgent in L^t.
Step 1: $Evaluations \leftarrow 0$; Initialize the agent lattice L^0: Generate
        a permutation randomly and assign it to $L_{i,j}^0(P)$,
        $L_{i,j}^0(SL) \leftarrow True$, perform MCD($L_{i,j}^0, 1$), compute $L_{i,j}^0(E)$, and
        $Evaluations \leftarrow Evaluations+1$, where $i, j=1, 2, \ldots, L_{size}$; update
        $CSAgent_{Best}^0$ and $t \leftarrow 0$;
Step 2: Perform the competition behavior on each CSAgent in
        L^t: If $L_{i,j}^t$ wins, then $L_{i,j}^{t+1} \leftarrow L_{i,j}^t$; otherwise, $L_{i,j}^{t+1} \leftarrow Child_{i,j}$,
        compute $L_{i,j}^{t+1}(E)$, and $Evaluations \leftarrow Evaluations+1$;
Step 3: Update $CSAgent_{(t+1)Best}^{t+1}$; If $(CSAgent_{(t+1)Best}^{t+1}(SL) = True)$, then
        perform the self-learning behavior on $CSAgent_{(t+1)Best}^{t+1}$,
        where $Evaluations$ has been updated in Algorithm 3, and
        go to Step 5; otherwise, go to Step 4;
Step 4: Perform the mutation behavior on $CSAgent_{(t+1)Best}^{t+1}$, compute
        $CSAgent_{(t+1)Best}^{t+1}(E)$, and $Evaluations \leftarrow Evaluations+1$;
Step 5: If $CSAgent_{(t+1)Best}^{t+1}(E) \geq CSAgent_{Best}^t(E)$, then
        $CSAgent_{Best}^{t+1} \leftarrow CSAgent_{(t+1)Best}^{t+1}$; otherwise,
        $CSAgent_{Best}^{t+1} \leftarrow CSAgent_{Best}^t$, $CSAgent_{Random}^{t+1} \leftarrow CSAgent_{Best}^t$, where
        $CSAgent_{Random}^{t+1}$ is randomly selected from L^t and is
        different from $CSAgent_{(t+1)Best}^{t+1}$;
Step 6: If $\left(CSAgent_{Best}^{t+1}(E) = 0\right)$ or $(Evaluations \geq Evaluation_{Max})$, then
        $s \leftarrow CSAgent_{Best}^{t+1}(V)$ and stop; otherwise, $t \leftarrow t+1$, goto
        Step2.

4 Experiments

Since any CSP can be equivalently transformed to a binary CSP [4], binary CSPs are used to test the performance of MAEA-CSPs in this section. The test suite[1] was used to compare the performances of 11 available algorithms in [5]. It consists of 250 solvable instances. They are divided into 10 groups according to their difficulty, p={0.24, 0.25, 0.26, 0.27, 0.28, 0.29, 0.30, 0.31, 0.32, 0.33}.

Three measures, the *success rate* (SR), the *mean error* (ME) and the *average number of evaluations to solution* (AES), are used to measure the performance of MAEA-CSPs. Four parameters of MAEA-CSPs are set as follows: L_{size}=5, p_c=0.2, p_m=0.05, and $Evaluation_{Max}$=100 000. We perform 10 independent runs on each of the 25 instances belonging to a given p value.

4.1 Comparison Between MAEA-CSPs and the Existing Algorithms

Reference [5] has made a comparison among 11 available algorithms, and the results show that the best four algorithms are H-GA.1 [2], H-GA.3 [2], SAW [3], and

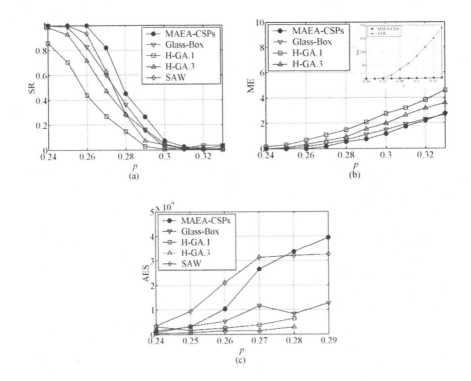

Fig. 2. The comparison between MAEA-CSPs and the four existing algorithms

---

[1] http://www.xs4all.nl/~bcraenen/resources/csps_modelE_v20_d20.tar.gz

Glass-Box [1]. Therefore, in this experiment, we make a comparison between MAEA-CSPs and the four algorithms. The comparison results are shown in Fig.2.

The SR of MAEA-CSPs is the highest among the five algorithms, and it achieves to 100% for p=0.24~0.26. Fig.2(b) shows that the ME of MAEA-CSPs is also the best among the five algorithms. Since the AES is not statistically reliable when the SR is very low, only the AES of the instances whose SR is larger than 10% is plotted in Fig.2(c).

4.2 Parameter Analyses of MAEA-CSPs

p_c and p_m are increased from 0.05 to 1 in steps of 0.05. MAEA-CSPs with the 400 groups of parameters are used to solve the 10 groups of instances. According to the SR, the 10 groups of instances can be divided into 3 classes. The first class includes the instances with low p values, that is, p=0.24~0.26. Since the SR for this class is higher than 90% for the most of the parameters and the ME is very low, the graphs for the SR and the AES are shown in Fig.3. The second class includes the instances whose values of p vary from 0.27 to 0.29. Since the SR for this class is lower, the graphs for the SR and the ME are shown in Fig.4. The last class includes the instances with high p values. Since the instances in this class are very difficult, only the graphs for the ME are shown in Fig.5.

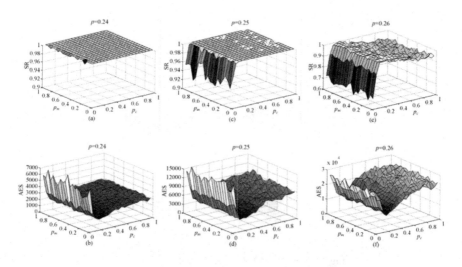

Fig. 3. The SR and the AES of MAEA-CSPs for p=0.24~0.26

As can be seen, p_c has a larger effect on the performance of MAEA-CSPs. For the first class, the SR is higher than 90% when p_c is larger than 0.1. Although the AES increases with p_c when p_c is larger than 0.2, the AES is smaller when p_c is in 0.1~0.3. For the second and the last class, the results are similar, namely, when p_c is in 0.1~0.3, the SR is higher, and the ME is smaller. Although p_m does not affect the performance of

MAEA-CSPs obviously, Fig.4 and Fig.5 show the SR is a little higher and the ME is a litter smaller when p_m is small. Therefore, it is better to choose p_c from 0.1~0.3 and choose p_m from 0.05~0.3. In addition, although the performance of MAEA-CSPs with above parameters is better, MAEA-CSPs still performs stably when p_c is larger than 0.2. It shows that the performance of MAEA-CSPs is not sensitive to the parameters, and MAEA-CSPs is quite robust and easy to use.

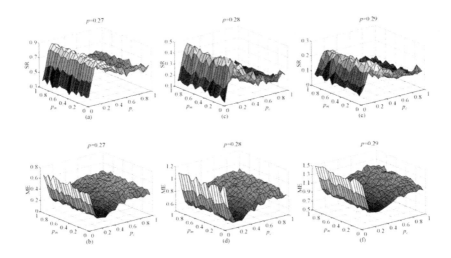

Fig. 4. The SR and the ME of MAEA-CSPs for p=0.27~0.29

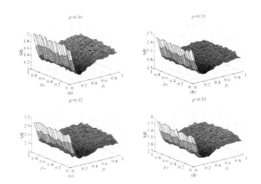

Fig. 5. The ME of MAEA-CSPs for p=0.30~0.33

5   Conclusion

In this paper, multiagent systems and evolutionary algorithms are combined to form a new algorithm to solve constraint satisfaction problems. Parameter analyses show that MAEA-CSPs is quite robust and easy to use. It is better to choose p_c from 0.1~0.3 and choose p_m from 0.05~0.3. The comparison between MAEA-CSPs and four well-defined algorithms, H-GA.1, H-GA.3, SAW, and Glass-Box, indicated that MAEA-CSPs outperforms the four algorithms.

References

1. Marchiori, E., Combining constraint processing and genetic algorithms for constraint satisfaction problems, in *Proc. 7th Int. Conf. Genetic Algorithms*, Bäck, T., Ed., 1997, pp.330–337.
2. Craenen, B., Eiben, A., and Marchiori, E., Solving constraint satisfaction problems with heuristic-based evolutionary algorithms, in *Proc. Congress Evolutionary Computation*, 2000, pp. 1571–1577.
3. A. Eiben and J. I. Van Hemert, SAW-ing EAs: Adapting the fitness function for solving constrained problems, in *New Ideas in Optimization*, Corne, D., Dorigo, M., and Glover, F., Eds. New York: McGraw-Hill, 1999, pp. 389–402.
4. Rossi, F., Petrie, C., and Dhar, V., On the equivalence of constraint satisfaction problems, in *Proc. 9th European Conf. Artificial Intelligence*, Aiello, L. C., Ed., pp.550–556, 1990.
5. Craenen, B. G. W., Eiben, A. E., and van Hemert, J. I., Comparing evolutionary algorithms on binary constraint satisfaction problems, *IEEE Trans. Evol. Comput.*, vol. 7, pp. 424–444, Oct. 2003.
6. Zhong, W. C., Liu, J., Xue, M. Z., and Jiao, L. C., A multiagent genetic algorithm for global numerical optimization. *IEEE Trans. Sys., Man, and Cybern. B*, vol. 34, pp. 1128-1141, Apr. 2004.
7. Tsang, E., Foundations of Constraint Satisfaction, Academic Press, 1993.
8. Tan, S., et al., Automatic image enhancement driven by evolution based on ridgelet frame in the presence of noise. Lecture Notes in Computer Science, Springer-Verlag, Berlin Heidelberg New York, 2005.

Author Index

Lecture Notes in Computer Science

For information about Vols. 1–3327

please contact your bookseller or Springer